AF096707

PROJECT MAVEN

PROJECT MAVEN

A Marine Colonel, His Team,
and the Dawn of AI Warfare

KATRINA MANSON

W. W. NORTON & COMPANY
Independent Publishers Since 1923

Copyright © 2026 by Katrina Manson

All rights reserved
Printed in the United States of America
First Edition

For information about permission to reproduce selections from this book, write to
Permissions, W. W. Norton & Company, Inc., 500 Fifth Avenue, New York, NY 10110

For information about special discounts for bulk purchases, please contact
W. W. Norton Special Sales at specialsales@wwnorton.com or 800-233-4830

Manufacturing by Lakeside Book Company
Book design by Anna Oler

ISBN 978-1-324-12331-6

W. W. Norton & Company, Inc., 500 Fifth Avenue, New York, NY 10110
www.wwnorton.com

W. W. Norton & Company Ltd., 15 Carlisle Street, London W1D 3BS

Authorized EU representative: EAS, Mustamäe tee 50, 10621 Tallinn, Estonia

10 9 8 7 6 5 4 3 2 1

CONTENTS

A Note on Sources   vii
Prologue   1
Introduction   3

PART ONE: FIND

1. Old Wars   13
2. Tilting at Windmills   22
3. We Do What We Want   29
4. They Call It Algorithmic Warfare   43
5. The First Mavenites   55
6. Relaxed About Fury   65
7. The Colonel and the Math Whiz   76
8. Somalia   86
9. Moral Outrage   100

PART TWO: FIX

10. The Algorithms Have No Clue   119
11. Harbinger of Doom   134
12. Arms Race   145

13	Daddy Karp	155
14	Palantir, Palantir, Palantir	169
15	Palantir Splits the Team	184
16	A Striking Operation	194
17	Data Hell	204

PART THREE: **FINISH**

18	We'll Find It and We'll Strike It	223
19	Nobody Knows Targeting Better Than Trey	238
20	Kill Chain	246
21	Ukraine Fights Back	261
22	Tens of Thousands of Targets	275
23	We've Drunk the Kool-Aid	287

PART FOUR: **FEEDBACK**

24	Machines Shouldn't Kill People	309
25	Trump's Robots	319
26	The Winchester House	335
	Epilogue	351
	Acknowledgments	355
	Notes	357
	Index	398

A NOTE ON SOURCES

This account relies on multiple, lengthy conversations with more than two hundred Project Maven insiders and opponents.

I'm grateful to the people who agreed to go on record for this book and to those whose names are not written in the text because they consider their information too sensitive.

As I note in chapter 12, Project Maven records are not subject to release under the Freedom of Information Act and, as a program of record run since 2023 by a US intelligence agency, its budget is also now classified. In addition to reviewing documents that form part of the public record, I have also reviewed multiple private emails and other messaging formats, internal defense and company documents—including nonpublic memos, reviews and slide decks, nonpublic algorithmic score assessments, personal photograph collections, contemporaneous notes, travel records, and sketches—to corroborate and cross-reference the timing and recollections of multiple individuals with varying experiences and perspectives. I have tried to understand and explain the experience, motivations, and intentions of each person I interviewed or who appears in the text, whatever side of various fraught divides they stand on.

The full history of Project Maven, and the past, present, and future of America's effort to send AI into war will rely on records and stories that may still come out in the years to come. This is the best I could do today. Any errors or omissions are my own.

PROJECT MAVEN

PROLOGUE

HE PASSED THE EVENING as so many millions of Americans do. At home, watching Netflix. But Drew Cukor was doing it slightly differently. Sitting at a desk, with a notepad and pen beside him, he just kept rewinding. He must have scrutinized parts of the show eight times over.

The US Marine colonel, transfixed by a new military mission, was watching a 2017 documentary about an ancient Chinese board game focused on strategy and control soon after it came out. Lee Sedol, the eighteen-time world champion from South Korea, was in the hot seat. The twist was Sedol's opponent: AlphaGo, an AI program developed by Google subsidiary DeepMind. Go had always been the holy grail of AI challenges. Far more complex than chess, the game has more possible board configurations than atoms in the universe.

Cukor was absorbed by Sedol's growing confusion and panic as the world champion came up against an unexpected move from a machine initially perceived as dumb and formulaic. The unorthodox play, hailed by AlphaGo's backer as "a moment of exquisite algorithmic ingenuity," handed AI the winning advantage in the second game and floored Sedol, who ultimately lost four games to one. That evening, the colonel watched a human at the top of his game discover the humiliating limits of his own mind.

To Cukor, the lesson of AlphaGo was that a human alone would never be enough. A human plus a machine would always be better. That had profound implications, he felt, for his own gruesome field: war. He already knew the future. Cukor needed to bring AI to the battlefield. He wasn't the only significant viewer: he heard China's president Xi Jinping watched it too.

INTRODUCTION

WHEN I FIRST MEET DREW CUKOR he has little in the way of easy smiles. It is mid-2024, and I have already spent almost a year trying to convince him to speak with me. I wait for him after work in the lobby of the towering New York office of J.P. Morgan, the bank where the retired Marine Corps colonel is now leading the transformation of artificial intelligence for chief executive Jamie Dimon.

Upstairs, Cukor offers me a bottle of water. He takes nothing for himself. We sit directly opposite each other at a booth in the emptied office café. And I watch the former intelligence officer decide if he wants to talk to me after all. My main task seems to be to meet his stare, which is not an entirely straightforward undertaking. It is clear the only person being interviewed is me.

He barely lets me write anything that first meeting. I remember one bit by heart. "War is terrible, war is terrible, war is terrible," he intones, holding my gaze and giving voice to a universal chorus.

His hair is tight-cropped. His demeanor stiff. His opinions uncompromising. But he softens into his passions: the catastrophe of the Dieppe Raid in 1942, the failures of the US military to match firepower to intelligence, and Project Maven, the effort he led at the Pentagon to put AI at the heart of how America makes war.

Over the months to come, I learn that Drew Cukor is a leading historical figure in a war that hasn't happened yet. That seemed to be what almost everyone to do with Project Maven thought, whether they feted or hated him.

Launched in 2017, Project Maven ostensibly aimed to use computer vision to sort through thousands of hours of drone footage taken across Asia and the Middle East. But I would learn that Cukor and his backers always intended to use AI for much more than surveillance; from the outset they wanted to target people and objects with the help of AI.

The problem with war, Cukor told me, had always been the humans. "They're materially corrupt, inefficient, and they get tired." And when they die it affects the campaign, he went on brusquely. He believed humans could do better with the help of machines, and that AI could pierce the fog of war.

More immediately, Cukor wanted to fix the bureaucracy he felt repeatedly let down America's fighting forces overseas, bring intelligence directly into combat operations, recognize the value of software over hardware, and test emerging technology in real wars. "We were really good at killing people, but it didn't get us very far," one caustic former member of Joint Special Operations Command (JSOC) told me. Cukor wanted America to stem the flow of mistakes that saw US forces accidentally killing civilians, allies, and America's own troops. He wanted America to reconceive what victory looked like: not destroying the enemy, but defeating it.

And he wanted to use government defense spending to tether a nascent commercial AI market to America rather than let it go off in search of customers in China. It was the government's role to help the venture capital community monetize their investments, he wrote in one draft 2017 paper I reviewed.

The colonel's relentless advocacy for Maven "monetized" startups and the ingenuity of AI researchers who usually spent their time writing esoteric academic papers. He recruited Big Tech companies better known for online shopping and office software: Amazon Web Services (AWS) and Microsoft now deliver algorithmic warfare for Maven.

Palantir Technologies, now celebrated as an insurgent force in the S&P 500 and one of the most valuable (and, many analysts and short sellers suggest, overvalued) American companies worldwide, was on the way out of the Defense Department when it won its first Maven contract. It arguably owes its rebirth to Drew Cukor.

Scale AI, the data labeling company in which Meta has taken a 49 percent stake, could say something similar about its own rise. AI chipmaker Nvidia was there from the start. Anthropic and OpenAI are suddenly plunging into defense work as they seek a commercial home for their generative AI platforms. Even Google, whose workers protested involvement in "the business of war" when they discovered they were part of Project Maven in 2018, now embraces national security work. Notably, Apple has said no.

Whatever form America's pursuit of AI warfare takes today, and in the future, it will owe something to Cukor. Alex Karp, the billionaire chief executive of Palantir Technologies, the CIA-backed data analytics company that soon joined the project, would later describe him as the "founding father of AI targeting."

CUKOR THOUGHT IT WOULD take twenty years to remake the US military and mainline AI. For the first five years, he encountered controversy and resistance as he pushed Congress to fund new types of firepower to help America cling on to the apex of global power. He infuriated doubters and critics, often from within his own team, never mind among civilians who balk at the prospect of *Terminator*-style global extinction.

Cukor's iconoclasm filtered down to his team. Many "Mavenites" saw themselves as maverick renegades within the Department of Defense. They carried themselves with a tech startup's insouciance in the heart of the button-down Pentagon. But Mavenites also reflected the overconfidence and anguish of a military superpower that repeatedly ran up against its own limits and flaws. I would learn the team dynamic was a constant rollercoaster as Cukor bulldozed a path to his dream at personal and professional cost.

One of Cukor's most controversial decisions was to push the US military to use minimally tested systems in hot wars. The colonel always argued getting AI on the battlefield, before it was ready or reliable, was the only way to improve it and develop the trust and know-how of a new generation of fighters in using it.

He was exacting in his every demand. Cukor was a noun, a verb, an adjective. "To Cukor" connoted tremendous hours, tremendous pursuit, tremendous invention, and tremendous intensity. "Getting Cukor-ed" meant having to pursue the same yourself, at his instruction.

"He was like a mastermind," one person told me, desperately reaching for the words to describe the power Cukor could bring to bend the bureaucracy—and the people in it—to his will. "I don't know what it is," they stumbled on. He was admired and feared. "It's a mental thing."

There is nothing unusual about me, Cukor would say. Meanwhile, everyone I spoke to for this book told me otherwise. *Have you spoken to Cukor?*, they'd ask. *You need to speak to Cukor. If Cukor would speak to you, it would be good. You can't tell this story without Cukor.*

THE RISE OF AI WARFARE speaks to the biggest moral and practical question there is: Who—or what—gets to decide to take a human life? And who bears that cost?

Cheerleaders argue that AI and the automation it makes possible will save lives. They claim algorithms bring a precision to decision-making that will limit civilian and friendly-fire casualties. They argue AI-empowered systems could deter conflict with China—or help win World War III, in which automated machines will putatively run combat at a pace faster than humans can understand.

Detractors think AI has already led to civilian deaths, will spread uncontrolled destruction, and potentially hasten the end of the world. Still more think the claims made for AI war tools are grandiose and the truth will be more prosaic, suffering from problems of rickety infrastructure, adoption, and trust. Pragmatic supporters argue an incremental mix of humans and machines will forge that trust.

The problem with many theories about what AI will do to warfare is just that: they remain theoretical. I wanted to go in search of the specifics. I wanted to tell the story of the people making AI warfare a reality, and of the US military members actually using it. What was inside the black box?

Ten years since Cukor started his effort, the AI decision-making systems developed under Maven, and some of the Pentagon's eight hundred other AI projects, are used on the battlefield. Maven Smart System (MSS), a software platform that develops targets with the help of AI, is now deployed in every branch of the US military and all over the world, incorporating more than 150 data feeds and the work of more than fifty companies. NATO started using a version of the system in the spring of 2025, and I would learn in October 2025 that ten NATO members were lining up to use it for their own militaries.

Maven has already sped up the pace of war. I learned from an official at the National Geospatial-Intelligence Agency that with the help of computer vision the US went from being able to hit under a hundred targets a day to being able to hit a thousand. In combination with large language models (LLMs) integrated into the Maven platform, that number has risen fivefold to five thousand targets a day.

The AI algorithms developed under Maven now deploy in submarines and in space operations. They're in subsea sonar systems belonging to America and two of its closest intelligence allies (the UK and Australia) designed for nuclear deterrence. They're fielded on autonomous drone boats. I learned AI targeting systems live in at least two highly secretive systems—one aerial and one aquatic—that could surveil, select, and kill targets entirely on their own, intended for the defense of Taiwan.

The US will have to define carefully the relevant use cases, guardrails, and doctrine if it wants to stick to the Geneva Conventions and avoid shooting civilians and its own allied forces.

I STARTED WRITING about the future of war after I became the US foreign policy and defense correspondent for the *Financial Times* in 2017—the same year Project Maven started. As the US reckoned with the rise of China, I watched a global powerhouse humbled by poorly equipped enemies in Afghanistan and Iraq attempt to embrace AI as a shortcut to sustaining global military dominance.

The first Trump administration's 2018 national defense strategy predicted new commercial technology "will change society and, ultimately, the character of war." Four years later, after I became a Bloomberg correspondent covering emerging tech and national security, the arrival of chatbots and AI agents only accelerated this shift. Under the second Trump administration, the Department of Defense has reemerged as the "Department of War" devoted to AI and autonomy, under a secretary who wants to make it easier to acquire weapons and free US forces from "overbearing rules of engagement."

Three experiences also drew me to write about AI's potential impact on future war. First, I've had Jan Bloch rattling around my head since 1999, when I sat down to read the yellowing pages of the Polish banker's 1899 book. The English translation of his work was retitled for what turned out to be a hopeless question: *Is War Now Impossible?* He was exploring whether lethal weapons produced at industrial scale would make war obsolete. He suggested that mass-produced rifles and other new technologies wouldn't make for decisive wins, swift wars, palatable killing. It would make for stalemate, long wars, horror. He didn't quite prophesy four years of trench warfare and 8.5 million combat deaths starting in 1914. But nearly. More than a century later, would the potential calamity of sending AI into war make great war impossible, or would Bloch be proved wrong once again?

Second, I spent a dozen years as a reporter covering business, investment, and politics in multiple African countries. I saw the impact of violence in countries from Sierra Leone to Somalia, and logged the distance between policy and reality. When it came to AI warfare, I wanted to know how theory on high would match reality on the ground.

Third, I carry with me the memory of a journey I took on a military plane back from Afghanistan in 2009. The British soldiers beside me told me about the friends who had just been killed in combat. They showed me the explosions they couldn't stop watching on their phones. And they told me they desperately wanted to leave the military but were trapped by contracts they could not escape, and now felt equally unable to survive civilian life because no one would understand them. In that moment, they felt bound to death. Whatever worse terrors wars visit on civilians and enemies, I also cannot shake what it does to the people sent to fight. Could AI alleviate the burden and suffering of war?

Project Maven sits at the intersection of colliding trends: America's rising insecurity about its place in the world, a technological revolution forcing AI into almost every aspect of life and war, fraught civil-military relations in the world's most powerful democracy, the dominance of Big Tech, China's growing military and technological ambitions, and all-encompassing surveillance made possible by ubiquitous sensors and commercial software.

The next ten years are still waiting to be written. Russia's invasion of Ukraine has upturned military expectations. The Pentagon's deadly strikes against boats in the Caribbean are graying the boundaries of the rules of war and underline the ease of declaring war at a remove. US military commanders say China is rehearsing for the military takeover of Taiwan. Rival superpowers are arming for conflict. Campaigners argue afresh for AI redlines. And a new generation of venture capital–backed Silicon Valley leaders is chasing defense contracts, talking up the superiority of the West and the appeal of AI-enabled killing with newfound braggadocio.

National security strategists now worry that no country can win a war without AI. The UN's aim to ban lethal autonomous weapons that select their own targets with the help of AI by 2026 is a lost hope. And yet AI remains a narrow, faulty tool with considerable limits to its usefulness and reliability that the US military is still discovering.

AI warfare can go wrong. And it is already here.

PART ONE

FIND

1

OLD WARS

"We've killed more people on Office than you'd ever imagine."

THE DOOR GUNNER let the .50-caliber machine gun rip into the night. The blacked-out helicopter was on its way to Kandahar. Among the people aboard now keeping needlessly silent was US Marine Corps Sergeant Dave Spirk.

The juddering of the 50-cal got Spirk's attention. "It was quite a moment," said Spirk. "Sporty for sure."

Beside him was Captain Drew Cukor, a Marine Corps intelligence officer.

"It was me and Drew. Our thighs were tactically one," Spirk told me.

The two men had begun their journey several weeks prior on a warship anchored in the North Arabian Sea. They were part of Task Force 58, an effort led by Brigadier General Jim Mattis to insert the first conventional forces into Afghanistan. The 9/11 attacks were little more than two months in the past, and the United States wanted to topple the Taliban and defeat the al-Qaeda jihadis they harbored. Afghanistan had a strong track record of fending off foreign incursion.

Mattis planned a series of raids aimed to "create chaos" as part of the US takeover attempt. The job echoed his own call sign "CHAOS," a sardonic appellation from an earlier job: "Colonel Has Another Outstanding

Suggestion." The goal was to prevent the Taliban, already getting hammered by US bombing raids in the north, from regrouping in the south around their spiritual home in Kandahar. Spirk and Cukor had each flown four hundred miles inland over Pakistan's Balochistan province to reach their waypoint: a remote Afghan airstrip in Helmand province. That was twice as far from the sea as usual Marine operations allowed, but Mattis wasn't much interested in doctrine and nor were the troops. He was about to send the amphibious wing of the military to invade a landlocked, mountainous country.

They arrived a couple of days apart at a spartan desert camp originally set up for Arab sheikhs eager to hunt bustard, a heavy-bodied bird. Renamed Forward Operating Base (FOB) Rhino by Army Rangers, there was no water supply but there was a long, single dirt runway. Air Force pilots improvised precarious nighttime landings on the dirt with the help of night-vision goggles, another departure from the norm. In those jumpy early days at FOB Rhino, US forces received wounded US and Afghan personnel, fended off attacks against the camp from Taliban fighters, and one night shot at a lone camel for fear it was carrying explosives.

Leaving FOB Rhino some days later, the two men were on the first wave of helicopters leading a Marine Expeditionary Unit mission to seize Kandahar Airport in southern Afghanistan. They flew packed in tight, shoulder to shoulder under a sea of infantry gear, listening to the 50-cal blazing into the dark. But this helicopter was a person short. At Captain Cukor's command, a bulky computer unit took the place of a lance corporal.

Spirk remembers it well: "I carried the damn thing."

The precious computer was meant to help Cukor find and assess targets, detect threats, help plan missions, and brief the commander. Loaded on the computer were his tools: Excel, Word, Google Earth, and PowerPoint, and some in-house military software none of them liked.

"It was my brain," Cukor said about the computer. But it was also close to useless.

"It didn't do anything, it was no good to us," Spirk remembered.

As soon as Cukor had landed in FOB Rhino, he set up the computer and a map. His job was intelligence in a land that would prove unintelligible to the new arrivals. He was the ground intelligence officer for the 26th Marine Expeditionary Unit (Special Operations Capable), responsible for feeding intel into planning and operations. Spirk was the intelligence analyst ("which means I could read and write").

He looked up to Cukor, who'd made Spirk something of a pet project after they met two years earlier, pushing him to work harder and harder. "I was an enlisted guy, he was an officer, but he never ever treated me like that," says Spirk.

Cukor often dished out unconventional advice, including—a few years later—that Spirk should leave the Marines and go do something else. Many Marines stay in, but Cukor thought Spirk could stretch himself as a defense official outside the confines of being enlisted. He would have more impact, faster. "Drew was a little bit different." (Spirk would eventually become a civilian leader at the highest echelons of the Defense Department, as its first chief data officer.)

The US conjured its Afghanistan invasion plan from scratch, but had years' worth of intelligence to draw on. There were logs of sites used by the Taliban in the past. Caves where the Taliban might be hiding out. Physical features that might help or hinder friend and foe. Tipoffs about weapons stashes. They could cross-check all this with new information—mapping signals showing up from phones or electricity when rogue lights showed up in the desert, or mining information from phones, computers, and paperwork harvested in raids, or reports from detainee interrogations whose treatment would become the subject of crisis in America's vision of itself. They could overlay details of Taliban attacks and other incidents to learn more about America's latest enemy.

But Cukor's tools could make little sense of all this information. Each piece of software was separate and the system was slow. Al-Qaeda targets were listed in long Excel tables. Lines and circles in PowerPoint could attempt to draw out connections between people. Word was for analysis write-ups. Google Earth was for zooming in on the geographic location of an object. Cukor would copy-cut-paste his way onto a slide deck to sum it all up to brief a senior commander. Nothing about these tools translated to war.

"Microsoft is where data goes to die," Cukor told me. "You might as well put a tombstone on it."

An Army artillery officer later summed up the effectiveness of going to war with target lists compiled on software designed to type up quarterly reports. "We've killed more people on Office than you'd ever imagine," he said.

It's not that Cukor didn't think data wouldn't save lives. He wanted information more than anything, but what he had was nowhere near ready to be sent into battle. His brain could hold only so many connections. Even if he could approximate a web of information for one potential high-value target on a digital document, there were several thousand more al-Qaeda suspects to think about. They had no way to divine the patterns of war, of the death traps being laid for them across the country. They were missing so many insights. And each connection they missed had sometimes deadly ramifications for US forces on the ground.

"It's just criminal," he tells me tightly, remembering it nearly twenty-five years later.

Eight Americans had already died in the first weeks since the US started official operations in Afghanistan on October 7, 2001. At least 2,465 would die before the US withdrawal twenty years later, with 20,149 wounded. Suicides would surpass it all: 30,177 among the US troops who fought in the wars that followed 9/11. Mattis had called the Marine insertion "Operation Swift Freedom," a name quickly amended by his military seniors. This would be a long war. Enduring.

US attacks started going wrong from the very earliest days of the invasion. Maybe they were mistakes, maybe it was the unavoidable fog of war that every military strategist had warned about since Clausewitz. Maybe it had something to do with battlefield tech not cut out for battle.

On the morning of December 5, a group of US special operators and Afghan forces were on a mission to install Hamid Karzai as leader of a post-Taliban Afghanistan. They were taking intermittent fire from a group of Taliban hiding inside a cave on the far side of a bridge on the outskirts of

Kandahar. A US combat air controller among the group mistakenly misdirected an attack to his own location after changing the battery on his handheld device, a Precision Lightweight GPS Receiver. Turning it back on automatically reset the coordinates for the strike to his own position, rather than the Taliban position he'd just carefully calculated. Responding to the call, a B-52 bomber released a 2,000-pound bomb right over the US position, killing three Americans and several Afghans fighting with them and wounding dozens more Americans and Afghans. Karzai, among the wounded, was evacuated with the others to Camp Rhino for medical treatment.

Not that Cukor noticed: he had his head down, "intense but unflappable," buried in the intelligence. "I was the guy on the computer trying to build the next package for the next operation with very primitive tools."

When he and Spirk landed before dawn at Kandahar Airport, enemy attacks probed the north side of the facility. And Spirk found himself lugging the computer unit into the airport as some of the first conventional forces poured in.

The US had built the airport in the 1960s, erecting its low-slung repeating arches against the mountain backdrop long before the Soviets arrived, long before the Taliban. By the time the Marine duo deplaned, the airport already bore the signs of war. The windows were blown out and the airfield was a wreck too, pocked by craters from US attacks, soon filled in by Seabees, the US Navy's construction battalions. The rose garden was quickly repurposed as an impromptu ablutions block, as new Marine arrivals supplemented the smattering of Special Operations forces already there. Soon US troops would be giving their living quarters tents names like "Jihad Motel" and putting up crude cartoons of "Santa Taliban" to keep them going through the holiday season.

Dangers lay inside even this minimal cordon of security. On the morning of December 16, the Marines suffered their first serious injuries when three were hurt by a land mine exploding at the airport they had just commandeered. One managed to take off his belt, turning it into a makeshift tourniquet and tying it around his femoral artery. After he was evacuated he lost his left leg below the knee, and was awarded the first Purple Heart of the war. It would be a sign of things to come.

America's high-tech weaponry on land and sea, in air and space, surpassed the rudimentary weapons of the Taliban by far. But the Taliban knew their operating environment and they had information networks that actually worked. America's ability to make sense of the information coming off their intelligence systems lagged far behind their abilities to collect the data. The most sophisticated US munitions still ended up underpowered compared to the Taliban's low-tech weapons and local knowledge.

It's not that no one had tried to address the problem. Every service had their own intelligence analytic software and processes to help them log and make sense of incoming information and target lists. They spent hundreds of millions of dollars on rival defense contractors to develop them. Billions, sometimes. Cukor was blunt about their usefulness. "They were crap," he said. "Just horrible."

Sometimes they were too complex for the middle of combat. Other times they simply stopped working, blinking back the blue screen of death. Sometimes they wouldn't turn on. One internal word processing software called "Enable" was just as quickly dubbed "Disable," because it would crash after three pages.

It was all Cukor had to store the data that might make or break a mission—and its troops. One time he lost everything on his hard drive and he wanted to go home. Spilt data was worse than spilt milk. "I just wanted to cry," he said.

Trying in vain to make sense of the terrain and threats, US troops went back to basics. Cukor and Spirk fixed a paper map up on the wall, marking up layers of acetate over the map by hand. It was a more efficient way to find, log, and warn of threats than relying on the computer.

It was a story that would be repeated elsewhere: America's munitions might have become precise, but the information that fed them was not. Pilots taped physical maps over their knees to help them navigate from the cockpit. Operators on missions cross-checked the locations of targets and landmarks as if they were arranged on a clockface, or a phone keypad, exchanging imaginary times and numbers in the fray of battle. Back at base, they'd put cutout circles up on the wall, physically segmented into pizza slices of time. A day. A month. They called them "pattern wheels." Operators could mark when attacks happened to try to discern their adversary's habits. Sometimes the only intelligence tools available would be printouts and highlighter pens. Even though it was useless, Cukor and Spirk never abandoned the server. They dragged it with them for the duration of their mission as US forces expanded their military footprint in the first months of the invasion—first to the Bagram Air Base in the northeast of the country, and then to the capital Kabul—for one reason only: they were told to.

Cukor at least found one reliable use for the computer server. "The box made the room warm," he deadpanned. The windows had all blown out in the fighting, and the air could cool quickly. "I always took that box because it was good as a heater."

THE WAY CUKOR THOUGHT ABOUT IT, he had a normal Marine Corps career until the Twin Towers came down. "All of a sudden we're in these wars," he told me in one of what became our weekly afternoon talks over the course of more than a year.

He grew up under the warm sun of 1980s Los Angeles, super skinny and super physical, with a group of young people who thought they might be incinerated in Reagan-era nuclear war at any moment.

The military had been a good place to get an education, and becoming "a soldier of the sea" felt at first a little like being in the Boy Scouts. By then it was 1993 and the Berlin Wall had come down, history had ended, and nuclear war was sliding into a cartoon threat of the past. They could enjoy themselves.

When Cukor got his first military assignment at nearby Twentynine Palms combat training base in Southern California, it meant running around the middle of the red-hot Mojave Desert for a couple of years. Fun, hard, absorbing, freeing. "For me, it was perfect," he said.

He found camaraderie in the Marine Corps. It was far smaller than the Army or Navy, and often insular too. The service had an expeditionary mind-set. It prided itself on its flexibility, opportunism, and aggressive speed. Marines went in first and took ground, then handed it over to the Army to hold. Marines had to be fast fighters.

Cukor asked to work in intelligence and was one of only two people he said made the grade. Intelligence to him was a small and somehow maligned function—unlike in the other services, Marine Corps intelligence officers rarely rise to the highest ranks. Since 2000, only eight Marine intel colonels have made it up to major general and four to lieutenant general. Cukor set up an intelligence section at the training base, and then ran up a $30,000 phone bill in 1993 because he was using dial-up internet to get unclassified news sources on his laptop. That act alone burned through the unit's intelligence budget allocation for the year.

When it fell to him to name areas of the unforgiving, sprawling, scrubby base for military exercises, he'd turn to his wife Kirsten for inspiration. All she had really ever wanted to do was to write and think, and he paid homage to her literary favorites. He named one training area Austen. Another Byron.

They met in the winter of 1993, the year after he joined the Marines, through Kirsten's housemate, a mutual acquaintance who invited Cukor over specifically to meet Kirsten. The housemate worried Kirsten would be annoyed at such a blatant setup. Cukor said to go ahead. "He just has this certainty," Kirsten says.

As she approached the living room she could hear Cukor appraising her books and wondering out loud what she saw in *Ulysses* by James Joyce. Kirsten always figured she was born without the gene that gives respect for authority. She went through the door ready to be sassy and argumentative. He was down on his knee in front of the shelves when she showed up.

He looked up at her and smiled. "Oh man, I'm done," Kirsten remembers, her twenty-one-year-old self instantly falling for the twenty-five-year-old, against her better judgment.

"That's her," he told himself.

When she discovered he worked in Marine Corps intelligence, she instantly quipped: "But isn't that . . . "

"... an oxymoron," he completed, just as quickly, delighting her that he even knew the word, never mind that he could laugh at himself.

She couldn't put him in a box. He was a Marine with a low opinion of the Marines. He liked *Phantom of the Opera* and *Les Mis*. He was wired for war but children trusted him. She was attracted by his tumble of contradictions. "There's no way there's another human on the planet like this."

She learned his favorite book was *Don Quixote*, the 1605 novel in which the tragicomic and misunderstood hero pursues a doomed quest for an idealized version of the world that does not exist, forever trying and failing to save the world and right wrongs. "He's such a Don Quixote-ish kind of person," she says. She would get paralyzed in theory, but he was all action. "He's willing enough to be wrong that he'll actually try to solve a problem."

He was due to deploy to Okinawa in a few months' time, but Kirsten figured if it only lasted two weeks between them it would still be the most interesting two weeks in her life. On the wall of her bedroom, she had a poster of her favorite movie, *The Philadelphia Story*. It included the name of the director, George Cukor. "Wouldn't it be funny if that was my last name," she remembers confiding to herself the evening they first met. They married within four months.

Three years and one child later, Cukor was writing his master's thesis on the need for "urgent reform" of the Marine Corps intelligence enterprise at the Naval Postgraduate School in Monterey, California. He was obsessed by what he saw as the failure of Marine ground intelligence and how ill-equipped it was for the emerging threats of the twenty-first century.

Drew Cukor had his own solution, but his vision relied on a way of seeing that did not yet exist. He was about to embark on a seemingly doomed quest.

2

TILTING AT WINDMILLS

"Someone forgot to involve intelligence."

DREW CUKOR'S 1997 THESIS faulted the Marine Corps for being "centralized, hierarchical and slow." He excoriated his service for barely changing a thing since World War II. The US might have won the 1991 Gulf War, he wrote, but Marine intelligence against the Iraqi Army "failed." The Americans ended up fighting blind, attacked many times on the ground, with no idea where the enemy was.

The Americans bludgeoned their way to victory with the help of overwhelming firepower, but with intelligence practices so inept that Cukor saw no sign they'd win against a bigger, tougher foe. The Iraqi Army's repeated reliance on surprise nighttime assaults—along with furtive retreats—outclassed the Americans on the front lines. Simple, predictable enemies were a feature of the past. "Military victory would need to come through military competence, not from sheer superiority of men and material as it had in previous conflicts," wrote a thirty-year-old Captain Cukor.

The Marines were meant to fight wars of maneuver. The service was designed to move forces swiftly into place, to hit an enemy's weak points with such ferocious speed and impact, often from multiple directions, that they simply gave up fighting. These tactical maneuvers are designed to avoid costly wars of attrition, but they come with the heightened risk you could end up shooting at your own troops.

In Desert Storm, however, Cukor argued that America's Marines had fought according to an outdated idea: "movement to contact." Cukor explained it to me like this: you have no idea where the enemy is, so you just keep moving and once you slam into the enemy you've found them and that's when you just start fighting. The "movement to contact" approach relies on—and risks—masses of manpower and machinery and tends to result in heavy casualties, something he argued Americans simply wouldn't stomach today or in the future. Instead of swift and tactical maneuvers, the Marines had been deployed as if they were a giant army. Things got messy and mistakes were common. They sometimes fired on their own people. In the Gulf War, more than half of all Marines killed (fourteen of twenty-four) died by so-called friendly fire, mostly from air-dropped munitions. Another ninety-two Marines were wounded.

Cukor identified problems with the movement of information. New intelligence flowed to command headquarters rather than the tactical commanders on the ground—but intelligence was useless, Cukor argued, if frontline troops didn't have it. Critics recorded other intelligence deficits: troops had small-scale maps showing so little detail that it was hard to plot out their advance. Sometimes they'd be rifling around for the right map, each one curled tightly around tubes jiggling around the back of a truck.

Cukor was railing against Marine Corps practices at a time the Corps itself had begun to acknowledge it needed to do better. In 1992, the Senate directed the Corps to fix its intelligence problems and the House started issuing reports. But a plan—named for Marine Corps Lieutenant General Paul Van Riper—to improve the role of intelligence had yet to be implemented by the time Cukor was writing in 1997. Cukor thought that even if the Van Riper Plan were successful, it was attempting only "incremental, peripheral" change mostly focused on training and access to promotion.

Chinese general and military strategist Sun Tzu posed a fundamental question to aspiring military leaders more than 2,500 years ago: Where is the enemy? Cukor had watched the Marine Corps fail to answer this fundamental, basic, life-or-death question over and again. He wanted something "vastly different" from the status quo. He wanted wholesale change—to turn data into knowledge and action. He wanted to know the precise location of allied and enemy forces on a map, and for the map to have meaning. And he wanted "a collaborative partnership" with industry and academia to achieve it, combining human and machine, thinking and networks.

"White dots fused with an understanding of the enemy and validated by other collection assets can contribute to a highly accurate battlespace picture," he wrote in 1997.

Making real those white dots would become the drumbeat of his life.

Writing in 1997, he sketched out threats that made this effort so urgent. They couldn't wait fifteen years to go from starting something to delivering it. He argued that surprise would be central to any Chinese effort to take over Taiwan, describing missiles as easy to conceal, difficult to track, and "perfect weapons of surprise." Taiwanese forces could be overwhelmed and annihilated by a sucker-punch missile attack and shock troops. "Within 24 hours, Taiwan could be in PLA [People's Liberation Army] hands," he warned. "Caught by surprise, America would not have time to put adequate forces in theater."

Cukor didn't think of himself as a prophet. "These were things that were just common analysis. What eventually becomes a strategy and a pivot to China was so obvious—it just took forever to pivot the machine," he told me.

The 1990s saw a spike in US-China tensions, much of it over Taiwan, provoking diplomatic and military crisis. In April 2001, US and Chinese planes collided midair over the South China Sea. The US pilot conducted an emergency landing on Hainan, a Chinese island, and the Chinese took the sensitive spy plane apart. Things would have gotten worse between the US and China, Cukor says. It's just that 9/11 got in the way.

Pivoting the machine didn't just mean a new direction in US foreign policy. To Cukor it meant the Pentagon. The forces might focus on battle, but the bureaucrats focus on budgets. There were always three budgets underway—the one being spent, the one going through Congress, and the one coming up next. Budgets were enormous, easily outstripping every other defense budget in the world put together, but somehow it was never enough to run America's war machine. Things could grind slowly in the world's biggest office building. Each year each component resubmitted a budget for the next five years of inflexible spending. Congress tended to strictly allocate cash in five separate categories, known as its "color," spanning procurement, research, operation, personnel, and construction, each with its own use-by date.

The result was a confusing morass that rewarded the savvy over the sensible. Services, such as the Army and Navy, equip their own forces. Some of the Pentagon's budget was classified, with military intelligence funded through a different stream. It was so complicated there was an entire university devoted to defense acquisitions, and so lucrative that companies and Congress alike jostled over the distribution of jobs in constituencies that came with big contracts. All this, and the Pentagon had never passed an audit.

Information technology fell between multiple stools. The Pentagon had spent little time thinking about how to buy bytes rather than bullets. The pace of software innovation and updates ran too fast for the budget cycle, and paying for repeatable annual licenses rankled some.

Cukor was determined to reorientate budget processes to win future wars. That would mean software to blend intelligence into operations, folding in analysis, wits, and guile to the deadly delivery of might. A new holy grail of battle started to emerge: a single digital grid to combine intelligence, communications, and operations. The idea was to detect absolutely everything on land and sea, air and space with ubiquitous sensors, and then figure out what it all meant in one comprehensive screen, all in real time. An aspirational single pane of glass to clear up the fog of war—and, one day, a brain to fight the war too. The American war machine wanted omniscience, omnipresence, and omnipotence.

"The human element . . . slows down processing time," Cukor complained in 1997. A vast network of sensors could ultimately connect directly to the weapons themselves. This would be sensor-to-shooter warfare, Cukor wrote, and it would reduce the human element to "minimal."

At the beginning of 2001, Captain Cukor was ensconced at Camp Lejeune in North Carolina with a new job: intelligence officer of the 26th Marine Expeditionary Unit, II Marine Expeditionary Force. Kirsten was pregnant again, and her husband would be heading out for a mission two weeks after their child's birth. He was due to go float around the Mediterranean. On September 11, 2001, their third child was born. And Cukor's mission changed instantly.

THE WAY CUKOR SAW IT, the US started out winning in Afghanistan but then lacked the wherewithal and intellectual depth to understand that they had started losing. Then they invaded Iraq, too. When he was sent to Iraq as part of the spring 2003 US invasion force with a Joint Special Operations Task Force, it was with a different unit but the same tech. Still PowerPoint and Excel. Nothing improved. They powered through five hundred kilometers and captured Baghdad within two weeks, but had no insight into the city once they got there. They knew even less about Tikrit, which they were commanded to seize next with no maps and an impossible task: locate the nonexistent weapons of mass destruction. No one predicted the level of civic instability, even though it was entirely predictable. In the first two months alone, fifty-six Marines were killed.

The US turned to big conventional forces with "very primitive intelligence tools" and fell hostage to "American bias, and hubris maybe," Cukor told me. Nearly 4,600 American troops would be killed. Upward of 186,000 Iraqi civilians died. Washington, Cukor said, put too little mentorship into supporting an inclusive government in Iraq. For Cukor, America's war in Iraq was a blur of continuous mediocrity and frustration.

By the time Major Cukor went back to Anbar province in Iraq in April 2005, 456 Marines had been killed in the Iraq War. Cukor was there as a fusion officer for the II Marine Expeditionary Force, a job that was intended to direct and deliver data and clarity to the epicenter of the war. But everything was going wrong. The unit kept missing vital—and sometimes fatal—information. There were dozens of daily skirmishes. The deaths kept coming. During his tour, 189 Marines died. That November, the Marines would be accused of the worst massacre of civilians in their history. Marines allegedly shot dead twenty-four unarmed Iraqi civilians in three homes and near where a roadside bomb killed a Marine in Haditha, a town northwest of Baghdad. One three-year-old girl was later photographed lifeless and bloody on the floor, the number 12 drawn on her cheek in red pen, her hand near the outstretched arm of her dead five-year-old sister. Cukor was on watch duty that day and saw the reports come in. He says the Marines didn't bury it and studied what happened. (This is disputed.)

"Then it's 2006 and 7 and 8 and 9, and it's just a mess."

The insurgency "was not going super well" and Cukor felt sick about rudimentary tools. Intelligence felt like an afterthought even when it came to supporting Iraq and Afghanistan's own security services. The US helped train Iraq's security services in logistics, artillery, infantry, aviation, and special forces. "Someone forgot to involve intelligence," he said. "I was like, 'What the heck is going on?'"

He developed a program in Iraq and Afghanistan to answer this deficiency, named Project Legacy, that drew on civilian policing methods. "Drew is one of the most innovative thinkers the Marine Corps has had in twenty years," Matt McKnight, a young Marine Corps officer in Anbar at the same time, told me. Cukor was known, McKnight said, for maneuvering inside a bureaucracy allergic to change. Project Legacy would later become the subject of public controversy over its methods and funding, with multiple investigations into alleged bad behavior by contractors and overcharging. By then Cukor had transferred to the Pentagon. That meant more bureaucracy.

"I probably lost ten years of my life in stress and pain," he said.

And then came November 9, 2010.

The Marines had taken over Sangin district in southern Afghanistan only two months earlier. They patrolled the southern provinces so extensively these became known "affectionately" as Marineistan. British troops there before them suffered heavy losses in the district: a third of

Britain's war dead. I visited Helmand province the year before: tracing the perimeter of a forward operating base without an explosion going off counted for success. You couldn't leave the shadow of its walls. More than a thousand US combatants had been killed so far in the war. Now, two or three Marines were dying every week.

That November day, a twenty-nine-year-old Marine Corps platoon commander was on foot patrol through the Sangin Valley in Afghanistan's Helmand province. A makeshift landmine exploded. He was killed.

This one cut especially deep. First Lieutenant Robert Michael Kelly was the son of John Kelly, a three-star general who commanded Marine forces in the US. Robert Kelly enlisted as a Marine, and later commissioned as an officer and made his way to platoon command. In a public note, the highest-ranking American military officer to lose a child in Afghanistan said his heartbroken, proud family called on well-wishers to redirect their prayers to the rest of Robert's platoon in Sangin. They weren't victims, Lieutenant General Kelly said, over and again. They were doing what they wanted, what they believed in. By the time the tour ended a few months later, Robert Kelly would be one of twenty-five Marines from the same battalion who died. More than a hundred had limbs amputated. It was happening again and again, same places, different people.

Cukor knew Lieutenant General John Kelly directly. He worked for him just months earlier, when Kelly was commander of the Marine Reserves. He had worked under Kelly before in Iraq too, when he fielded Project Legacy. "I have all of this anger in me," Cukor said. "I had gotten pretty close to him."

Cukor had been fighting these endless wars since 2001. He was tired of not being able to connect the dots. Of new arrivals having to relearn the enemy every six months. Of no one remembering who lived where, or who took what road when. Of unnecessary death. Of death.

Lieutenant Colonel Drew Cukor was unbelievably pissed off.

3

WE DO WHAT WE WANT

"My plan is nothing short of deploying Palantir to the entire Camp Leatherneck."

AMERICA BROUGHT $8 TRILLION and dominant firepower to the two-decade fight it called the "global war on terror." But the US military toiled, largely unsuccessfully, to find an answer to the cheap improvised explosive devices (IEDs) that killed and maimed its troops. IEDs accounted for two-thirds of all American casualties. The Marines used thousands of heavy and expensive mine-resistant ambush protected (MRAP) vehicles to protect their troops, the result of a painful, delayed, and ongoing fight at the Pentagon and in Congress over production, design, and money. But the MRAPs came with drawbacks. They insulated the Marines from the local population and kept them on roads. There still weren't enough. And they wouldn't save anyone on foot. What Marines really needed, Lieutenant Colonel Cukor figured, was information.

When First Lieutenant Robert Michael Kelly was killed by an IED in November 2010, Cukor was a handful of months into his new job at the Pentagon. He had a useful title—analysis and futures chief at the Intelligence Department of Marine Corps Headquarters—a good boss, and the task he wanted: to fix intelligence.

He thought he had the perfect solution: a data management platform made by a company named Palantir Technologies Inc., a Silicon Valley startup founded a few years earlier that was already working extensively with the US intelligence community. Cukor wanted to field the software to the Marine Corps base in Helmand to help pinpoint the next roadside bomb: "My plan is nothing short of deploying Palantir to the entire Camp Leatherneck," he told me.

Cukor's faith in Palantir went back to spring 2009, when a junior Marine Corps officer named Joe Larson wrote to the Silicon Valley startup to introduce the company to Cukor's work.

Captain Larson, an intelligence officer completing his law degree at Stanford during reserve duty, interrupted his studies to deploy on active duty to Iraq in 2008. He was bowled over by Cukor when he met him at the Camp Fallujah intelligence center: he thought him "brilliant." Cukor would later come to regard Larson as "my right arm."

And now Larson was bowled over by Palantir too. His job had been to track down high-value al-Qaeda targets. But he, like Cukor, despaired of "the failure" of defense intelligence systems to do modern, complex data analytic work. "My analysts spent many hours in Iraq doing a 'Ctrl-F' search in a windows folder to find relevant documents for building their network analysis, and that just isn't the way to do business," Larson emailed Cukor. He had previously discussed Palantir with him at a Marine intelligence conference. Marines "may need to look" at this new company of Stanford computer engineers to tackle the Afghanistan intelligence problem, he continued.

And so began Larson's key role in bringing Palantir to Cukor and Cukor to Palantir. Larson and his wife, lawyer Ann Marie Rosas (who is now general counsel at Anduril Industries), would three years later both go on to work for Palantir. The upstart startup would develop its own insurgent take on the long-standing "revolving door" that defines America's defense-industrial complex.

Larson also sent Palantir a paper penned by Cukor. In it, Cukor explored the best way to create and analyze a terrorist network database for connections and clues. Raking over America's failure to thwart the 9/11 bombers' plot, the 9/11 Commission Report had highlighted the need to do exactly this. Inside disconnected US systems had been clue-fragments of multiple links among a few men training in American and European flight schools since 2000 and a Saudi construction millionaire who long dreamed of felling American skyscrapers the same way Lebanon's towers once burned.

In his January 2008 paper, Cukor and his co-authors tested out how to digest and learn from 183 public texts about al-Qaeda. One approach had human intelligence analysts simply read them. The other tried an emerging branch of software named network text analysis, which sought to extract meaning by finding linked words among masses of text.

Cukor's study didn't come down on the side of the machine alone—the software missed things humans picked up. But humans also missed things the machine picked up. And the machine took far less time and fewer people. The authors argued for both: "an integrated human-centered automation support approach."

The network text analysis software they assessed was developed under Cukor's co-author Kathleen Carley, a computational social scientist at Carnegie Mellon University who was already working with the CIA to automate intelligence analysis. In combination with other software she developed, her network text analysis program would hunt individuals at the center of any web and predict the impact that arresting—or killing—that person might have. She modeled how soon the network might reconstitute itself without them, and through whom.

For some, amalgamating and analyzing data en masse like this presaged a terrifying future. Automation machines could propose unwitting individuals for death in war and help build an overpowering domestic surveillance state. For others it would be a smart way to keep America safe.

Cukor's and Palantir's efforts overlapped, and Palantir began to consider whether it might find a lucrative route into the Marines through Cukor.

Jason Payne, a Palantir employee, knew of Larson through a friend. When they met for lunch and a demo, they discussed Cukor's work, approach, and role. In an internal company memo following their lunch, Payne flagged Cukor to the rest of the company.

"LTC Cukor is trying to shake up Marine intel," wrote Payne. "Joe thinks he could be a conduit to get us to the right folks."

Palantir prided itself on spending time and energy stitching together an influence network to drum up customers and contracts. "We have no salespeople working for us at Palantir," co-founder Alex Karp crowed in a 2009 television interview. But the company was no less aggressive for lack of a sales staff.

"We were trying everything we could as a small scrappy startup to get in front of the actual operational users," a former Palantir employee who worked on US military outreach told me. "Not procurement teams, not bureaucracy; but people who actually had skin in the game."

Cukor had done his homework on Palantir. It was founded in 2003 after PayPal investor Peter Thiel wondered if the emerging solutions that helped his financial payments company combat online fraud could be applied to broader efforts to rifle through sensitive, jumbled government and enterprise data.

Thiel decided it could: he named his new company after the "seeing stones" in *The Lord of the Rings*. The office vibe mimicked Silicon Valley—bean bags, video games, free meals—but catered to a distinctly different clientele. In its first few years, investors worried Palantir had no retail product and repeatedly turned down funding requests from Thiel and his former Stanford buddy Alex Karp. But then they got a tip: Go see the CIA.

By 2005, In-Q-Tel—the CIA's independent tech investment arm established in 1999—invested $2 million into Palantir. In-Q-Tel's savvy intelligence professionals provided other value: they introduced companies on their roster to the CIA as a customer who might influence, shape, buy, and use their software. Susan Gordon, former principal deputy director of national intelligence who had helped start In-Q-Tel when she was at the CIA, told me she thought their Palantir investment was one of the two best they ever made. (The other was in the company that came to be Google Earth.)

The CIA and the military meant new territory. Akash "Aki" Jain, one of Palantir's first five engineers, had never worked in defense or intelligence. He spent the first five years at the company figuring out the product and things as basic as a facilities security clearance to learn the bureaucratic "secrets" of working with the national security community. The company developed a particular selling point: it liked to spend as much time as possible with the people who would use their systems. That came with three main benefits: the company could shape a bespoke platform around the customer's needs, fix things as they cropped up—and mold a captive audience. Years later, Jain would try to explain Palantir's way of sorting data to me. "It's like two thousand Lego pieces in six bags with no manual: How do we create the Death Star?" he said, referring to the planet-destroying superweapon in *Star Wars*, before correcting himself. "Or something more positive: Falcon 9."

By the time of Larson's outreach, Palantir was deeply woven into America's counterterrorism spy wars. Its platform could enable analysts and targeters "to get ahead of the terrorist, to see patterns the terrorist doesn't realize they're giving off," Karp claimed in the 2009 interview.

He taunted competitors and seduced users directly: "Compare us to what you have."

The Marine Corps had almost nothing at all.

Palantir's hard sell to Larson worked. He wasn't just impressed by the company's software, he was "blown away," Payne reported up his chain in the company memo I saw.

For Larson, the software couldn't come too soon. He was due back on active duty in the summer of 2009 to train more than one hundred new Marine intel analysts and was set to deploy with an intelligence battalion back in Afghanistan in 2010. As units turned over, old intelligence reports were often lost or siloed, with data locked up in static documents no one looked at again. That was one reason behind the bitter line that the US fought the Afghanistan war twenty times over, for each time a new unit came in. After seeing Palantir, Larson told Payne he was "depressed" at the thought of having to teach all the existing substandard tools for making links between data. One of those was Analyst's Notebook, made by i2, an intelligence analysis software company that would be bought the following year by IBM. "He_hates_ANB," Payne emailed his colleagues, with such emphasis it surely revealed a little delight. Users complained ANB was susceptible to confirmation bias and couldn't call up profile metadata with ease. Errors could be treated as facts, with no way to check why someone had drawn a certain conclusion.

Palantir tagged not only the data itself but every single link between the data too. The system recorded an audit trail noting which analyst from which agency linked a person to a phone or place. It logged the report the insight was based on. It kept track of relationship changes. Every conclusion or hunch could be inspected, audited, interrogated. The system could investigate if conclusions resulted from circular reporting that led back to a single potentially erroneous data point. It could ingest multiple data sources. Anyone with the requisite security clearance could make and log changes across multiple agencies and locations. McKnight, the Marine intelligence officer who'd struggled with rudimentary tools in Iraq and who advised Palantir for a few years, described the product as a really large-scale indexing capability that allowed intelligence operatives to tie together disparate databases in the same secure environment.

Cukor flew over to the West Coast and visited the company's Palo Alto headquarters the next year. He liked how it could make sense of jumbled data: "I knew it was special."

The system impressed him because it could handle difficult things that looked simple. "My name is everywhere and misspelled constantly," Cukor tried to explain to me. Basic systems would struggle with spelling and data entry errors in nouns like Dru or Dew or Dude. But by baking rules into the system, Palantir might recognize Dru Cooker and Drew Cukor as potentially the same person, especially if other links connected them.

Working with the Palantir system, intelligence analysts examining phone call transcriptions, interrogation reports, captured documents, notebooks, memories, and mistakes might start to realize different elements of information might, if combined, lead to a single person. Perhaps only three sentences in a ten-page intelligence report—or interrogation transcript—might be relevant for a specific query. The system might find that the same person was popping up in Mosul, Kabul, Riyadh.

The approach was also highly controversial. The company was in bed with US intelligence agencies at a time when American security personnel were pursuing targets all over the world, detaining people at black sites, gathering so-called confessions of unclear value and unclear veracity, and feeding information sometimes extracted under conditions of torture into software systems. What if Dru or Dew or Dude were Mohamed or Mohammad or Muhammad? It might be possible to conflate the world's most popular name inaccurately with the wrong person. Civil rights activists frequently accused US airlines of barring people from flights on the sparsest of information, sometimes just their name alone. "Flying while Muslim" became a sardonic reference to the pursuit of dragnet security based on stereotypes instead of evidence of any real risk.

The immense volume of data that systems such as Palantir's could sweep up, hold, and analyze also threatened to expose much of what there was to know about a person. As the FBI and homeland security agencies developed their use of Palantir, some Americans feared such systems would subject their civil rights and personal lives to tools developed for foreign wars, long before aggrieved NSA contractor Edward Snowden would spill the beans on extensive domestic data collection by his spy agency in 2013.

It wasn't only civil rights activists who worried about Palantir. Detractors in the Army would later argue the official program, DCGS-A (Distributed Common Ground System—Army), was capable of doing much more than Palantir could. DCGS-A did the hard work of collecting the data, one senior Army acquisition official would say in 2013; Palantir just displayed it nicely. DCGS-A was like the whole iPhone while Palantir was just a single app, the official pleaded to a *Politico* reporter, saying the Army was "somehow" not getting the message out fully. (Cukor reckoned DCGS-A was horrible. Sometimes it wouldn't even open. You couldn't hand data over to a green incoming team.)

In 2010, the company that produced the analytic tool that so riled Larson accused Palantir of industrial espionage on an epic scale, via a "years-long scheme of deception." Shyam Sankar, who joined Palantir in March 2006, illegally accessed and copied parts of i2's software, a lawsuit alleged. He allegedly misrepresented who he was, corresponding with i2 through personal and university email accounts and a company registered to his parents in Florida. He attended i2's customer video conferences under a fake company name, Xoom, the lawsuit went on. Palantir copied screen icons representing people, places, weapons, and planes, and used its unlawful access to learn how to export and ingest i2 data into its own system, the lawsuit alleged, developing a technique that could help new customers switch. (Palantir said at the time that i2 had "unclean hands," arguing its rival used information about Palantir's products obtained through a company acquired by i2.) The pair would settle out of court in 2011—reportedly with Palantir paying i2 $10 million—and say no more about it.

Cukor, who read this lawsuit and was familiar with other informal complaints against the company, described Palantir's early sales practices to me as "absolutely pernicious." He said the Pentagon was not used to such aggressive overtures from startups. But after his visit to the company he decided he was desperate for their tech. He faced obstacles. Palantir's platform was neither accredited nor was it a "program of record," the holy grail of defense contracting that meant Congress had bestowed approval and funding for it, without which you couldn't buy or supply the platform to US troops.

Change could take a long time. Cukor's boss Lieutenant General Vince Stewart once related a story to Peter Dixon, a Marine captain who had just started working for Cukor. When Stewart was a young officer, he put a request in for a new project. The request took so long to work its way through the system that it eventually came across his own desk when he was a much older one-star general. Rather than sign off on his own request, Stewart canceled it. By then, it was obsolete.

All members of militaries run into bureaucracy at some point, but not all of them make it their life's mission to reform one. Cukor wasn't one for waiting. The US was nearly a year into a 33,000-strong troop surge ordered by President Barack Obama and registering heavy casualties. Cukor's efforts to shake up the system would be tested: more Pentagon employees work in acquisition and procurement than in the entire Marine

Corps. The Marine Corps is the smallest of the US military services, and its intelligence component was smaller still, and humble. "A basket case," Cukor put it to me with his usual tact. Cukor helped orchestrate a formal request from the top US military intelligence officer in Afghanistan, Army Lieutenant General Michael Flynn, designed to bring in Palantir. Flynn had excoriated the state of intelligence at the start of 2010, arguing it was only "marginally relevant," obsessed with being secret rather than effective, and shared too little to be useful. "The real intelligence hero is Sherlock Holmes, not James Bond," he and his co-authors quoted in approval. Then in July, Flynn put in an emergency request for new analytical tools known as a JUONs (Joint Urgent Operational Needs).

"It's the most powerful document a commander can launch," Cukor told me, explaining it leapfrogs the usual procurement hierarchy. "The Pentagon is required to pay attention to it."

Flynn's emergency request specified a solution that, without naming any company, described Palantir software so closely that senior Army officials would later say it was "clearly ghost-written by a Palantir engineer." In fact, I learned it was ghostwritten by Cukor and some of his team.

Dixon, the Marine intelligence captain working under Cukor, had worked with In-Q-Tel and the State Department, and was now working for Cukor on Project Navigator, an effort to understand the potential threat from future tech, like drone IEDs. But that was the future. The Marines were losing three people a week right now. Dixon knew what that meant. His unit had experienced terrible losses in Afghanistan during his tour: "Without battle-tracking software we were knocking on compound doors not knowing if the previous group of Marines had been met with gunfire or tea," he recalled to me.

Dixon visited Google and Palantir and then he and Cukor drafted the parts of the emergency request that detailed requirements for new intelligence processes and analytical tools. Cukor had an uncanny way of convincing people that Palantir's software was valuable, a former Palantir employee recalled. "We jammed that thing through," said Cukor, who still remembers typing in the draft.

The Army point-blank rejected the appeal. The urgent need didn't apply to them, they argued: they were already spending $550 million a year on DCGS-A, the $1.6 billion predictive intelligence-gathering network. Cukor decided that if the Army wasn't going to act, the Marines would. He was fed up, he told me, with the people in charge and the decisions they made. He started making some of his own. Cukor figured he didn't have much of a career ahead of him in the Pentagon to squander. "I probably am a little too audacious," he offered, as if advancing a subtle or somehow debatable point. The thoughts in his head nevertheless ran big: he half-wondered if he could be fired, or sent to jail.

"We were all fueled by the knowledge that we would be saving lives on the battlefield," said Dixon. "But none more than Drew Cukor, he was relentless."

Cukor set about hunting for money. The Pentagon's budget is massive: $691 billion in 2010. But Congress pre-allocates every dollar, and finding spare money—even in wartime—can be the stuff of bureaucratic impossibility.

Cukor by then "knew all the tricks" of acquisition and caught wind of a plus-up. That was Pentagon-speak for extra money granted by Congress, thanks in part, he said, to Palantir lobbying. The additional money was allocated to a research and development lab for irregular warfare called the Combating Terrorism Technical Support Office (CTTSO).

The curious office spent its millions on everything from sniper comms to armor for dogs. Cukor had previously turned to it to fund Project Legacy; he knew the team well. Congress often sent a little extra money its way, but in 2010 the plus-up was huge: $36 million extra on top of a baseline $82 million.

But CTTSO couldn't get anyone to touch it. Despite Flynn's appeal, the Army turned down the office's extra money. Cukor didn't hesitate: "I grabbed that money because CTTSO didn't know where to put it." In the final budget week for that year, the office reprogrammed $10 million "in view of the urgent needs assessment."

Now Cukor just needed a contract. Under conditions attached to the money, he could do sole-source expedited funding. ("This boring stuff shapes my life," he said. "Money, contract, statements of work—they are three legs of the stool.") So when the Army Research Laboratory put out a statement of work on Cukor's behalf, it was turned into a public solicitation that only one company could meet: Palantir.

Next Cukor wanted the new system out with operators in the field. If his boss, General Stewart, had known his plan, Cukor figured he would have said no. Stewart would have told him to deploy only what the acquisition commands ordered. But the acquisition commands weren't ordering him to deploy anything. Lieutenant General Kelly's son's death in November motivated Cukor to act. "He had no idea what I was doing," Cukor told me. (It was not possible to check this; General Stewart died in 2023.)

Cukor got clearances for two Palantir engineers and they took their $20 million servers into Afghanistan on a C-17. The team hauled themselves and their equipment to Camp Leatherneck at the end of February 2011. But no one wanted to connect the IT systems and there wasn't enough power. Palantir didn't wait for a solution or an invoice: the company sent a $100,000 generator.

When still nothing happened, Palantir appealed directly to Major General John Toolan, commander of the 2nd Marine Division, whose troops had recently deployed to Afghanistan. Toolan had earlier taken a meeting with Palantir, astounding his aides. He came to the rescue again when no one would turn the system on, curtly reminding his team to use Palantir. On March 21, 2011, the system went live.

"This thing lands in Afghanistan and it is immediately a hit," Cukor triumphed. Within an hour, fifteen people were using the software at Delaram, a forward operating base in Nimruz province. Training started apace. Within eight weeks, it was up to 250 users among thirty different subgroups. The Marine Aircraft Wing used it to plan air routes and assess threats to helicopter landing zones. The biometrics team built forensic IED networks with fingerprints from people the US interrogated. Biometrics captured on phones and laptops linked to information from other reports. Even the meteorologic operations cell started analyzing whether a certain kind of weather or phase of the moon coincided with conditions picked by bomb-makers.

US Army soldiers started flying into Camp Leatherneck to see for themselves. That created more pressure at Army headquarters to roll out the same. Even the Brits came over, and wanted to know what the hell the Marines were using.

That's about when General Stewart got wind of what Cukor had done. Stewart started taking heat for deploying an unapproved system. One time an Army colonel showed up, furious, at Cukor's office when he wasn't there, saying the Marines had gone against the program of record and they needed to fall back in line. "I didn't understand this world at all," said Dixon. DCGS-A wasn't even due to arrive with the Marines for three years. "I had just helped save more lives putting Palantir into Afghanistan than I had leading infantry outside the wire. Seeing a colonel come and not talk in terms of supporting the men and women but supporting a defense company was so shocking to me."

It was the beginning of an intense hatred among some army personnel for Palantir: many worried that the company was expensive and arrogant, and that it trapped their data. Cukor felt some heat himself. Stewart was mad but he "wasn't super mad," Cukor recalled. He figured his commander liked that the Marines were out in front. "This is the Marine Corps; we do what we want," Cukor told me, suddenly sounding kind of giddy. "All of a sudden it's cool to be in the Marine Corps that's using Palantir. The politics around this was fundamentally the Marines had the balls."

On a small scale, Cukor had scored his first win. He began to experiment with bolder ideas to turn around Marine Corps Intelligence. What if they fed more than intelligence information into the data management system? What if they plugged in the results of military operations? What if he poured in new data every time US forces did a raid? Cukor liked to think of information "pulsing" its way to the front—constant drip-feed updates of the latest data into real-time calculations.

Once they started feeding in operational data from Afghanistan, "people's minds blew up," Cukor recalled. It turned out US forces had raided one place twenty times already. "We had no idea. OMG we look like idiots."

Palantir kept up its pressure campaign. In August 2011, a Palantir worker sleuthed out that Jim Mattis, the four-star Marine Corps general in charge of Central Command, was going to visit the Marines' Memorial Club in downtown San Francisco. The Palantir employee showed up and shoved a letter in his hand advocating for the company's software. "I had one second to get something to him," the worker recalled.

It would take time—and acrimony—for Palantir to embed themselves. First came an explosive 2013 session in Congress, in which a congressman accused a four-star Army general of blocking Palantir to the detriment of US forces. Then came an astounding 2016 lawsuit in which Palantir successfully sued the Army in order to be able to sell to it.

"Palantir gets a black eye but wins and wins ugly," is how Cukor described it. "I'm just sitting smiling; my name isn't on any of this stuff."

But he would still hanker after more wins of his own. Palantir couldn't solve the entirety of what he called the analytic problem: So a battlefield commander or a general at headquarters could answer big questions—Where is the enemy? What on earth is going on? What was the point of the battle?

"We always see a past that was predictable but not predicted," went one sorry mantra that informed a Cukor effort to shape a new approach to global geostrategic intelligence in 2011. McKnight, the Marine Corps intelligence reservist, said Cukor led the Marine Corps in thinking about future intelligence concepts. But to Cukor, it was all small wins so far. He wanted a big one.

"A nation not prepared to wage existential war because it is rare faces catastrophe," concluded the draft geostrategic analysis he commissioned about future threats the Marines would need to address.

New tools were still emerging. Cukor would argue in a 2014 essay that the modernization of military intelligence analysis "will largely determine the success of US forces on the battlefield." But one new tool in particular was coming. Cukor was taking his first steps toward a new type of warfare, one that the movies—and academics, diplomats, politicians, military historians, and generals—had long worried would imperil the world.

"It's now the age of AI," he recounted to me. "And I'm asking where the fuck is the department's AI?"

4

THEY CALL IT ALGORITHMIC WARFARE

"I know this is now going to be a knife fight."

WHEN DREW CUKOR GOT WIND of an informal but influential group called the Breakfast Club, he immediately wanted to join.

It was mid-2016, and he had just got back to the Pentagon, working in the Office of the Under Secretary of Defense for Intelligence, but still despaired that intelligence wasn't viewed as mission critical. "Like a flower in the room," he'd tell me. US combat personnel would ask for some intel the same way they'd ask for the weather report, and then to Cukor it would be: "Now go sit down and shut up while we do some business here."

The Breakfast Club was a twenty-strong group that had met every two weeks since late 2014 in the conference room adjoining the deputy defense secretary's office. No breakfast (or coffee) was served. The initiative was named for a previous Pentagon working group which during the Cold War had strategized ways to counter the Soviet Navy. Its reincarnation forty years later devoted itself to the problem of beating the United States' newly looming foe: China.

"It was trying to bring back that magic of getting smart people around the table to focus on specific operational challenges," recalled Greg Grant, who convened the club. "It was almost like the deputy's brain trust."

Cukor got himself in through a typically entrepreneurial "self-invite." He got permission from his new boss, a kindly and thoughtful three-star Air Force general named Jack Shanahan who'd flown more than 2,800 hours in reconnaissance aircraft, bombers, and fighter jets and who now worked as a director for defense intelligence at the Office of the Under Secretary of Defense for Intelligence.

The meetings were meant to scout out new tech, vet potential use cases, and figure out how to overcome the latest military threats from China. Such threats, in the group's view, were aimed deliberately at US vulnerabilities. They spanned Beijing's development of hypersonics and space weapons to stealthy aircraft like the new J-20 and anti-ship ballistic missiles better known as "carrier killers." These were designed, the US feared, to destroy America's ten-strong active fleet of aircraft carriers, which for decades had projected US power into the South China Sea and the Taiwan Strait. Instead of floating weapons launchpads, these $13 billion ships might now be sitting ducks.

Departmental planning meetings often languished in anonymous offices and off-site buildings, but the Breakfast Club meetings took place just down the hall from the secretary of defense. They surfaced new ideas, thought about how to pull them off, and set the agenda for quarterly discussions held by top US defense and intelligence leaders focused on the same issue. Cukor loved having the chance to think big.

"It's cool, this is a big deal," he remembers. "We're just chin-stroking."

The chin-stroking came thanks to Robert Work, the deputy defense secretary since 2014, who would often poke his head in the room or join them at the table. While the defense secretary is the face of America's military engagement with the rest of the world, the deputy secretary role is often the most consequential in the building—responsible for running the Pentagon and its future plans. Work was constantly wanting to push the group in the direction of artificial intelligence and autonomy, Grant recalls.

"The US has never tried to match any adversary tank for tank, airplane for airplane, ship for ship," Work explained to me. "It's always looked for military technical superiority."

He worried that China had started closing the technological gap with the US. Work was a former Marine Corps artillery officer who had his eyes fixed toward World War III. He grew up during the Cold War and was driven by fears that, while America had been fighting wars in Iraq and Afghanistan, China had spent fifteen years learning the US military's weak points. America's idea of deterring war rested on acquiring and maintaining global "military dominance." Size was never going to be enough—by that measure China and Russia would always win. It needed something else.

Retiring as a colonel in 2001, Work never fought the terrorism wars. After he left military service he was director of studies focused on future warfare at a Washington, DC, think tank that published papers with titles including "The Promise of Directed-Energy Weapons" and "Rethinking Armageddon."

For Work, the US needed to be first at inventing new weapons of war to stay ahead. The US was first to develop nuclear weapons, and the only nation to detonate them in war. Advanced nukes were America's answer to Soviet mass mobilization. The Soviet juggernaut put a million men on railroad security alone. But the US figured technology could substitute for numbers. Better weapons would compensate for smaller size, went the argument. A bomb delivered at arm's length was also far cheaper than sending in a large conventional force. The number of US nuclear warheads peaked at 31,255 in 1967.

But Russia soon caught up with the US, producing its own "tactical" nuclear warheads and pugnacious theories about how and when to use them. By 1978, Russia was producing more warheads than the US and indicated no intention of slowing down.

So the US leapfrogged the USSR again in the 1980s, this time pivoting to develop long-range guided missiles. These could be dropped by stealth planes to deliver overwhelming strike force right on target thanks to GPS satellites run by the US Air Force.

Work wanted the US to retain aggressive first-mover status for each new era of warfare, a "never-ending" effort to provide the US with a steady stream of next-generation weapons. To critics, that sounded like warmongering, but Work thought of it a different way.

"Commanders who don't go after it are putting their people at risk," he would say.

He worried the US was far too slow to react. He likened the US approach to planning, developing, and fielding new weapons as equivalent to serving up a meal of badly defrosted vegetables after they'd been trucked across the country. "Sometimes it's too old; doesn't taste good," he said in a speech.

He wanted the Pentagon to come up with new ways to make it harder for Beijing to catch up. His answer was autonomy, and AI was the way to get him there.

"AI was just a means to the end of autonomy," he told me.

To Work, autonomy meant two things. The first was making better and more relevant decisions faster with the help of AI and human-machine collaboration. That meant autonomous mission control and targeting, situational awareness, course of action development, cognitive electronic warfare, predictive maintenance, and logistics. He called this "autonomy at rest." The second was taking humans out of weapons platforms altogether to create a new force of swimming, flying, driving unmanned systems. This was autonomy on the move.

"I'm telling you right now, ten years from now if the first person through a breach isn't a fricking robot, shame on us," he said in public remarks in 2015. "We can do this."

Under Work's stewardship, the department began wringing its hands over autonomy, which was arriving all over the country except in the Pentagon. Work thought the Air Force, Army, and Navy services responsible for equipping and training its people were dragging their feet. He recalled that the armed services hadn't wanted new tech like GPS either, which underpinned the development of guided munitions. Back then, Deputy Defense Secretary William Perry overruled the services. Now Work thought it was his time to do the same. A Defense Department study published in June 2016, the same month Cukor returned to the Pentagon, warned the US military was far behind the commercial sector on autonomy. IBM had Watson, Google was pursuing self-driving cars, and, despite funding AI research dating back sixty years, the Pentagon had nothing much at all.

Recent leaps forward in AI called for "immediate action" on military autonomy, the report argued, saying it would reduce the number of US troops in harm's way, increase the quality and speed of their decisions at war, and enable entirely new missions. "Advances in AI are making it possible to cede to machines many tasks long regarded as impossible for machines to perform," concluded the report authors, who included experts at Amazon, Google, and IBM.

But the services didn't budge. One reason was money. While the Pentagon budget was huge in 2016, there was a dearth of money available for new spending. Many also balked at handing over to machines life-and-death decisions that carried the weight of the law. Bringing autonomy into command and control in military operations and combat was "one of the most contentious applications," the experts' report acknowledged. Far harder than delivering the technology, it said, would be getting anyone to trust it.

Many of America's top commanders had never heard of machine learning and had no idea what AI meant. And, besides lacking talent and willingness, the Pentagon had none of the practical ingredients on hand needed to make AI work at scale—cloud, computing power (known as "compute"), and machine-readable data. Almost nothing inside the gargantuan Department of Defense lived in the cloud. Even though the internet was a Pentagon creation, and the DoD spent $38 billion a year on IT, it was slow to catch on to the meaning of the cloud for war.

And although every weapon system, satellite, and computer produces data—the feedstock of AI—it was bundled away on old slide decks and hard drives in computers and cupboards that sat as good as discarded. No one was bringing together original raw data to share, sort, or understand it.

The scientific advisory group's final report argued the best way to get the US military colossus to adopt AI and autonomy would not be through diktat, but through experiments, pilot programs, and prototypes that had a chance of taking hold. So far, the Pentagon had invested only in "small niche projects in an uncoordinated fashion," according to a 2017 draft document I read. These had "little chance of meaningful transition to the warfighter." Work wanted to build confidence and case studies so US fighters themselves changed their minds. And he wanted it clear he intended AI for combat.

"I do not want it to just be about intelligence," he told me, describing the steer he gave the Breakfast Club. "I want to have some type of direct warfighting applications."

The club's convener, Greg Grant, had hit it off with Cukor from the moment he arrived, describing him as "super interesting" and "simpatico." One day Cukor showed up to the Breakfast Club with an idea.

Cukor didn't want to do a two-year study. He didn't want to do a prototype. He didn't want a test. He wanted to do "a thing." Something real. In 2014, he had written an essay, "Operate to Know," focused on "actively" hunting the enemy. Now he had a program, he announced, that would make a great demonstration. It would illustrate to the military services the importance of AI, and why they should partner with leading-edge AI companies, which didn't usually work in defense and national security.

"I was like, oh my God, Bob's going to be so excited," Grant recalled.

Cukor pitched Bob Work in his office with Grant alongside him: he wanted to use AI to help analyze drone data.

The Pentagon was drowning in a deluge of footage: America's drone wars had taken off in the belief that remote combat could save troops' lives. From flying fifty-four battlefield drone sorties in 2001, the number had reached close to eight thousand in 2010. By 2016, the US would be spending $3 billion flying drones. The US was flying seeing machines over lands far from America, in Afghanistan, Iraq, North Africa, East Africa. More than thirty countries had developed their own armed drone programs. US drones weren't just watching lives, they were taking them too. Mostly al-Qaeda, Taliban, and ISIS. But, sometimes, dozens of civilians at a time at weddings, hospitals, markets, roads, farms. Sometimes the US apologized, and sometimes the US covered it up, and sometimes the US didn't even check.

Drones might produce pictures and videos, but the US couldn't process or make sense of them. The intelligence, surveillance, and reconnaissance (ISR) data recorded in 2016 was so overwhelming it was equivalent to eighty full years of video footage. Frank Kendall, a senior Pentagon official for acquisition and technology, was appalled to learn during a visit to Baghdad that no one was even looking at it. And still the numbers of drones were growing. It was a war factory without a workforce.

Cukor's pitch to Bob Work suggested that AI could help do the work of processing, exploiting, and disseminating (PED) the information contained in drone feeds. That could take the load off humans who were tackling only a fraction of the work. In Mosul, Cukor said, the US had five airmen staring at intelligence feeds for twenty-four hours a day. The incredibly laborious process wasted their brainpower, morale, and time. "Why can't we have algorithms do this work?" Cukor asked.

In years to come, I would hear those in Cukor's orbit say he could sell salt to slugs. His pitch certainly convinced Work. "He was super psyched," Grant remembers. Work dispatched him to dummy up a demo.

Cukor started visiting Silicon Valley companies and the world of emerging tech. Here lay the best talent in artificial intelligence, rather than with big traditional defense companies such as Boeing, Booz Allen, Lockheed Martin, and Raytheon that were known as "prime" contractors.

Much of the thinking behind tomorrow's AI wars would be inspired by Cukor's visits to autonomous car companies: Tesla, Waymo, Uber. At Uber's plant in Pittsburgh, he found a dozen cameras mounted on a car could process everything around them and help the vehicle (mostly) avoid cars, people, and other obstacles. He saw AI plastered over billboards in Silicon Valley. He read a mountain of AI papers. One startup named Clarifai offered a public portal to run images through AI on the open internet. AI would describe the picture by coughing up contextual words. The team ran a picture of Bob Work himself through the portal. "Military," the portal shot back. "Leader."

Now the Pentagon just needed algorithms of its own. Cukor turned to Dave Spirk, the Marine who'd helped him carry his computer around Afghan airfields.

Spirk had left the Marines, left his senior intelligence job at Special Operations Command and taken a side swerve, investing in a local Tampa coffee roasting company Buddy Brew, named after the owners' dog. On a very part-time basis, Spirk was still advising the Intelligence Systems Support Office, part of the Air Force. Then Cukor showed up in Tampa: "Hey Dave, I need you to go full time."

In March 2017, Spirk went to San Francisco to see IDenTV, a startup founded four years earlier by Mohammad and Amro Shihadah, and for $120,000 got them to build a computer vision model for some unclassified video pulled off a drone and shipped it back to the Pentagon. The AI could put a white dot on the screen every time a specific object showed up. The makeshift demo was a lot better than an immobile list on PowerPoint and elicited wows.

Work's reaction was immediate. *Go do this*, he said. Now they just needed money. Grant cast around for budget and found an existing project that Congress had already agreed to fund. Will Roper, a senior defense expert, was running a program initially so secret he couldn't even tell his wife about it. He had been reading up on public Microsoft research on machine learning and had managed to strike up a project to use computer vision providers by defense contractors to help pick out objects on still satellite imagery. "The Pentagon had tried automatic target recognition programs many times, and none of them had worked very well," Roper told me. Now, with the help of masses of data, more computing power, and advances in research, maybe AI could understand more about the battlefield than human operators and act as an expert—or maven—to guide the human. At Bob Work's request, Roper allocated his hard-won Congress-approved money and name of his effort to Cukor's new project.

"I ended up with all of Will Roper's money, which he was always mad about," Cukor says with light glee. (Roper has always given loud support to Project Maven.) Cukor started briefing Congress and told me he went rummaging for funds in an obscure corner—a pot of money kept in reserve to cover the ups and downs of fuel prices. "Bam; I got money," he said. Eventually, Project Maven would become the Defense Department's largest AI investment to date, starting with $40.8 million for the first year, according to a document I reviewed, with much more to come.

Work signed the project into existence in a memo outlining its scope on April 26, 2017. General Shanahan would be the director. And Cukor got the most consequential job of his career: chief of Project Maven.

The memo itself said the new team would start with applying AI, big data, and deep learning to drone feed footage focused on the fight against ISIS. The new team would also consolidate all the department's algorithm-based technology initiatives related to defense intelligence and eventually address "other defense intelligence mission areas" that went unspecified. The aim was to provide "actionable intelligence" and insight at speed.

The catchy name might have been poached from Roper's effort, but it was the official, dry, bureaucratic title of the project that most appealed to Cukor: "The Algorithmic Warfare Cross-Functional Team."

That was not for ease of saying it (some were soon wildly translating AWCFT as "aw shit"), but because of what it implied for the Pentagon as a whole: AI was going to war, and the whole of the Department of Defense was going to send it there.

The popular conception of AI at war focused on Skynet, the fictional artificial intelligence system at the heart of the *Terminator* movies. The US military creates Skynet with the help of a defense contractor, but when an Air Force lieutenant general activates the system, it turns out it has become self-aware, takes over the nuclear controls, and an hour later launches a world-changing attack against the entire human race, killing three billion people instantly. "A war between man and machines," intones the hero. Some US military officials were just as scared of this possibility as moviegoers; other US military officers would soon be emailing around clips from the movie in approval.

The Maven memo was careful to delineate its scope. Joe Larson, the Marine reservist and lawyer who Cukor called back to work on Maven, told me Cukor would have "winced" had anyone described Maven as a targeting program. It simply wasn't that, Larson insisted. Intelligence systems only, with no connection to any munition on a weapons platform, and strictly for research and development.

But the official name of the project appealed to both Cukor and Work precisely because it made clear the ambitions for it: algorithmic warfare.

I asked Grant if targeting and offensive weapons strikes were intended to be involved as part of Project Maven. "Yah, of course," he told me. "It's not like we're doing it for kicks. The goal of the intel is to take out high-value targets." The ambition to do "targeting" appeared in a 2017 draft Maven memo I saw.

Speaking to me years later, Cukor made no bones about it either. He always had operational targeting in mind, he said. He had always wanted to bring intelligence into the business of operations and Project Maven was the way. Their efforts went to the essence of targeting: AI would help select and prioritize targets and match the appropriate response to them.

To Cukor, that ambition could be encapsulated in one simple idea that turned on a complex recipe: a white dot on a screen that came preloaded with an accurate coordinate. General Shanahan made the broader intent public: "The Department of Defense should never buy another weapons system for the rest of its natural life without artificial intelligence baked into it."

Another person who joined the new Project Maven team described their own vision for the contribution they expected AI to make to warfare: "We're gonna fucking kill people all the time."

Bob Work had a vision as well. "I really do believe it will lead to a revolution at war," he told me of Project Maven's effort to push AI into battle. Combined with autonomous drones on land, sea, and air, AI would change war forever, he argued.

Looking back, Work told me sorrowfully that if he could rewrite the memo to make clear the focus on AI-enabled targeting and weaponry, he would. "I have taken a lot of heat," he told me. The text's narrow focus on the intelligence side, rather than an explicit focus on developing AI for the combat operations side too, may have backfired, he went on. "We were trying to make sure that we started on the right side of Congress," he said. "But it's totally fair to say that it probably caused confusion inside the department."

From the outset Maven was in fact always about precisely the thing that would most scare its detractors: creating a revolutionary targeting application with the help of AI to compress the kill chain between the time from seeing a target to hitting it: Find, Fix, Finish.

This and other elements of the project would inspire internal departmental friction. "Anything in DoD is a competition for money," Grant told me, even more so when it comes to a program pushed by the deputy and focused on such new territory. "People either want control over it or they want to kill it."

The idea was that, with the authority of Bob Work behind him, General Shanahan as Project Maven's three-star director would give Cukor "top cover," and Colonel Cukor as project chief would go make it happen.

"If it wasn't for Drew Cukor, Maven would never have existed," said Grant.

Work thought Cukor was uniquely equipped for the technological and bureaucratic battle to come. In his view, Cukor combined in one single person the best of three very different worlds within the Department of Defense. He'd fought wars. He was technically savvy. And, perhaps most important in the sclerotic world of the Pentagon, he knew contracting and acquisition processes down cold. "He was a bureaucratic ninja like no other," according to Work.

Whenever you hit a problem, the deputy defense secretary told Cukor, just come into my office. "You will fire anybody that resists you," Cukor says he remembers Work telling him.

"So now the game begins," Cukor told me.

Cukor's ambition ran big. "AI will transform war," he said in a 2017 presentation to defense officials and potential commercial partners, according to slide notes I reviewed. AI would be able to outthink enemy commands and run trillions of simulations about what an enemy might do next and the way a war would go.

One arresting slide contained a clip from the 2012 Bond movie *Skyfall*—the high-speed chase scene where Daniel Craig races around an Istanbul marketplace on a motorbike. As the clip runs, a widely available algorithm identifies objects in the fast-moving video, pinging up boxes labeled "person," "umbrella," "horse," "chair." If the commercial sector could already do this, the Pentagon needed to catch up.

The Defense Department needed to function more like a software company than a weapons factory to deal with the "tsunami" of data, Cukor argued, citing ominous stakes: making the switch "may determine how successful we are in the next war."

Making regular updates to the DoD's software code could end up mattering as much as training and morale, he argued. "I believe Google updates their core software several times a minute ... we're lucky to update meaningful processes once a year," read his presentation. "Algorithms at the end of the day may be the key variable in how we win."

He wanted data processed "instantaneously" to bring target, weapon, and operator together and called for an "algorithmic revolution" to get there. In Cukor's vision, as with Work, AI was a route to automation. Cukor knew it came with risk: AI would make mistakes and get things wrong, and it would take time to figure out how to use it, he warned. "There will be false detections and the need to correct algorithms will be constant," he said. "We are still figuring out the golden utilization of AI."

Cukor dared to assail the very workforce whose support he needed. Recasting the Pentagon as a software company "is a massive and painful undertaking" that would be filled with cultural, organizational, fiscal, and policy "pain." Software culture would require new people, he goaded. The pace was "not suited for everyone," his presentation continued, warning some would lose their jobs. "No doubt there will be some who are currently part of our major weapon systems who will need to move to other more stable tasks such as fifteen-year-long lead physical projects ... trucks ... ships ... airplanes."

However delicate the wording in the memo that launched Project Maven might have been, Cukor was seizing on the controversy. "I know this is now going to be a knife fight," he recalled to me.

The effort to put AI at the heart of how America makes war had begun.

But Project Maven had almost no people. Besides a barebones crew, the Pentagon wasn't going to staff his project.

5

THE FIRST MAVENITES

"I don't know what the fuck AI is."

COLONEL DREW CUKOR set about building a team that would come to regard itself as a scrappy and subversive cult working from within the traditional walls of a stodgy, aging Pentagon. Before long they took to calling themselves "Mavenites."

Besides Cukor himself, Project Maven had only one other officer on the Pentagon payroll. Air Force Lieutenant Colonel Garry Floyd came with the call sign "Pink," thanks to his last name. Pink developed a lasting interest in AI when he worked for the eavesdropping National Security Agency in Alaska, and Lieutenant General Jack Shanahan brought him over to be Cukor's deputy in June 2017.

Delivering the future of war would require more than one paid officer. "We had to stand this thing up out of thin air," said Cukor. He turned to who he knew would get things done: Marines.

Marine reservists were unlikely to know anything about AI. But they were cheaper than contractors, Cukor knew how to call them up, and he wanted people he could reliably set on a task and trust they'd figure out how to achieve it.

Cukor had little respect for the Pentagon. He joked the building's onsite pharmacy made most of its money from selling statins and remedies for cracked teeth. He brought in younger, more aggressive officers better suited for frontline combat than riding cubicles in the world's largest office building. Cukor had followed some of their careers for years. He rated them hardworking thinkers who wouldn't stop until they succeeded, and now he gave them license to break norms. This risky choice would make Cukor's legacy on Maven—and taint his career.

Eric Schmidt, executive chairman of Google's parent company Alphabet who urged the DoD to pursue autonomy, was deeply skeptical about the Pentagon's ability to turn things around. "You absolutely suck at machine learning," he'd told Tony Thomas, a four-star general who led Special Operations Command and who immediately wanted to throw Schmidt out of the vehicle, according to a *New York Times* report. But I was told Schmidt put his faith in Maven. "These guys are going to pull this off," Schmidt judged. The reason? "Because you've got five pissed-off Marines who aren't just going to do it the same way."

How Marines get things done relies on a central tenet: commander's intent. Commanders assign tasks to a subordinate "without specifying how it must be accomplished." On this, Cukor was no iconoclast: he gave the team he assembled almost no direction at all.

The Marine Corps tended to endorse fury as a way to get things done. "Marines may use anger to accomplish missions with energy and focus," reads one official service document. "Properly channeled anger can motivate us to work toward goals and it alerts us that something is wrong and we need to respond."

Cukor hand-selected his initial group of Marines precisely because they were going to rage against the way things were done and get them changed no matter what stood in their way. Some Marine Corps reservists who joined later would be amazed at the freedom Cukor afforded them. One new recruit struggled to get used to casual attire in the office and using everyone's first name. He had never been able to address a colonel, much less give his view, without first asking for permission to speak freely. A Marine Corps colonel was usually responsible for at least two thousand troops. "It's a big deal just to be standing in front of this guy," the reservist told me.

Cukor insisted he needed "big personalities," not least, he said, because he wanted them to tell him when he was wrong. "We train them to be like that, we reward them for being like that," he told me. Some I spoke to agreed strongly that Cukor hired big personalities for the team and disagreed just as strongly that he wanted to be told when he was wrong. Others said Cukor always onboarded negative feedback about Maven in order to improve it.

Among the first big personalities he called on was Captain Colin Carroll, a Marine Corps intelligence officer aged thirty-two. Carroll had worked under Cukor for years. He had been a deception planner focused on befuddling China about US military intentions in the Pacific, and most recently had been in Iraq focused on ISIS. Cukor thought Carroll was impressive and smart and wanted him to join Project Maven as a reservist. "Come do this for six months," Cukor said. "We need a deception plan."

"I don't know what the fuck AI is," Carroll said. And then. "Sure."

Joe Larson, the Marine Corps reservist, encouraged Carroll to join. It was going to be awesome, he told him. But Carroll soon realized he'd signed up to deceive China about something that didn't even exist. Project Maven had barely one piece of paper to its name. There was nothing to protect.

Carroll, who had deployed to Afghanistan three times, figured Cukor just called on people "who in his past life he had abused and who could put up with it for twenty-four-hour days for seven days a week," he once told a podcast host.

Carroll and Cukor shared one quintessential quality: they got things done. Some on the team thought they were so similar, other than being separated by fifteen or so years, that they came to see Carroll as a "mini-Cukor."

Carroll was "super smart and driven" and wants all the right things for the country, another person who'd known him for years told me. But the person had more to add: "There's a line from *The Big Lebowski*: 'You're not wrong, you're just an asshole.'"

Carroll was certainly less interested than Cukor in gentle tactical persuasion. "I'm just a general asshole," he's announced on multiple podcasts, to multiple people and to me. He was also, he said, "competitive as shit" with his twin brother, who went into the Navy while Carroll picked the Marines. When Carroll finished basic officer training, his instructor told him she wouldn't be surprised if he ended up on the front page but she wasn't sure if it would be for getting a medal of honor or for getting court-martialed. "Ma'am, I'll take that as a compliment," he told me he'd replied. In the years to come, he would be fired twice from the Pentagon—first by the Biden administration in August 2021 and again four years later by the second Trump administration. Carroll would suggest in podcasts he was accused of being abrasive in the first case, and falsely accused of leaking information to the media as part of an internecine Pentagon power play in the second. Spurred on by his new experience with AI and his shock at the parlous state of the Pentagon's tech, Carroll would slowly become an advocate of autonomous weapons—the killer robots that so worried campaigners and military ethicists. He would land at Anduril for several years in between firings and in 2025 would start his own company devoted to building a $15 million autonomous plane.

But first Carroll would spend more than two years on Project Maven, later describing it as "a one in one hundred" team. He loved Maven, he loved working, and he loved Cukor. He prided himself on being passionate and effective. "It takes a certain type of person to wake up every morning knowing that the day will be spent waging an uphill battle against the frozen middle, in the basement of the Pentagon, surrounded by bureaucrats who work at best six-hour days and exist to say no—and still enjoy the camaraderie and the slog," he later posted to his old teammates. He mixed truculence, brainpower, and appeal, often in a single sentence. I put it to the person who'd known him for years that Carroll could be quite charming. "So is a sociopath," came his answer. (When I later asked Carroll about this, he considered the definition carefully and told me he doesn't think he is a sociopath; he cares too much, he said. "I don't know why I am the way I am," he said.)

On his first day at Maven, Carroll walked into Room 1 Bravo 855, a secure, windowless enclosure on the ground floor of the Pentagon.

Carroll had an undergraduate degree in aerospace engineering from the US Naval Academy and reckoned it might help. And he'd read a single, fifty-page Boston Consulting Group guide to AI, which informed him that to make a working AI system they needed data, contracts, and integration. That was pretty much everything he knew about the brave new world. As Carroll recalls it, he looked questioningly at the men standing around him in Maven's new digs. But they just stared back, unclear what was happening.

Among the group was Sy Poggemeyer, a recently promoted Marine Corps captain aged twenty-seven whom Cukor had first met in 2010 when Poggemeyer was just starting out as a Marine. As he did with Carroll, Larson, and others, Cukor took a close interest in Poggemeyer's career: he thought him a brilliant, dedicated, hardworking, and promising officer. Poggemeyer had been in the game longer than his rank suggested: he could remember trundling around amphibious assault ships at the age of four, jumping on the bed of his Navy captain stepfather's stateroom on the USS *Peleliu*. Two decades later, Poggemeyer would serve on the same ship and under the same motto: "Peace through Power." In between, he studied abroad in St. Petersburg in 2011. And in 2014, when he was a second lieutenant, he, Cukor, and two others published an award-winning paper on reforming intelligence.

Others on the team would come to think Cukor sometimes looked on Carroll and Poggemeyer almost as his sons. But even if that were the case, the two would never approximate brothers. The dynamic between them would track closely the ups and downs of Maven, eventually coming apart at the seams.

Carroll took it upon himself to assign jobs. As far as he was concerned, Poggemeyer—at five years his junior and less advanced in his rank—discounted him from taking a leading role. Then there was Mike Rhoads, a Marine Corps artillery officer who in 2012 had been shot in front of Carroll in Afghanistan, who Carroll recruited to the team. Carroll announced Joe Larson should do contracts since he was a Stanford lawyer. (In an office short on titles and hierarchy, Larson didn't end up doing contracts, but became Cukor's unofficial chief of staff.) Dave Spirk, who helped put together the first demo for Project Maven and was still down in Tampa, was the diplomat getting the special operations community on board.

What was left? Data and integration. Rhoads would do data, with Poggemeyer soon in support. Carroll himself took integration, focused on engineering, meaning how to roll out AI to the troops in working systems. Carroll promptly sent himself on a fact-finding mission to visit drone bases around the country and within a couple of days was absolutely convinced on one thing: there was no way AI was going to work. He threw himself into it anyway.

TO CUKOR, there was no difference between the pace of work in the battlefield and the pace of work at the Pentagon. "Why would I be different when I'm in Kandahar than when I'm here in the Pentagon?" he asked me. "My argument was always the pursuit of technology is warfare."

Cukor wanted a skinny operation—he thought staffing and administration costs should take up less than 10 percent of Maven's budget—far less than the usual 20 to 30 percent overhead that he considered a bloated waste of taxpayer money. He told me he ended up at 7 percent. Soon a microcosm of Marine reservists, Pentagon civilians, and a clutch of companies led by ECS Federal as the primary contractor would all be working on Project Maven.

Jack Shanahan, the three-star general who saw his key task as providing top cover for Cukor, described him as a hard-charging, bullheaded disrupter who does not take no for an answer and breaks a little glass.

"Everything we were doing was different and new enough that somebody, somewhere was guaranteed to be upset at him at some point, including people on his own team, because he was brusque," Shanahan told me. "It was such a violent shakeup to the status quo he's going to make people angry. You're shaking up a system in which he is not taking no for an answer."

The pace would be too much for some; Cukor just exhausted people. He worked eighteen-hour days, rarely appeared to eat, and seemed to expect everyone else to do the same. He was infamous for double or triple-booking himself and was regularly meant to be in two different places at once. "Apparently he never sleeps," a contractor emailed colleagues in 2011. He doesn't get grumpy, his wife Kirsten told me. He just suddenly stops. "Boom and he's down. It's like, oh, his battery just wore out." The team made jokes about pub mix, crunchy salty snacks that counted as the mainstay food in the office. For liquid sustenance, Cukor relied on caffeine-free Diet Mountain Dew. No booze. Not even coffee.

Cukor came with mystifying quirks. If it rained, he might announce it was a good day for photosynthesis. "Should be fun," went another Cukor tagline the team soon took to mean nothing of the sort. He made them all stand up for long morning meetings and worked so hard he had no time for voice mail: he hadn't checked it since 2005, according to his recorded phone message. Send a text.

Vendors and team members alike could expect calls and visits late evenings and on weekends. Carroll said he stopped setting an alarm because Cukor would reliably call him every morning at 6:30 a.m. Even today, if Dave Spirk's phone rings before 7 a.m. or after 8 p.m., his children automatically say: "Say hi to Drew Cukor, daddy." Members of the team back from a six-month deployment in Afghanistan, sometimes their first-ever deployment in bewildering and potentially dangerous conditions, could expect little in the way of backslapping or thanks even on their first day back in the office. Cukor was never personally insulting and he didn't much raise his voice, but he could be withering toward failure. Some detected he got their names wrong when they messed up, rewarding failure with anonymity and irrelevance. He changed his mind about what the project needed next so often that some workers ached that their toil was destined for the shredder. Some felt humiliated by him.

"Some of the people that worked for him in Iraq said I'll never work for that fucker again," Shanahan related to me. But he said they'd soon relent: "Dammit he's a great boss I'm coming back and working for him again," the refrain would go.

A person who found him "Elon Musk-esque" put it this way: "You did not really want to work for him, except that the types of things you were doing were so important either to national security or saving lives, or finding bad guys. Even if you hated him, you still saw the brilliance and you wanted to learn." Another team member wondered if they might have Stockholm syndrome: "I felt it was the most important thing in the world to work on."

Joe Larson considered Cukor not only brilliant but highly capable, technology-oriented, unafraid to take risks, and innovative too. Dave Spirk admiringly held him responsible for his own career. A variety of other assessments were more lurid. According to Brian Ward, who joined as a contractor on Maven with ECS Federal in summer 2018, Cukor was "a psychopath." Sometimes he added a key qualifier: a psychopath "in the best way."

Cukor tended to wave all this off. He simply liked to say they had a dizzying number of things to get done, and fast. "At the time it was kind of annoying, but then now I think back on it, I miss it to death," said another former Mavenite. "Because it kind of made a lot of us what we are now."

CUKOR ARRIVED AT the Marine Corps in 1992 without a father, siblings, or much money. All Marine officers are made in Quantico, and there for the first time he met several other young men from fatherless families. "Usually single moms, usually lower economic class, and usually lower opportunities," he says of many of the intake. "Suddenly we're all the same. We're all modestly aggressive and we know we're working within a system and we want to follow rules because we appreciate them," he says. "I was not exceptional."

He grew up in a small apartment in a poor part of Los Angeles in public housing with his young single mother in Reagan's 1980s. She worked as a bookkeeper and then a schoolteacher, and the way he remembers it, they weren't particularly close. His father's side were Hungarian Jews who landed in midtown New York, and a small branch eventually headed out to Southern California. "He wasn't much in my life. He kind of walked away," Cukor says.

Around the age of four, a friend of his mother said she could take him to church on Sundays. His mother "was just happy to see me gone probably on Sunday so she could rest from a hard work week," he says. He started going every Sunday. It was the Mormon church. The friend just kept taking him. "She was lovely."

When he was eight, the youngest age at which The Church of Jesus Christ of Latter-day Saints will take conversions, Cukor got baptized. Converting also meant a future without tea, coffee, alcohol, or tobacco, all prohibitions he observes. When Cukor was fourteen, his mother converted too. He was eager to leave home at eighteen.

"It's not like it's the most loving family, she was working hard and I was doing my thing," he says. "Just a normal single mom trying to be helpful, and probably a very arrogant son who didn't want so much help."

He talks through this recollection of teenage claustrophobia and estrangement in a soft, steady voice. "So you launch, you just go."

I figure he must have turned over this calmly recollected version many times before he tells me this, long over a year after we first meet, although I learn that perhaps he hasn't said it out loud much after all. Kirsten later tells me he is a private person.

In the absence of funds, he figured his best chance of further education was the Reserve Officers' Training Corps (ROTC), a program that sponsors university education in exchange for signing up to serve as an officer in the US Armed Forces. "The military in the 1980s was really the only way to go to college, honestly, for a lot of us."

In 1987, he interrupted his time at the University of Southern California, losing his military scholarship, to do his two-year stint as a Mormon missionary. The Mormons sent Cukor to Panama. He would be spreading an intrinsically American faith: church members rely on the Book of Mormon, which says Jesus visited and blessed the people who lived in ancient America. Run by General Manuel Noriega, Panama was in the crosshairs of Washington's efforts to stamp out Communism and the drugs trade and return the military to its barracks.

Cukor wore the white shirt and went about as a twosome with a Panamanian missionary who spoke no English. "You truly are on your own," he says. "You get used to that independence."

Panama could have made for fallow territory for the Mormons: the United States, in part because of the US military presence, was not popular. US troops were also largely confined to base for fear of sparking political violence. Cukor's remit, by contrast, was to go all over. "I had no clue," he says. "We loved it."

He got a taste of instability too. Missionary program leaders pulled him out as tensions rose in the run-up to the 1989 US invasion to depose Noriega. He started over again in Costa Rica. He accustomed himself to sudden change and developed techniques to inspire the people he met. "I certainly was no Paul from the New Testament but I did my part. But the conversion part is really a miracle, because, like, honestly the church has a lot of baggage."

He was "moderately successful" at chalking up conversions. "It was just a gentle message." A nineteen-year-old Cukor had discovered his own powers of persuasion.

6

RELAXED ABOUT FURY

"I want to reduce the non-American population."

THE MOST SOME NEW MAVEN RECRUITS KNEW about the Silicon Valley ethos they were supposed to imitate was the HBO comedy television series of the same name that sent it up. The show's conceit focused on a group of young developers trying to develop a software platform "they feel will change the world while trying not to get run over by rampant dueling billionaires with massive egos and scores to settle." Some Mavenites watched with expectant glee, hoping to see themselves in the series. Some also shielded their eyes from its main intention: to "skewer the people that think all tech is good for us." The satire was saturated with financial crises, firings, scheming, and AI. Mavenites were soon gorging on all four.

The group united as insolent contrarians, proud of their windowless Pentagon basement. A small bottle of water soon arrived on a desk in the office labeled "Federal Budget." The joke was that anyone from Maven walking past it could drink from it; the project was guzzling cash. Other office paraphernalia included a toilet paper roll with Vladimir Putin's face on the sheets.

The office dynamic immediately started turning heads in the rest of the Pentagon. "People are loud," recalled one person who worked there. "There's some yelling going on, you don't know if they're mad at somebody or they're just on the phone speaking louder 'cause everybody else near them is also loud. It's just a very dynamic, crazy environment."

Sometimes Colin Carroll would be yelling at Joe Larson. "Joe, you're like fucking dumb," he'd be saying to the Stanford-educated lawyer as they stood at the whiteboard. "This is the dumbest shit." New starters at ECS, the prime vendor, would be shocked, unsure of whether to enter. Old ECS hands would reassure them: "Ignore them; they always yell at each other like this. They still love each other."

It wasn't always love in the office. Air Force Lieutenant Colonel Garry "Pink" Floyd, the only other active-duty officer on the team, outranked everyone but Cukor. He modeled himself as the grown-up in the office. Carroll stood out to some as too abrasive for his own good, and Pink would counsel him to slow down. Carroll seemed to some as irretrievably bent on insulting other Pentagon officials; Pink would say those officials were all likely to outlive any single reservist called up to work on the Maven team. It might be better to work constructively with them rather than steamroll or alienate them if they wanted Maven and AI adoption to take off.

But Pink's own veneer of parental composure could disintegrate as well. One day he overheard Carroll complaining loudly about the "fucking Chair Force." That was the derogatory term Carroll used to suggest the Air Force was sedentary when it came to getting going on Maven and AI, or anything much at all. (Carroll said it so often that his phone autocorrected to Chair Force.) Pink saw red. Boiling point. He stormed across to Carroll. "You have no idea what you're talking about! I'm tired of cleaning up your messes!" he yelled. "We are never going to get anywhere with you bullying these guys!" Pink slammed down the snack container he had been holding, which promptly exploded. Bits of food went flying over the floor, reaching the tops of lockers. Pretzelgate was underway.

Carroll stood up. An office standoff. "Hey, first off, I hate the Air Force, but I hate the Marines more and I'm a Marine."

They were eyeballing each other. Larson stepped in. Then, not for the first time, Major Carroll issued Colonel Pink, his uniformed senior, a command: "Go fucking clean all this up."

Cukor was unmoved by office explosions and didn't seem too

interested in hearing complaints. But he was constantly offering personal counsel to irate individuals on his team who'd show up at his door with their frustrations. He knew how hard he worked them, how constantly exhausted they were, how much everyone was up against, how explosive it was in the Maven back office. In this particular fracas, it was Cukor who later took on the real cleanup job: counseling Pink and Carroll they were two strong personalities who had to work together. Outbursts weren't productive, he told Pink. I know the pace is frustrating, he told Carroll. "We got the mission done in spite of our own internal foibles," Cukor told me, glowing that Maven did ultimately deploy to the Air Force, Army, and Navy.

Besides, Cukor was generally relaxed about fury. "There'd be these flareups, right? And I'd be like, go back to work, man. Like, this is how we do tech. We have flareups, we work through it, we apologize, and we move on," he recounted.

He sort of liked it that way. "Did I have a culture where people could tell Colonel to go pound sand?" he said, referring to himself. "Yes. That allowed for that flexibility." To Cukor it was just the military culture—no different in office corridors than in combat. "These are wonderful minds. Let's let them do their thing."

That didn't work for everybody, though, and some couldn't stand the pace or the dynamic in the office, especially as the OG Mavenites who'd served together in Afghanistan and Iraq left for new jobs and were replaced with a looser-knit and less aggressive team. Cukor would eventually go through six rounds of people. "A copy of a copy of a copy," he'd say. Others saw tensions as the necessary cost of an innovation office, part of a certain glamor that appealed to them in the first place.

Word spread through networks of reservists. Some would try to get transferred onto the hectic team while others would be recruited through friends of friends. Interviews could be perfunctory. Carroll, who recruited many Mavenites in the first two years and says he and Larson "built an awesome team", sometimes ran interviews for as little as fifteen minutes. They were often over before they started: "You've got a clearance, you're a reservist, when can you start?"

Jaim Coddington, a Marine reservist recruited by Sy Poggemeyer partway through 2018, was among Maven recruits who were startled, and eventually amused, by precisely how little training Maven's new starters would get on this complex and little-understood technology. Coddington tried to capture it in a cartoon he sketched.

"Go do AI!!!" ran a speech bubble in one cartoon delivered to a bewildered new starter. Another cartoon made fun of the way Cukor tended to squirrel himself away in his office "working like a madman" at inhospitable hours. As Mavenites lined up outside Cukor's office to clear their latest slides at 7:13 p.m. on a Friday evening, Cukor would keep them waiting, head down on his own work. " . . . Well I'm sorry they can't have their early weekend; it's only 7," went the tagline.

While Cukor always wore his uniform, his reservists didn't. The team adopted startup culture fashion and started wearing gym pants and white-soled sneakers. One Mavenite recalled feeling embarrassed in an interview by his smart Pentagon attire, which he ditched the moment he got the job.

Some of Maven's newer recruits looked up to the fashion sense and verve of two foundational members on the team—Carroll and Poggemeyer. But others found the situation uncomfortable, somehow untenable. The Pentagon was not Palo Alto.

In a building of rules and an institution of uniforms and hierarchy, wearing Lululemon, Allbirds, and three days of stubble helped erase the divide between military officer and tech worker, but it also ranged somewhere on the spectrum of unorthodox to sacrilegious. As easily the smallest of all the services, derided as unintelligent "crayon-eaters" by rival services, Marines used outward discipline as one way to distinguish themselves.

Clothing came with a "concept of operations." New uniform regulations, signed off in May 2018 by the Marine Corps Uniform Board,

Opposite: Cartoons by team member Jaim Coddington depicting workplace culture at Project Maven during 2018. The cartoons spent a couple of years pinned on the team's office noticeboard.

carried stunning specificity, embossed with a seal. A single allowable visible tattoo could stretch from the second knuckle of the index finger to the first knuckle of the pinky finger and no further. On the lower leg, a single tattoo must fit within the space of the Marine's own hand "with their fingers extended and joined with the thumb flush against the side of the hand." (Note: "The measurement will be from the base of the palm to the tip of the fingers and from the outside of the thumb to the outside of the palm.") Each crease of fabric, shine of boot polish, and length of facial follicle came with rules. Commanders (like Cukor) were meant to ensure such regulations were "appropriately disseminated and enforced," while broad enforcement came down to "the good judgment of Marines at all levels." It was all "a matter of tradition and pride, and of good order and discipline." Rules meant values. Values meant respect. And it wasn't clear if Project Maven was showing any.

By the end of 2018, a distinctly Maven take on a classic Christmas carol would be penned in Cukor's honor, capturing the never-ending vibe at the end of the second year:

"On the twelfth day of Maven, Col Cukor gave to us / A Friday evening phonecon / Deployment with no notice / Disgrace for drinking coffee / A red eye to the Valley," went the final verse of the song. "Demands for 100 million / Shame for Joe's Detritus / TERRRRRRRRY MITCHELLLLLL."

(These were references to Joe Larson, whose mess had grown quite a reputation around the office, and a person whom Maven members considered unaligned with their vision whom I could not track down.)

"Shit for leaving early / 4 marine reservists / A crude algorithm / 8 impromptu meetings / 103 slides to brief."

One new recruit who described the office vibe as "controlled chaos" looked up in particular to Kelly Martin, an Air Force lieutenant colonel who would replace Pink when he left midway through 2019. Besides Cukor, she was the only other active-duty officer on the team, and thanks to her Brooklyn father she could curse as well as any of them. She was a mother of three at home, and now her duties appeared to extend into the office: she ended up as "mother of the misfit boys."

The Project Maven team on the steps of the Defense Department in a June 2019 "unapproved" photo shoot. In the front row, from left to right, is Lieutenant Colonel Garry "Pink" Floyd, Major Joe Larson, Colonel Drew Cukor, Major Colin Carroll, Lieutenant General Jack Shanahan, Gregory Christ, Captain Sy Poggemeyer, and Jamie Kovarna. Directly behind Gen Shanahan is Brian Ward. "We'll see whose book it ends up in first," read the June 17, 2019, email from Major Stuart "Death" Wheeler sending it around.

She was kind, determined, clever, and tactful. "She was just a phenomenal officer," said the new Maven signup, who'd suffered at the hands of a terrible lieutenant colonel in the past. At Maven, others joked to me she played the role of "heat shield," protecting members of the team from Cukor's wrath or disdain.

Recruitment interviews conducted by junior officers in Maven's early days would not have passed muster with employment lawyers. "Are you married? Do you have kids?" went one popular line of questioning. Once a pregnant woman showed up and the Maven representative wondered if he even had to go through with the interview. A certain compassion laced the unfeeling, unfair, and illegal questions. Whoever got the job was about to work hours wildly inhospitable to family life and travel at no notice for uncertain lengths of time to war zones. New recruits could show up and by their second day would learn they were being sent to Afghanistan. Some who started out married at Maven left Maven unmarried. One divorcé, who blamed the breakdown of his marriage on Maven, told me it was like the Apollo program—America's 1960s push to get to the moon that was popularly held to be the driver for a spate of divorces. The Mavenite said multiple marriages failed during the heady early years of the project. No one wanted to be responsible for that level of personal attrition.

And then there would sometimes be a third and final question: "Why do you want to work here?" One celebrated answer, according to two people familiar with the matter: "I want to reduce the non-American population." The applicant got the job.

PROJECT MAVEN PLANNED TO get AI onto three different types of drone platform. That required training separate algorithms on data from different sensors for each drone type.

Tactical drones—such as the small drones made by Aerosonde, or the ScanEagle drones made by Boeing—tend to be flown, controlled, and analyzed by Special Operations units out in the field. They would fly fairly low, at 1,000–20,000 feet, close to the troops who controlled them, to help them, see ahead, or keep watch as they carried out an operation.

Medium-altitude long-endurance (MALE) attack drones that flew at 25,000–50,000 feet above ground included MQ-1 Predators made by General Atomics and were piloted from control rooms back in the US—in California, Nevada, Virginia, and elsewhere. They could carry weapons and produce video footage in multiple formats. The full-motion video feeds, or FMV, came embedded with details such as time stamps and other metadata. These could enable an operator to view a video feed overlaid on top of a map; measure distances between objects as well as height, area, and traveling speeds; and came with specific playback rates for reviewing footage. Cameras were limited by narrow fields of view, however. Military operators likened these to looking through a soda straw.

Some of the bigger weaponized drone systems, such as MQ-9 Reapers from General Atomics, could carry Hellfire laser-guided missiles, laser-guided bombs, and GPS-guided bombs known as JDAMs. These were equipped with synthetic aperture radar that could "see through" clouds, bomb smoke, dust, and the dark, and made them a platform of choice in what the Defense Department described as all-weather hunter-killer operations. By early 2016, the Air Force fleet of more than two hundred MQ-9s had flown nearly a million total flight hours. They could produce "WAMI"—wide-area motion imagery—that took in an entire town, city, or landscape when equipped with a new bug-eyed sensor called "Gorgon Stare" whose hundreds of cameras collected and stitched together huge reams of footage. Different drone sensors could record in infrared to pick up heat signals, particularly useful at night and when looking for people, vehicles, factories, and fires. Each system and sensor would need a different algorithm to learn what to look for, based on differing angle, altitude, scope, video format, playback speed, and more. Getting hold of data to train the bespoke algorithms in the first place would be more than half the battle.

The job of securing data, the lifeblood of artificial intelligence, fell initially to Dave Spirk. In April 2017, he hitched a ride in Joe Larson's Honda Accord to a small drone program office run by the Navy and Marines in Maryland. The duo's mission that day was to get hold of battlefield drone video imagery so it could be labeled, to teach algorithms how to recognize objects that appeared in footage produced by different drone platforms.

When Spirk and Larson got to the Patuxent River drone office, called PMA-263, the Maryland team welcomed them with open arms, pointing them toward the data store. Opening a closet, they found a mess of hard drives all stacked up and collecting dust since the days the Marine Corps were deployed to Afghanistan in 2010. That was years before the arrival of the advanced drone platforms and sensors they were meant to be using.

Driving home, the two men were worried. "That data we saw was no good," Spirk lamented to Larson. "We gotta find a way to get data."

Then Spirk had an idea. He called a special operations friend who'd just got back from Djibouti, headquarters for the secret US drone operations in the Horn of Africa that were constantly capturing fresh reams of video footage.

"Hey, what do you guys do with all of that video that you were taking over Somalia?" Spirk remembers asking. The answer: It's all with the Navy SEALs.

That's when Spirk set about converting the shadowy world of Special Operations Command (SOCOM) to embrace AI. It began with another phone call: "Can I pillage and plunder some of your video?" He got the thumbs-up from then-Rear Admiral Tim Szymanski, commander of Naval Special Warfare Command. "Have as much of it as you like," he told him.

On May 9, Spirk got a huge data dump from Group 10, a new part of Naval Special Warfare set up in 2011 to run drones and provide specialized surveillance to Navy SEAL teams carrying out missions on the ground. Spirk tracked down more. He felt as though he were becoming a data dealer, hunting out chunks of precious imagery from the Navy SEALs, hounding anyone he could for video. That first year, much of Maven's data arrived thanks to Spirk's ad hoc outreach to his friend back from Djibouti.

The special operations community saw itself as the vanguard of new tech and new approaches; they were receptive to the gospel of Maven. Soon SOCOM commanding generals were turning up to lunchtime sessions on AI, taking courses on AI at MIT, and coming around to the iconoclastic notion of fielding AI directly into the battlefield. In 2018, Spirk would become the first director of AI and chief data officer for SOCOM, bringing AI to some of America's most secret, deadly missions, and two years after that for the whole of the Defense Department. General Richard Clarke, who took over SOCOM in spring 2019, would later say he wanted SOCOM to become the first AI-enabled command.

Drone footage provided by Group 10 brought more challenges. Some was monotonous, showing arid land and not much else for hours on end from above. Deadly raids brought different problems. Sharing classified imagery with AI experts, many of whom were unused to the realities of military raids and violence, and several of whom were not even US citizens, would be unthinkable to the Pentagon—and illegal.

The team planned to take the secret drone footage and chop it up into more manageable chunks, into three-second or five-second clips of useful scenes. Project Maven evolved a plan to pass these clips onto data labelers to identify relevant objects and then pass on part of that labeled feed to AI experts to train an algorithm how to pick out these objects. A separate portion of the footage would be retained to test the resulting algorithms.

Cukor secured one of his many unlikely breakthroughs when Deputy Defense Secretary Bob Work signed a decree saying Project Maven was exclusively allowed to input full-motion video classified "Secret" through the team's own process, then sanitize it and share the resulting declassified imagery with commercial companies whose (possibly foreign) computer scientists would carry no security clearances.

Project Maven teams would scrub out frames that contained strikes, gory images including casualties, or operational battlefield information such as location, date, or any specific event details that could help an enemy, cause a public stir, or create other problems. "We didn't want anybody traumatized," Spirk told me. Even in bite-size chunks, the drone imagery flicking through at dozens of frames per second would require intense work to label each individual frame accurately, highlighting what could be dozens of people and vehicles in each scene and each frame.

Commercial Pentagon vendors were dominated by big defense companies such as Booz Allen or Lockheed Martin or Northrop Grumman, which dominated defense contracting. But Cukor wanted faster, smarter companies with cutting-edge technology and less bureaucracy to create the algorithms. He wanted a new defense-industrial base equipped for the future. Drew Cukor wanted to create America's first AI defense contractors.

7

THE COLONEL AND THE MATH WHIZ

"It's always, like, a divisive thing, like working with the military or not."

COLONEL CUKOR KNEW America's future AI defense contractors wouldn't be behemoths the size of Lockheed—at least not yet. But one of the first companies he set about wooing had only recently started out with a head count of precisely one: a New York math whiz hungry for revenue who knew nothing about working for the military.

Matt Zeiler was an AI starlet. He started his company in November 2013. Within a month, it delivered the top five most accurate models for image recognition in the world, beating tech stalwarts including Google and others. Zeiler hadn't even made his first hire yet: he did it himself.

Zeiler grew up in central Canada, outside Winnipeg, the son of a doctor and a nurse. Like his brother, he was meant to go into medicine. "They all wanted me to go to medical school," he told me. He did a year of premedical studies. But something kept snagging. Physics and calculus just felt more interesting to him than biology and chemistry. He switched to engineering science, trying to decide between computing or nanotechnology for his final year. Someone on his dorm floor at the University of Toronto showed him a video of a flickering flame, and he was gobsmacked to hear it was completely generated by artificial intelligence.

The student, it turned out, was studying under Geoffrey Hinton, one of AI's leading lights who uncovered the wiles of neural networks.

Zeiler wrote an undergraduate thesis under Hinton investigating how to deploy AI to quantify the curious head-bobbing of lovestruck pigeons. Back then Toronto was one of only four schools looking at deep neural networks and image recognition (the others were New York University, University of Montreal, and Stanford).

Zeiler was hooked. After his investigation of pigeon courtship, he headed to New York, to study under NYU's AI godfathers—Yann LeCun and Rob Fergus. In 2012, while he was still there, a new breakthrough at an annual computer vision competition called ImageNet suggested for the first time that it might be possible to deliver AI detections with error rates far lower than one in four.

At the time, Zeiler was working on an AI model that rifled through street-view images to find each house number. Pinched for money, he was interning at Google Brain, Google's deep learning project, which had only just got going itself in 2011. One day, Jeff Dean, Google Brain's co-founder and leader, asked him to present his research to four hundred people, and Zeiler realized his approach was working better than anything Google had. Zeiler thought about joining Microsoft, Apple, or Facebook—meeting Mark Zuckerberg in one interview—but ultimately declined what Google told him was the largest-ever graduate offer and instead started his own company, Clarifai, in November 2013. Three weeks later, his brand-new company swept the board at the ImageNet computer vision competition with the top five most accurate models.

Zeiler's victory led to a groundbreaking 2014 paper that did something more important. It edged closer to visualizing how convolutional neural networks actually work, confirming for the first time what the greats like LeCun and Hinton had been saying for a decade or more: that deep learning mimics the way the human brain learns.

Zeiler tried to explain the real-life version to me. "Like, the first layer of your eyes kind of filters things into different orientations like a vertical line, a horizontal line, different angles, lines, maybe small things like circles, et cetera," he was saying, as he gazed on the machine that won him the contest. Covered in Perspex for posterity in his office, it's a plodding mass of plastic and wires surviving under a layer of dust.

"And then the next layer is thought to have, like, started composing them into like Ts and rods and cones and that kind of stuff. And then eventually at a third or fourth layer you'll see an eyeball feature emerge, a finger, whatever. And then, you know, a couple layers above that you'll see, like, oh, this is a face or a person. And so my paper showed that these neural networks actually learn in that exact same way."

Neural networks—a form of deep learning—rely on multiple layers of differently weighted artificial neurons. These function much like synapses, the sluice gates of the brain that determine when and what information is important enough to pass around the body's nervous system. Convolutional neural networks apply this idea specifically to images.

Zeiler used his own money to buy his first server. His first angel investment funded two more. The three machines, packed with $60,000 worth of GPUs—specialist electric circuits or computer chips that can process pictures and video with lightning speed, store code and memory, and perform many calculations at once—overheated his small apartment in Manhattan's financial district. Each machine had to go on a different circuit because of how much power they drew. He had one in the living room, one in the kitchen and one in the bedroom. The GPUs drew so much heat he had to open the windows even in winter to avoid working in a sauna.

His first customers were stock media companies, travel companies, food. His biggest client was a wedding blog. They wanted to tag their stuff so people could find their content easily. He built a wedding-specific model—it could recognize things like "cake," "veil," "bride," "groom," and "suit." The more images they ran through models, the more they paid.

To train the neural network models, much like the way we train our own brains, the company fed the models information, and then compared the model's prediction to the ground truth. Was the image it declared to be a cake really a cake? If yes, hey presto, eureka, let the people eat cake, send over the invoice. But if no, you comb through every mistake in a process called backpropagation, then update the parameters attached to the filters, so the algorithm will be a little bit better at predicting "cake!" the next time it sees three concentrically smaller tiers. "You go through millions of examples iterating like that, improving the parameters over and over and over."

Zeiler had no personal experience with the military, but he grew up nursing a certain fascination for it, honed from watching World War II movies with his grandfather. One friend went into the Reserves, and Zeiler was amazed at the endless dedication to routine and pain. "They had to, like, stay up for multiple days in a row awake and do that laying in the mud and it's crazy stuff," he recalls. "It just sounded brutal." But none of this particularly dented his broad view. "Like, obviously war is bad in general."

The way Zeiler saw it, if technology was going to be used at all, it should be used to prevent future wars.

Cukor proposed a meeting, swapped his cammies for service charlies, and boarded the Amtrak with Joe Larson. It was the first time Zeiler held a customer meeting on a Sunday. Zeiler, still in his twenties, was amazed to see the pair encumbered by military backpacks on the streets of New York, but he thought they "had their shit together." An hour into giving his demo, Zeiler remembers Cukor's commanding interjection: "We definitely need you to be part of this program."

But there was a catch, which involved Clarifai and the others remaking their approach. Clarifai was going to have to swap training data from wedding paraphernalia for images of warfare. Details were sparse. Cukor said he wouldn't let any government data leave government systems. That meant Clarifai would have to remote in via a virtual private network to the clunky, slow, secure government network known as SUNet. As clouds go, SUNet was risible, but it counted as an unclassified but protected working space, hosted on government data centers, where AI startups could train their models.

Zeiler was appalled: he was used to working in his own speedy high-capacity computing environment. Cukor knew what he was offering was going to be hard and less performant. "It's government grade, which means it's not very good," he said. But he didn't want to hand over "exquisite" government data on a hard drive and risk getting hacked or suffering some other embarrassing leak. Cukor already knew, as he would say at a conference that July, you don't buy AI like you buy ammunition. His solution was unorthodox by Pentagon standards, but offered a novel way of working with commercial software companies: the government would own the underlying data and pay a recurring annual license to use the algorithms, while Clarifai would retain its own intellectual property but would not retain access to its own model. "We'll pay you to use your tech but you cannot take it out—it's the government's," went Cukor's stiff pitch. "I didn't want any of our tech to land in China."

Zeiler's concession came with grumpy caveats: "Just know it'll be slower and dumber."

Cukor barely heard him. The government didn't pay much, but Cukor figured it was amazing money for Clarifai given the company was still starting out. And he demanded speed after all: "Every ninety days you gotta give me a new algorithm."

Cukor was appealing to Zeiler's competitive nature. Besides Clarifai, the Maven team assembled three other AI startups for Cukor's algorithmic war factory that summer. He wasn't going to hire just one company; he was going to hire all four and set them against each other. This, he thought, would birth a new world of AI researchers within the Pentagon's orbit. These were, multiple people told me, IDenTV, the company Spirk enlisted for a demo earlier in 2017; Pilot AI, a Silicon Valley startup led by Jonathan Su; and Xnor, a Seattle startup founded in 2017 by Mohammad Rastegari, and led by AI professor Ali Farhadi as chief executive, which was ultimately bought by Apple in 2020. All were initially reluctant about doing work for the military.

"They all took time; nobody came over immediately," Cukor recalled. "It was not an easy decision for them. Once you do defense work you get stained."

Cukor thought bringing them on was necessary: he couldn't go to the big traditional defense companies because they simply weren't good enough in this new field. But every single startup he spoke to was skeptical about doing military work. "I was trying to build a sophisticated technology program in an age where most people who built that technology were not very favorable," he said. That was Cukor's way of referring to widespread negative views of America's drone wars. At the July 2017 conference, he blasted out his public appeal to sell AI to the Pentagon, and then flew to Honolulu shortly afterward with a group of defense officials to scout out talent at the top annual research gathering on computer vision. Each of the AI companies he tried to enlist, he told me, "had a Matt." To Cukor it was worth the effort to persuade them: "These are guys that are literally building some of the greatest tech on the planet at the time."

One of those "Matts" was Joe Redmon, who won plaudits at the conference the year before for his work on computer vision that could quickly recognize images, an award-winning milestone for the field referred to as "You Only Look Once."

Redmon was thrilled that computer vision could, thanks to these new approaches, detect objects within twenty milliseconds, down from twenty seconds in recent memory, and all using code and the source models he had made free and available on the open internet. AI could now track threatened gorillas, find cancer, and explore the ocean in ways it never could before, he enthused to a live audience.

Redmon worked at Xnor, which secretly signed up to work on Project Maven, although he wasn't on that team. In February 2018, Redmon would describe his horror at being approached at a 2017 computer vision conference in Hawaii by a US military official. "His technology worked at levels no one else had been able to replicate," said Cukor, who remembers making a beeline for the gentle, thoughtful computer scientist as he prepared to give a presentation at the conference. He was sitting on the floor, so Cukor joined him. "He was so generous with his time. I gave him my spiel; we want to be more precise and accurate."

Redmon remembers things slightly differently: he said in his talk that a military official (who Redmon described as an army general, though Cukor thinks he might have meant him) discussed a video of his algorithms being used to track and find vehicles on battlefield drone footage. "I was kind of horrified," Redmon later told the audience at a public TEDx presentation in India of his reaction at the time. US drone strikes had killed thousands of people across the Middle East, he said. "They've hit weddings, schools, torn apart communities. I couldn't imagine that my work could be a part of something like this."

He perceived in the US military's pursuit of AI an urge to be stronger than anyone else regardless of human cost. It reminded him of warnings he had read from Albert Einstein, who argued that a "disastrous by-product" of science and technology was the mechanization and depersonalization of our lives that led to an "abominable deterioration of ethical standards." Redmon talked about his fear that technology could became a tool of the elite, used not to equalize power but to maintain and consolidate it.

He would later say he gave up doing any computer vision at all because he saw the impact his work was having. "I loved the work, but the military applications and privacy concerns eventually became impossible to ignore." These views would strengthen over time. "The US is the largest state sponsor of terrorism in the world," he would post in June 2025. "Don't work on tech for the military."

Cukor had more success convincing others in the AI space to support the US military. For some the idea of working with the military appealed simply because it was such new and unfamiliar territory. To others, I learned, it felt more like diving into war movies than anything from real life.

Pilot AI became a team favorite because unlike traditional vendors they were always "so chill" about making changes that weren't itemized on the contract, said Kateryna (Katya) Volkovska, a project manager at ECS Federal, which managed the subvendors. "They were these crazy gamers from California. All they'd ever done in their life was teach AI and build cool algorithms."

Each startup would be given a new task and then graded on it ninety days later. If they didn't make the cut they'd be dropped from the program. "I'm a Marine," Cukor told me as if that explained it, or as if that were news to me. He went on. "If we need a helicopter we have three on standby," he said, channeling the military mantra of "two is one and one is none." "It's a mentality of no-fail. You need three Humvees because two will break. Always three backups."

Cukor liked to call it horse racing. Zeiler liked to call it unrealistic. Goalposts for judging the algorithms could change halfway through the task, and teams would be like "What the hell?" Evaluation metrics made no sense to the experienced AI practitioners.

But first Cukor needed his yes. As they initially discussed price over the phone, Zeiler started to hem and haw. Maybe it was a negotiating tactic. Maybe he saw there were uncomfortable moral questions to reckon with. Instead of seeking out brides with bouquets they would be looking for jihadis with Kalashnikovs. At least some of his brainy hipster New York workforce would have to swap cakes for bombs, and that might not go down too well.

Clarifai liked to say the company's mission was to accelerate the progress of humanity with the help of continually improving AI, and Zeiler's workforce was under the firm impression they refused work for the military or for pornography "because they didn't improve life." "It's always, like, a divisive thing, like working with the military or not," he said. "I knew that was gonna be, like, friction."

Cukor sensed the delay. He returned to New York one evening to meet for dinner. The colonel always conducted himself exactly the way Zeiler imagined a military officer: "regimens like to the tee." But Cukor also had, when he wanted, the charm of a good storyteller.

"Here is why this project is important," Cukor set about telling him over their meal. Zeiler remembers Cukor recreating the moment when somewhere near a river somewhere in Africa, a routine patrol boat left camp for its nightly run down the river. But they didn't come back in the usual time. People were starting to get worried. They were wondering if they should send another boat to go check on them. Eventually the group came back, their clothes tattered, ammo out, stressed out. They had just done battle not against enemy fighters but a rampaging herd of animals. "It was just like this perfect example where he is like, none of that was necessary. Like they could have saved those animals' lives. They could have saved like all the stress that the humans have to go through, could have saved all the ammunition and all the time if they just sent an AI drone down the river to do the routine surveillance," he told me.

Cukor told me he doesn't remember telling Zeiler such a story, and I strained to take this foundational parable seriously—that the military needed AI to stop themselves from needlessly shooting dead the pastoral animals on which nomadic people relied for their existence. That was hardly the driving idea behind Maven. Bob Work wanted to use AI and autonomy to outpace China's rising military prowess, not to save cattle in Africa. Plus, a surveillance drone without AI might have been just as helpful. I waited to see if Zeiler had more to share.

"I thought it was great because it's not like you're killing humans or like you're causing a war or anything like that," he went on. "There's a lot of tasks that the military has to do that we can start automating with AI and it's actually gonna save lives, not hurt lives."

I understood the broader point he was seeking to make, but still strained to take the specifics of this story seriously. I thought back to the Maven member who wanted to use AI "to reduce the non-American population." And the other who joined Maven thinking that with AI they could now "fucking kill people all the time."

But the military also undertakes offensive missions, I offered. Did he have his head in the sand about that? Zeiler said targeting decisions are tough but he thought AI could do a more reliable job. "It doesn't get tired, it doesn't get hungry, it doesn't get stressed out," he gestured, arguing targeting relies on lots of approvals and other steps, and that full automation would be a long way off. "Ultimately it can make better decisions." He didn't discuss targeting back at the office though. Cukor's yarn had given Zeiler just the (nonlethal) ammunition he needed.

"I love that story and that allowed me to take it back to the company as well and be like, guys, we're gonna be working on military contracts. There's a huge opportunity for this technology to do good and to ultimately save lives 'cause we're gonna do better decision-making more quickly using AI. And that helped, you know, get everybody in the company."

It didn't quite. "I shouldn't say everybody," Zeiler corrected himself.

"Most people in the company aligned. A couple people quit when I made it kinda a definitive decision."

Zeiler rationalized it was good the workers who remained felt the same way and were aligned to company values. Those company values had, via an unlikely war story set in the African brush, just taken a huge swerve.

In years to come, more employees would leave as more details emerged, but Clarifai would double down. Eight years after his dinner with Cukor, Zeiler's company would create more than a million algorithms, including for facial recognition and sentiment analysis, and would for a time proudly display on its website that it worked for Joint Special Operations Command, home to some of America's most lethal military units that carried out some of its most secret missions.

For Cukor, it was job done. He could now count Clarifai among his clutch of emerging AI primes. Whether he wanted to acknowledge it or not, Zeiler had just become an AI defense manufacturer.

8

SOMALIA

"Just get it in their hands."

LONG BEFORE THE ADVENT OF AI, Colonel Drew Cukor pored over the history of just how difficult it is to get the US military to do something new. Evolution comes in stages. First the idea, then building the technology to match. But the longest tail, he told me, is just getting people to use it.

"It took almost twenty years to rebuild the army at the end of Vietnam," he told me.

The Vietnam War blindsided leadership and population alike. America's bombing campaigns and the extensive use of napalm against civilians including children, the death of 58,000 American troops—and the ignominy that went with it all. Protest snowballed throughout the 1960s, delivering two million antiwar demonstrators to Washington, DC, in 1969 as military desertion rates rocketed alongside reported incidents of soldiers attacking their own officers. "Don't desert. Go to Vietnam and kill your commanding officer," went the advice of an outlet described as a West Coast news-sheet.

On top of overseas deaths and disfavor at home, the Pentagon leaders missed something else entirely: the shifting tides of military expansionism elsewhere. "They're like, 'What did the Russians just do? They've literally rebuilt their entire military while we were in Vietnam,'" Cukor related to me. "We were still on World War II stuff."

It took ten years for the US military to get their heads around the new challenge, and another ten years of fielding new equipment for operators to actually start using it. It took fifteen years to develop night vision. Cukor thought it would easily take twenty years for the military to figure out AI. The only way to speed up the clock would be to give US military personnel a chance to start looking at, touching, using, and playing with AI as soon as possible.

He didn't expect perfection. The first planes and cars came with problems and worked only under narrow circumstances. Failure would always be part of AI, he figured. That's where the learning would come in. But much like planes and ground vehicles, AI would become essential to war, he thought. He wanted AI to be tested and developed in war too.

"Just start," came the advice of one West Coast doyen of the new technology.

But the Pentagon preferred test conditions for good reason. A battlefield is not a lab; it is a place where human beings die. "Move fast and break things" might work in San Francisco, but in a real combat scenario, malfunctioning equipment could be deadly. AI getting something wrong could break the law and undermine the morals the US military claimed to embrace.

For Cukor, there was also the survivability of Project Maven to think about. If AI led to a mishap that harmed civilians or America's own troops, upset Congress, or irked the commanders he was trying to convince to adopt AI, it would risk his project's funding and prospects.

And yet to escape the world of academic papers and internal departmental defense research studies, Cukor felt he needed to make AI real. If they waited until AI was pristine, he figured they'd never use it.

"It's one of the rules of Maven, you always deploy before you're ready," recalled Matt Barchick, a Marine Corps reservist who would join the team for more than two years starting in April 2019. Cukor preferred to call it "field to learn." It was a version, in a way, of the idea he'd developed in his 2014 thesis: "Operate to Know." Both relied on constant action and constant feedback.

This sort of risky approach might invite questions even from supporters within the Defense Department, never mind their detractors. Officers who don't steer clear of controversy rarely make it to general. Cukor, who was approaching his thirty-year up-or-out retirement date, was beginning to wonder whether he would manage either. If things didn't work out for him, maybe they would work out for algorithmic warfare.

"I think there's a bunch of junior officers that should be motivated to want to actually raise their hands and participate," he said to me one day. "All too often, people don't wanna step outside their box."

He was inspired by a trio of controversial US military strategists who came before him. All three were passed over for promotion but managed to make lasting impacts on the US military. John Boyd was the unorthodox Air Force "ghetto" colonel who remade America's aerial onslaughts and sped up decision-making through a novel analysis of combat operations processes. Admiral Hyman Rickover was known as the father of the nuclear Navy. He helped develop the first nuclear-powered submarines and spent an unprecedented sixty-four years in active duty, including three decades as director of the US Naval Reactors office. Lieutenant General Jimmy Doolittle was the Army aviator who argued for pilots to trust not their senses but their instruments, led air raids against the Japanese after Pearl Harbor, and in the 1950s secretly advocated for the CIA to raise standards and yet also abandon "fair play" in the struggle against Communism.

"These were total mavericks," Cukor said. "Each of these communities has their fathers."

Would Cukor play that role for AI? One Mavenite remembers Cukor sometimes cast himself to the group as the next Rickover, the sharp-tongued Navy admiral who—according to one write-up from 1954—"spurred his men to exhaustion, ripped through red tape, drove contractors into rages," and managed to solidify a cult around himself, even as he made enemies in the bureaucracy. Cukor told me he wasn't anywhere near Boyd's level. "I am not even a tenth of that man." But another Mavenite called him "the next John Boyd," after all.

Cukor was looking for a total culture shift among operators on the front lines of war. He didn't expect warfare to change in twelve months, or even five years. He figured change would take time, infrastructure, investment, and sustained work.

"That's really what comes from this AlphaGo thing," Cukor told me one day. "You can't play these games anymore without this technology. You can't go and step on a modern battlefield without this, otherwise you are behind, you're not serving your country well, and if the other country has it? You're Lee Sedol. You can't win. You will lose."

That constant fear of losing ran constantly through America's top defense leaders, he said.

"The number one thing for all of us any time we step on a battlefield—I remember this twice in my little tiny career, Afghanistan and Iraq—is, are we going to be the first to lose? Like, catastrophically. Are we ready for this? The Marines aren't thinking that but the senior leadership are like, 'Do we have everything?' And that's why they plough so much in, because no one wants to preside over the force that's going to lose."

Success for Cukor on Project Maven—however patchy—would be about getting AI out into combat scenarios in support of some of America's deadliest missions.

"Just get it in their hands," went the mantra. "Honestly, the biggest push that we had was, let's just get this out to as many military officers as fast as we can so they can begin thinking, because this is going to be the future."

The team was already working fast. Within the space of a few months they'd got a slew of new tech companies signed up to work with the Pentagon, they were hoovering up data from US platforms around the world and they had started making algorithms intended to detect relevant movement and objects. But they had one more impossible deadline to meet: Maven's sponsor Bob Work wanted AI in the field by the end of 2017.

Colin Carroll, who had put himself in charge of integration and engineering, now had to get Maven into formations all around the world. But in those early days, the team got turned down. The US Air Force said no. Even the US Marine Corps turned them down. The services were worried, the Maven team figured, about undermining their own investments.

But as Carroll remembers it, none of these impossibilities mattered to Cukor. He conveyed the instruction to get it done to Carroll in typically flat terms: "Christmas."

CARROLL FIGURED Somalia was a good place for the first Maven algorithm to land. That was not only because the data they were training the AI models on came from there, which meant the algorithms would be likely to detect objects from the same area, but also because the operational risks of messing up weren't as high as in Afghanistan. Plus the Maven team had friends in the world of special operations—an old contact of Carroll's from the Naval Academy was a company commander with Marine Forces Special Operations Command out there. "I chose a friendly site in an area where it was more of a free-for-all," Carroll told me.

Special Operations Command (SOCOM) had already proved more willing when it came to sharing data. Teams from the most elite part of SOCOM, known as Joint Special Operations Command, or JSOC, were already in places where the US was "schwacking bad guys on a daily basis." And what they did often filtered down to larger conventional units.

"JSOC wanted AI on everything," according to one internally commissioned, nonpublic review of Maven I saw, attributing the statement to an anonymous government leader interviewed for the report. JSOC wanted to be, the interviewee went on, the first AI-driven organization. Maven "would go on to put AI in every single formation they had."

Carroll saw Somalia as the ideal vehicle to start the spread of AI across the entire force, but Somalia's ongoing civil war was decidedly low-tech. When I visited in 2013, I walked the beach surrounded by armed protection and dined on camel milk. I saw signs the capital city of vaulting white arches pockmarked by war was tentatively coming back to life after UN-backed African troops pushed out the jihadis in charge. But al-Shabaab Islamists gained new strength, and a twin bomb blast in the capital in October 2017 killed more than five hundred people, including three US citizens—the deadliest attack in the country's history. The Pentagon was upping air strikes and ground operations against al-Shabaab and Somalia's small but deadly ISIS chapter, which formed in 2015 and grew over the course of 2017. Armed with Kalashnikovs and trucks, they blew up themselves and hundreds of Somalis with homemade bombs. The setting could hardly offer heavier contrast with well-equipped US forces—or the sci-fi imaginings of *The Terminator*.

But the Maven team wasn't aiming for great-performing AI in Somalia. Success would be something much simpler: get the system up and going, and into the hands of the first real-life users.

"If that thing catches on fire, we're still declaring victory," Carroll told me, describing the attitude toward the first Maven fielding effort.

And did it? I asked.

"Maybe it smoked a little bit."

The main task was to get newly trained algorithms into the combat zone and then hooked into the military computer system that operators already used to view and analyze incoming drone footage. The Somalia team had four people staring at a screen showing drone footage on this system, known as the Tactical Imagery Production System, or TIPS. It was long, dull work for the intelligence analysts on site. The idea was—eventually—for algorithms to make their jobs easier.

The algorithms needed a system of their own to live in. Such a system would require specialist AI chips known as graphics processing units (GPUs), drive space, electric power, and some kind of cooling from systems in remote locations that were already overstretched.

The effort would also come with a bunch of red tape, including a six-month review process to check these systems were hardened against cybersecurity attacks. Cukor told me they never had a cybersecurity incident, which was hard for me to check. But they came close—the smoke to Carroll's fire.

On a work call with contractors, Carroll and others discussed how to integrate the algorithms into TIPS. During the call, one person from a different department started telling the Maven team they can't just plug an algorithm into the network.

The person started talking about "IA."

Carroll told the person he'd got his letters mixed up.

"It's called AI. I fucking know about it."

But Carroll's telephonic correspondent was indeed talking about IA—information assurance. That was the Pentagon's extensive rules-bound checklist to secure a software system against both physical and digital vulnerabilities so it didn't get hacked or otherwise mess up.

Carroll was by then googling frantically on the call.

"Fuck me," he remembers thinking to himself. "Like what is all this shit?"

He ended the call and began speed-reading about IA. He couldn't, it turned out, simply plug a new system carrying an algorithm into an existing system without getting it approved. But doing it right meant paperwork and process—maybe up to a year to get the golden security ticket known as "Authority to Operate," he was told.

Mavenites were not big fans of paperwork and process. The team was deep into developing what another member described to me as an "insurgency mentality." Carroll liked to think they were effective because they didn't properly understand the rules or impediments in the first place: "We weren't bogged down in our head."

Cukor himself had little interest in sticking to policy. Sometimes he'd check: Does a proposed action break the law, or just policy? If it was just breaking policy, he'd tell them to go ahead.

A few days later, Carroll figured he had read enough about the IA requirements to be dangerous.

"This is what we're gonna do," he announced. Carroll's idea was that instead of integrating the system carrying the algorithm onto the secret classified network on which America fights wars, he would instead put the algorithm onto the unclassified part of the network before any data traveled into the sensitive part. That would mean fewer hoops, meaning no need for inordinate security protocols and time. Data that beams down from the MQ-9 or ScanEagle drones to local antennae is encrypted, but it is unclassified until it enters the secret network. Carroll figured he would put Maven's AI on the system just before it got to the secret part of the network, so only the output of the system—known as the AI inference—would continue onto the classified network.

I could see it was an unorthodox solution, but what if someone hacked the algorithms? No one was going to do that, he told me, arguing they were still encrypted anyway. If experts began to question his approach he would rely on his Marine Corps charm: "Just shut the fuck up."

In late September, the Maven team did a test run for Somalia. They were at the Army's Fort A.P. Hill training and maneuver center south of Fredericksburg, Virginia, and the process just about worked: for the first time, they flew ScanEagle drones, pushed data through TIPS, and ran models against the footage.

Standards at the Soviet-built Baledogle Military Airfield, fifty miles northwest of Mogadishu, were altogether looser. Somali nationals sometimes rolled into the sensitive compartmented information facility area, or SCIF, an area meant only for classified information. Sometimes they'd even be on their cell phones while US teams were planning operations. "You could never do that in Afghanistan, Iraq. So we chose places like that where it would be less disruptive."

(A year later, al-Shabaab launched a massive assault against the base, prompting an hour-long shoot-out against about a dozen militants who detonated a 5,000-pound truck bomb—at the time the largest on the continent in history—shaking the entire base. The attackers tried again, with rocket-propelled grenades, machine guns, and assault rifles, but were eventually all killed with a combination of sniper, mortar, and drone fire.)

When the Maven team plugged the algorithm in that hot December morning in 2017, despite Carroll's earlier confidence, something still went awry with the internet protocol. "We fucked it up somehow and we just went to lunch," said Carroll.

Their return came with a jolt. They'd somehow managed to accidentally shut down the network. They'd been spamming the JSOC network with rogue bits of data known as packets, and the system had overloaded.

"We brought the whole fucking thing down for forty-five minutes," Carroll recalled. This was a dangerous operational failure. Had that been Afghanistan or Iraq, Maven might never have recovered.

When they finally got the system up and running, it signified that they had officially met Cukor's Christmas 2017 deadline for deploying the first computer vision algorithm to a combat zone overseas, all within eight months of starting Project Maven.

But no one in Baledogle was much impressed. The algorithm was meant to put a box around real-life things, naming what and where they were. But there were so many boxes around so many things it was overwhelming. The clunky, bright yellow line around a person was so thick it obstructed what was in their hands—it mattered whether they were carrying a bag or a Kalashnikov. Even the Maven folks admitted it wasn't working; to them it looked like a four-year-old had got busy with a fat pen over the screen.

AI boundary boxes were also calling out totally the wrong thing. Every twenty minutes or so, the MQ-9 would pan away from overflying the ground, shift angle, and scan the horizon to do a weather check. That's when things got sillier still. Screen alerts went up that there was a school bus in the sky. The algorithms were incorrectly mislabeling the clouds.

The special operators decided it was easier—and safer—to turn the AI off. So long as the boundary boxes were obscuring the scene, AI just seemed to be making things worse. Maven took the hint—they went back to the developers to narrow the line and make it fainter. Six times in five days that December, they updated the system in Somalia. They later began to experiment with displaying little dots instead of larger boxes.

They grappled over what to do about the clouds. Industry partners wanted to improve the algorithms, so they identified each and every pixel accurately. A better pipeline of images to train the model would help. Carroll wanted to fix the problem another way: just tell the model that it should never display a box around anything of a certain size and certain angle associated with the skyline. "Just never show it to a human," he said. The vendors considered that cheating; Carroll figured he'd simply made an adjustment so the model would be more accurate for the user.

Back at the Pentagon the flow of "ruthlessly candid feedback" seemed to slow to a drip: the special operators appeared to have stopped using the system altogether. If that got out to Congress, it could be bad for the project—its funding stream and any chance for it to improve. If it was going to work, Cukor decided he needed more help. Someone who understood what it meant to stare at a screen that played full-motion video, or FMV, all day long, day after day, eyeball-to-pixel with the horrors of war. Someone skilled in the art of gentle, consistent persuasion. Someone who could simply sit alongside special operators, all the time, and get them to turn on the AI and start using it.

A new order came down from Cukor: "Go find someone who knows FMV and has a bunch of tattoos."

JUSTIN GUZZARDO CAME WITH BOTH. His first tattoo arrived at age eighteen after a dare—a sweet little rubber ducky whose head was just visible above the line of his swimsuit. That was the year after his beloved adoptive father managed to dissuade him from following his passion for musical composition.

"He pulls me aside," Guzzardo remembers. The seventeen-year-old grew up in a family that loved music. His older brother loved the guitar; Justin went for piano. He was on his way to getting a scholarship to a Texas music school. "'Don't do music,'" said his adoptive dad, a former Marine sniper who had married his mom when Guzzardo was aged two.

The advice landed with Guzzardo. He was born on a military base, and remembers, somehow fondly, the go bag always being ready. But he never had any interest in the military for himself. His dad advised him against going into the Army or the Marines, but instead to pick the Navy or Air Force. That way Guzzardo would get a technical skill and encounter less struggle than he had.

But by the time Guzzardo was in Air Force training, America's secret drone wars were taking off, and his next few years were spent glued to screens out in California's Beale Air Force Base, calling out every detail he could see as the bombs rained down. His team got advice on how to stay focused on long monotonous shifts. Stay busy, do something sensory, and remember that for the people on the ground it was life or death. That was the fear factor.

Guzzardo developed his own cocktail for getting through fourteen hours a day glued to a screen. He'd go through a bag of sunflower seeds each shift, working his fingers and mouth, constantly topped up with Red Bull. Others on the team chewed gum or snapped a rubber band against their wrist. Anything to keep busy, and keep going. But mostly, he taught himself to keep asking questions. He went into his head, seeking out new details. Every question he asked himself, every detail he unturned, every minute step he took deeper into the picture and the scenario, was a way to stay awake. He kept his brain buzzing, sleeping for four or five hours, and then was back for more.

He started out as the "eyes on." It meant just that: you couldn't take your eyes off the screen, not even to turn a shoulder to sneeze. He'd call out what he saw out loud. Time, place, building, person, clothes, direction. And so it went on. Almost all day long, every day, for two years. A keen eye could spot the teal glow of muddy fertilizer as it dried on village rooftops, a possible explosive ingredient in improvised roadside bombs. He graduated up the roles. Next up was a year as a FMV screener, typing down furiously without looking what the "eyes on" said aloud. He'd plug into drone feeds from Predators, Reapers, U-2s, Gorgon Stare, and more. The information would end up in databases and slideshows, and sometimes the people he counted in the FMVs would end up in cemeteries. He worked with the conventional military and soon enough special operations. That was where everyone wanted to be. Scanning villages, rooting out IEDs, and searching for terrorist leaders. No room for error. More kinetic, as the bloodless military verbiage for live combat goes. In a single year, he worked on two hundred special operations assaults.

At first, the team cheered and high-fived when they got a bad guy. They were doing America's work. They loved their country and they were protecting it. But that started to change. A life is a life. The targets were people too. There wasn't so much to rejoice about, if you thought about it. A lot of death, on a relentless cycle. All told, five thousand hours tallied up to two hundred days and nights spent looking at emptiness or daily existence or violence, nonstop. In between came Kandahar in southern Afghanistan, Erbil in northern Iraq, and back to Beale. Looking for tiny black specks of people who might be about to attack a base, as they often did, or disappearing in the distance. The work was so hush-hush back then you couldn't talk about it at home, and you couldn't ask for help if it got to you. You'd be removed from service. You couldn't even whisper about a weakness in your armor. No one talked about the anxiety. Depression would get you fired. The skill was to sock down emotions.

Guzzardo was kind of okay at socking them down. But it was essentially an impossible ask of most people. He doesn't remember the first time someone on his team lost their life to suicide. He didn't like to think about it. Doesn't like to think about it. Friends and colleagues. He pushed it out of his mind. But he knows the count. Eight people. People would have breakdowns too. One seemed to go mad, right at the start of a shift. She got up on a table and started talking gibberish. She just kind of lost it.

Pretty soon, he had tattoos everywhere. A full right sleeve. A full chest piece. Half a sleeve on his left arm. Back rib cage. Pelvis. Mementos of friends lost. Little of him emerged unmarked. He struggled a little, felt less innocent, not so starry-eyed about the task his father had laid out for him all those years ago.

By 2014, the head honchos had started to realize the drone function was in crisis. The policy changed, all too late, and they were allowed to ask for help. A psychiatrist arrived on the floor. And a therapy dog too. But Guzzardo had decided to pack it in. He grew so interested in mental health he started to retrain as a nurse. And that was where he expected to stay for a while, until he felt the tug back to war.

Guzzardo told me only that he was deployed to east Africa as a defense contractor in 2018, refusing to be any more specific. But I learned from three other people that he went to the Baledogle base in Somalia where Maven had installed its first algorithm just a few months earlier.

He was the first in what would become a long line of people sent to the front lines for Maven on behalf of the project's prime vendor, ECS Federal. They were hiring for the first "field service representative," looking for candidates who could be "integrators, salespersons, and ambassadors"—a demanding jack-of-all-trades position. Guzzardo's potential new bosses wondered if he was too much of a lone wolf to fit in, but he came with niche capabilities, experience that enabled him to connect with screeners and other drone analysts who operated onsite. That was why Cukor had wanted to send over someone steeped in the language and minutiae of drone footage analysis. He figured only someone with Guzzardo's sort of experience and understanding might be able to talk US forces on the ground into swapping out exhausted screen-watchers like him for indefatigable AI. Guzzardo and his tattoos suggested that what really mattered when it came to launching robot wars was a wholly human quality—rapport.

But it was going to be difficult. Stuck in the middle row of a C-130 swinging in from the searing heat of Djibouti, Guzzardo had to steel himself to avoid passing out in front of all the Marines. When Guzzardo arrived at the site in Somalia after dark he might have remembered how to hunt on screen, but he still had almost no AI training. He found the algorithm not only turned off, but disconnected. It may as well have been in a closet. After about a week of introduction, his Maven hand-holders went home and left him to his plywood shack in the middle of the hot Somali scrub, whirring to the pulverizing sound of noisy air conditioning.

As a representative sent by a vendor, Guzzardo came with no rank at all. "I would have told a contractor to go pound sand," he said, imagining his fate. In the aloof and closed-off world of special operations, he wasn't sure if anyone would even eat with him come mealtime. "No one owes me anything." The only way to make friends, he figured, was simply to volunteer as free labor. He was back to being a screener.

The ScanEagle is a small tactical drone, operated not from big bases back in the US, but by deployed units on the ground. It needed two analysts watching the screen in the hot outpost, unable to move their eyes. Now they had Guzzardo's eyes too. After a few shifts helping them out, he convinced them to let him turn the AI back on.

Detection success was still low—more than half the time the

algorithms got its predictions completely wrong. But several possible tasks started to emerge. The AI could potentially help count people in a given setting such as the marketplace, or follow America's own convoy vehicles from above, or maybe even track people. These were all tasks that fell to humans, whose job it was to check for targets, as well as friendlies and civilians they needed to avoid hitting. The screeners eventually told him they wanted the AI to help them after all, but it wasn't getting hold of people and vehicles the way they needed. Guzzardo started to put parameters around its usage—it wasn't some mythical tool that delivered magic results, he tried to tell them. Its effectiveness depended on how they used it and what they wanted from it in a specific circumstance.

After a raid started, people would often scatter in different directions all at once. A single drone with a soda-straw field of view and two pairs of eyes following it struggled to keep track of where they all went. If everyone was armed, in the middle of a firefight, that made for an extremely dangerous moment: sometimes the people the US attacked just scattered, and sometimes they returned fire, or waited for their moment to launch a grenade. And one day the algorithm spotted a person hiding in the bushes who no one else on the screens saw by eye.

"Holy crap," Guzzardo thought. "This could actually work."

9

MORAL OUTRAGE

"I had a sense of how fucked this was."

INSPIRED BY ALPHAGO'S MACHINE victory in March 2016 over the best human go player in the world, Cukor had always considered Google's DeepMind as the global "A" team for AI. He wanted that team playing for the Pentagon.

If Cukor could bag them, it would mark a complete departure: Google had never done any classified national security work in its thirty-one-year history. Ahead of the company's 2004 public listing, the duo put into print the company's unofficial motto: "Don't be evil."

DeepMind was co-founded by Demis Hassabis, a child chess prodigy from North London who bought a computer with his winnings aged eight. As a teen gamer, Hassabis dreamed that AI could solve disease and fix the world's health problems. In DeepMind's early days, the company won backing from Peter Thiel and Elon Musk, but was soon ravenous for more computing power to develop its algorithms. Google resolved those problems when it snapped up the London startup for more than $500 million in 2014, making a big bet on AI as the technology that would one day rule the world.

Breakthroughs at the company started to come quick, including AlphaGo's victory over Lee Sedol in 2016. Hassabis seemed absorbed by the sheer intellectual joy of combining computer games with his pursuit of artificial general intelligence. But Cukor saw something else. The AI's unexpectedly creative and risky gameplay revealed to him the transformative potential of algorithmic warfare.

DeepMind was an unlikely target: Hassabis had signed a 2015 open letter against the perils of bringing AI into warfare, along with nearly five thousand AI and robotics researchers.

"Most AI researchers have no interest in building AI weapons," the letter read. "If any major military power pushes ahead with AI weapon development, a global arms race is virtually inevitable, and the endpoint of this technological trajectory is obvious: autonomous weapons will become the Kalashnikovs of tomorrow."

The reliable automatic assault rifle—created by young Soviet sergeant Mikhail Kalashnikov in 1947—became the world's most ubiquitous weapon, killing millions of people across the world in wars large and small. The contention behind the letter, signed by the Pope, the Dalai Lama, and Elon Musk, as well as Hassabis and thousands of AI researchers, was that AI would not create a new weapon so powerful as to be beyond the reach of use—like a new and terrifying variant of nuclear deterrence. Instead, it would create a ubiquitous killing machine. Their fears prefigured the arrival of millions of deadly drones in the Ukraine war, a fraction of which can kill autonomously, with more underway.

The letter's signatories were inspired in part by the Stop Killer Robots campaign. When it formed in 2012, several groups came together seeking a preemptive ban on the development, production, and use of fully autonomous weapons after one founder of the group read some US Air Force documents and realized the US military was already planning for their arrival. "We couldn't stop armed drones, but we could kind of stop the point at which you lose human control over the use of force," Mary Wareham, an arms expert at Human Rights Watch who became coordinator of the campaign, told me.

Sending "emotionless robots" into combat would make wars more likely and make it easier to kill, her group argued. Civilians would bear the brunt. The message caught on fast. Pakistan—which had suffered hundreds of unacknowledged CIA drone strikes that killed thousands of its citizens—became the first country to call for a ban. The UN took it up as an issue for discussion, and, in 2016, Iceland's *Alþingi* passed the world's first parliamentary resolution endorsing the call to ban fully autonomous weapons.

Perhaps more immediately consequential, thousands of computer scientists with AI skills refused to work on warfare. Military spending had supported Silicon Valley for decades. Now, the Pentagon scrambled to make defense work appealing to a new generation of tech workers put off by the drone wars and the 2013 revelations of the National Security Agency's mass spying programs. Worried by the trend, the Pentagon set up an outreach office in Silicon Valley in 2015 called the Defense Innovation Unit Experimental (DIUx) to bring in tech startups. Ashton Carter, the theoretical physicist turned US defense secretary, drove the initiative. But one man's hero is another woman's foe. Wareham didn't buy this appeal to techies by defense officials in jeans and sneakers. She saw it as an attempt to reframe an unconscionable AI arms race as a more palatable AI talent quest.

Hassabis also stuck to his unarmed guns: Cukor's outreaches to Hassabis landed flat. Hassabis didn't even respond to Cukor's entreaties, Cukor told me. "Those guys were high on their horse," Cukor told me. "They're very anti-military." There was just no way DeepMind was going to work on defense. They had the view, Cukor said, that artificial general intelligence—AI that could match or surpass that of humans—would destroy humanity. "They wanted to figure out AI and then trap it in a box." (A Google representative declined to comment.)

Cukor was striking out with what he saw as the next best option too: he couldn't get Google Brain on board either. Co-founded by Stanford professor Andrew Ng, a computer scientist who argued for the use of AI chips known as GPUs in deep learning, Google Brain captured Cukor's attention with its 2016 breakthrough on machine translation, when it scored a huge sudden improvement in translating Chinese into English. Jeff Dean, the engineer who led Google Brain, previously worked with Clarifai's Matt Zeiler and in 2013 hired Zeiler's former professor, AI pioneer Geoffrey Hinton.

Cukor wondered if he would have more luck convincing a more commercial part of Google to open discussions with Project Maven. Google Cloud, a separate new division of the company eager for business, was developing its own computer vision algorithms. The new cloud division had only two significant contracts in 2015 and was in search of two vital commodities: clout and customers. The company saw promise in a lucrative relationship with the Defense Department, and Cukor thought the company easily set the standard for commercial cloud compared to rivals like AWS and Microsoft Azure. If he succeeded, he would be snagging a powerhouse.

Cukor had two things in mind: first, he wanted Google to make computer vision models for Gorgon Stare. That was the complex, bug-eyed WAMI sensor that the US attached to Raytheon's large MQ-9 drones. The sensor produced too much information over the course of several hours to attempt to process by eye. Running algorithms against hundreds of video feeds at different angles to form a composite picture of an entire city over time presented a task so hard and so complex that he thought only Google would have the wherewithal to pull it off.

"Nobody had done this," said another person familiar with Maven's effort.

Second, Cukor told me he wanted Google to build the Defense Department its own Google Earth. It would function as a workflow tool for intelligence—and ultimately for targeting. This would go further than any other undertaking to bring intelligence into operations. Cukor wanted Google to build the Pentagon a cloud-based AI targeting platform.

Google's C-suite boasted at least one high-profile Pentagon fan: Eric Schmidt. As a UC Berkeley computer science graduate in the early 1980s, Schmidt took money from DARPA and attended Burning Man. He epitomized the tension at the heart of California: a sunny liberal outlook and economic growth built on the back of a Cold War defense-spending spree. In 2001, he had joined Google as chairman and chief executive to undertake "adult supervision" of the Google co-founders. (Schmidt initially thought "Don't be evil" was the "the stupidest rule ever.")

But even signing what Cukor referred to as Google's "B" team required careful diplomacy. At a defense conference in mid-July 2017, as Cukor was trying to snag the deal, he used a rare on-stage appearance to praise the man and the company to an audience of defense-tech experts: "Many of you will have noted that Eric Schmidt is calling Google an AI company now, not a data company," he offered suggestively.

The following month, US Defense Secretary Jim Mattis flew to California to visit Google headquarters. Mattis preferred ancient wisdom to newfangled ideas—he carried a tattered copy of Marcus Aurelius into battle in his rucksack—but under the influence of his advisers, Mattis was waking up to cloud and the need to enlist Silicon Valley. The Pentagon will get better at integrating AI advances into the US military, Mattis told reporters during his visit.

But a few weeks later, when it came to pushing play on Google's involvement in Project Maven, nerves were still running high among management.

Late one Friday afternoon, Google was still running the internal policy, legal and contracting gauntlet about the deal. Google Cloud's chief executive Diane Greene called Aileen Black, who ran public sector cloud for the company, as she raced across San Francisco International Airport to board her weekly evening plane back home to the East Coast. Black loved the US military. Her father was a two-star general. West Point and the Air Force Academy figured prominently in her family. She was one of the few in her family who hadn't gone into the military. For her, working on Project Maven was a chance to get closer to public service, do her duty, and protect America.

What's more, she was enchanted by Cukor. He would call her cell directly on weekends. She just trusted him, and she loved how hard he worked. He was focused, driven, ambitious. "I'd want to be in a foxhole with him any day," she told me.

He had portrayed Project Maven to her as a great opportunity for

Google to give back to the country. She was excited at the prospect of helping the US government.

Cukor also told Black about Lieutenant General John Kelly's son who died. She was convinced Maven would be about saving all innocent lives, and not just American lives. "It was to help identify at night that somebody's walking across a field carrying a sack of potatoes versus a shovel and a landmine," she told me. "I was proud and honored to be part of something like that."

No algorithm on a WAMI sensor would be able to identify so few pixels at such great distance. But there might be ways for AI to overlay, characterize, and identify movement, a process known as backtracking. As she raced through the airport, she tried to convince Greene on the other end of the call—impassioned, clearcut, out of breath.

"We're an American company; we should provide the best tech to our men and women out in the field protecting our rights," she panted on her way to the plane, according to a person familiar with the exchange.

She emphasized there were ethical guardrails, trying to persuade Greene to seal the deal: "It's on the books: there has to be a human involved and this project has nothing to do with autonomous weapons."

That has not stayed true. I learned during the research for this book that Maven is indeed involved in America's development of autonomous weapons. (See chapter 25.) The Pentagon's largely secret initiative to fashion, test, and deliver thousands of its own killer robots included efforts to incorporate Maven algorithms so drones can identify and select targets under their own steam—and kill. (I also learned that Maven ran Gorgon Stare algorithms produced by Google in combat in Afghanistan in 2018; see chapter 10.)

But that was seven years into the future. Terminator still didn't exist yet. Google signed.

Cukor had chalked up a new conversion. At the time, Google's new cloud division counted only four sales reps; the contract was a huge coup for them. Doubly so, since Google had no classified cloud for national security work at all.

"I was like all sad face," someone at Amazon Web Services (AWS) recalled to me, after catching wind that Cukor had picked Google to be Maven's partner. AWS already provided cloud for the CIA and intelligence community and had spent weeks chasing after Cukor for a spot on Maven, sponsoring an entire booth at a conference they heard he would attend just to have the chance to run into him and convince him, to no avail. "We're the ones with classified cloud."

It might have been Google's "B" Team, but Cukor luxuriated in the memory of snagging Google Cloud: "It was lovely."

Not for long. Soon a public furor over the Maven contract would tear Google's workforce apart and set back the Pentagon's efforts on AI and its relationship with Silicon Valley writ large. Google Cloud's chief AI scientist was Fei-Fei Li, a Stanford academic who arrived in New Jersey from China in 1992 at age sixteen, speaking no English and working weekends in her family's dry-cleaning business. As a computer scientist fifteen years later, she assembled fourteen million images across 22,000 categories as part of a mammoth campaign to build the very first computer vision training sets. "Flowers and nature and things—nothing I could use," Cukor dismissed. In 2017, she joined Google Cloud as chief scientist of AI/ML.

Aileen Black and several others on the team wanted to announce publicly the new defense contract with pride. The contract was the culmination of a five-month-long race among AI heavyweights. Google had beaten out AWS, Microsoft, and IBM and the news was "a really exciting great win," Black said in emails I reviewed. She also counseled that the agreement, subcontracted through ECS Federal, would in any case eventually leak and they should craft their own message. But Li worried about negative attention. In public, Li advocated for the careful, "human-centered" usage of machine learning. In private, she warned that weaponized AI would be "red meat" to the media to find all ways to damage Google.

"Weaponized AI is probably one of the most sensitized topics of AI—if not THE most," she wrote to colleagues in emails that were later leaked.

"You probably heard Elon Musk and his comment about AI causing WW3," she wrote. "Avoid at ALL COSTS any mention or implication of AI."

When I asked her years later about these comments, Li suggested her private comments had been misinterpreted. In an emailed statement, she told me that she didn't think Google's work on Maven really counted as advanced AI or machine learning but was rather "table stakes cloud products" focused on storage, infrastructure and data analytics. "I was extremely anti-hype," she wrote to me. "My email to the salesperson was very much about, 'if you're hyping it this much, it'll have consequences that you don't realize.'"

But Li's emphasis on "a vanilla cloud technology angle" hardly encapsulated Maven. "We can steer the conversation about cloud but this is a [sic] AI specific award," Black wrote back, unwavering, according to documents I reviewed. The award would mean $15 million for Google over 18 months, but as the program grew the expected spend was $250 million a year. Maven would also sponsor the company's efforts to do secure classified cloud defense work, she wrote in the documents I reviewed, describing such help as "priceless."

Black lost the argument. The company signed in September, decided against making the contract public, and a team from Google's Advanced Solutions Lab got quietly to work. The first key milestone would be to deploy Cloud AI models on the government platform for field tests, according to an October 13 weekly update from the product team I reviewed. The next week, a fifty-strong Maven team visited Mountain View for a two-day kickoff and made clear that Google's Cloud AI team was "the core" of Maven. The initial focus would be on detection, classification and tracking in city-wide images and video from WAMI, but the ultimate aims included scene recognition, event alerts and a "very important" user interface, according to customer-engagement updates I reviewed. A Google Cloud engineering team was "heavily engaged," visiting California's Beale Air Force Base at the end of October to meet drone pilots. Google received 251 images from US combat theater to help train models in mid-December, a number that rose to 5,000 labeled in-theater WAMI images by February 2018. Cukor told me that Project Maven ultimately shipped "tens of thousands" of expensive annotated data to Google, Cukor told me, adding that Google "of course" made AI computer vision models for Project Maven that subsequently ran in combat in Afghanistan. Google algorithms were on their way: they were already detecting vehicles that human labelers missed. Elsewhere the team was assessing cooling systems for the extensive infrastructure that carried the GPUs necessary to run the algorithms on WAMI footage from Reaper drones, and figuring out how to airlift all the contraptions in a big tractor trailer over to Afghanistan.

Not a drop of it in public, though. Four months into Google's secret work on AI for the US military, Google chief executive Sundar Pichai even told a forum, "Countries need to demilitarize AI."

But Cukor could see things were brewing. More than once, he told his boss Lieutenant General Jack Shanahan they had problems out at Google: "There's some people that don't like what they're hearing."

One of them was William Fitzgerald. By the time the Irishman from County Waterford became head of communications for Google Cloud AI, he had been at Google for nine years. He'd started out in sales, working his way up through roles focused on public policy in Pakistan and Hong Kong and occasionally accompanying Eric Schmidt on his jet and typing up his tweets. But he'd also read about America's drone wars on Pakistan, protested the Iraq war and read anti-war critic Noam Chomsky in college. He'd helped hide NSA leaker Edward Snowden in 2013 and cautioned against Google's Schmidt meeting Trump in 2017. He'd also watched Lieutenant General Shanahan court Google over months and seen how eager Google Cloud was to bag a defense contract that could lead to more Pentagon cloud work. When he got copied into an email one day discussing the Maven deal specifically, he started telling others at Google they shouldn't take the contract, he told me in an interview. And when they ignored him, he took more furtive action. "I didn't think it was a good thing for society or for Google," he told me.

Fitzgerald tried the legal teams, the policy teams, the ethics teams and one day he tried Meredith Whittaker. Hey, I know you do AI ethics stuff, went his outreach in late 2017. I'm one of the few people who's involved in a secret project and I'm really disturbed by it, went his gist.

Next came a stash of documents. Google was developing algorithms for America's illegal drone wars on an effort called Project Maven, Fitzgerald told her. *What do we do?*

Whittaker, who thought of herself as an "internal dissenter, an academic," hated the way the AI debate focused on sci-fi versions of the future. What AI was getting wrong right now was already a huge problem. Just a year earlier, Google's own photo app had confused dogs for horses. A month later, it was tagging dark-skinned people as gorillas, sparking condemnation and crisis within the company for embedding racism. She convened a 2016 meeting at the White House to highlight how integrating AI was already embedding bias and injustice into American institutions, and would only get worse. Throwing war into the mix was almost unthinkable.

With Kate Crawford, a Microsoft Research principal researcher, they founded the AI Now Institute, got a home for it at New York University, and started putting out papers on AI bias and mistakes. The AI Now Institute's 2017 report echoed calls for strict limitations on the use of AI in weapons systems. Whittaker hadn't known it was happening in her very own workplace. But as soon as she found out, she figured Project Maven amounted to "a very, very scary recipe."

She had read her Dwight D. Eisenhower and understood the military-industrial complex's misshapen incentives and its unwarranted influence on society. Like Fitzgerald and many US commercial tech workers, she fumed over revelations from Edward Snowden four years earlier about the National Security Agency's mass data collection.

"I had a sense of how fucked this was, and of the real dangers of yoking an unprecedented surveillance powerhouse like Google at a global scale to the incentives of the world's most lethal military," she told me. (This was the way the Pentagon bills itself, she pointed out to me.)

"I have a bit of vanity about my dignity," she drawled, somehow sincere and sardonic in one go. "I didn't want to stay and just greenlight that."

Fitzgerald and Whittaker moved their conversation onto Signal, the encrypted messaging app, and added a third person to the thread. The campaign had begun: before long, three people would multiply into three thousand.

Laura Nolan, a Google engineer based in Ireland who taught herself to code in the mid-1990s, separately discovered Google was planning to build an air-gapped system to run code and models for the Pentagon in the company's own data centers. Google leadership talked about the company's AI helping to save lives, but she didn't understand how a computer vision tool to identify targets would do only that.

For hundreds of years, as Nolan saw it, warmongers had always claimed safety as a reason for pushing forward with technological advancements in warfare—and yet horrifying battlefield and civilian deaths had continued. "People always say it will save lives. And they don't really specify how."

She couldn't shake the feeling that any work she did on the project or its apparatus would enable killing. She started having dreams about it. "I just couldn't live with it," she told me. She started to develop physical symptoms—gastric reflux "and all sorts of stress problems."

"If another attack like 9/11 occurred, what might the US do with Project Maven?" Nolan wondered. "Having this ability to generate these huge target lists, all it takes is some political will and, you know, you can go out and you can do thousands of strikes potentially, if you have the munitions to hit them."

But when she asked around inside the company, her colleagues just kind of looked at her "like I was a crazy person." Nolan started to feel like she was losing her mind, even as she was struck by moral clarity. "I can't work at an organization that's asking me to do work that I feel will just directly contribute to innocent people being blown up by missiles," she said. "It's just horrible."

Fitzgerald was adamant that nothing short of canceling Project Maven would be sufficient. Anything else—such as an ethics board or privacy review—would count as a "professional dodge," he told me. He coordinated a campaign, drawing in Google affinity groups, broader civil society players, and eventually, he told me, decided to leak information about the deal. On March 6, 2018, news of Google's work for Project Maven broke, saying the drone project had "set off a firestorm" inside the company. The public revelation made it easier for Nolan to breathe. But Fitzgerald was already signed off work for a nerve condition, something he later attributed to work stress, and he resigned from the company the same month.

Not much changed at first. The Pentagon refused to confirm Google's role and the company tried to bat away the worries. It was a "pilot" project, a Google spokesperson reassured. Flagging images for review was intended to "save lives" and was for "non-offensive uses only." ("Don't talk to us like we're stupid," thought Fitzgerald in response to this messaging from the company. "This is going to be used to kill people.")

Lifelong human rights campaigner Mary Wareham wrote to Google's leaders citing "concerns that this project could contribute to the development of fully autonomous weapons." AI-driven identification of objects could quickly blur into AI-driven identification of "targets" as a basis for the direction of lethal force, she wrote. (She was right, I learned: see chapter 25.) Project Maven could lead to machines having the capacity to determine permissible targets on their own, she argued.

Whittaker, Fitzgerald, and the third person had already typed out a draft petition by then, trying to work in every concerning point. The letter ran to two pages. "We believe that Google should not be in the business of war," opened the document, which the *New York Times* made public in April. Whittaker had no idea the letter was about to change her life and upturn the Pentagon's outreach to Silicon Valley. "Everything is quotidian until you look back at it and put it in a grand or historical narrative," she told me six years later.

Whittaker saw in the narrowly focused drone intelligence project an on-ramp to far greater enlistment of large West Coast tech companies into defense contracting. She saw deadly military applications coming within Google Cloud. She saw Eisenhower's warning flashing red. This was the arrival of Big Tech's new military-industrial complex. "There are many more Mavens," she said.

Senior defense leaders railed to me about Whittaker and her motivations, but she would be proved right about many of her predictions. More Googlers resigned, and the backlash grew so intense that in June Google announced it wouldn't renew its contract on Project Maven. A new set of company AI principles stipulated Google would not design or deploy artificial intelligence in "weapons or other technologies whose principal purpose or implementation is to cause or directly facilitate injury to people."

It marked a significant setback for the Pentagon. The whole point of the outreach since 2015 had been to encourage Silicon Valley back into the defense fold. Some in the Pentagon told me they saw China's hand in the protests at Google that accelerated through the year, and could hardly believe Googlers would "weaponize" advertising data but wouldn't support national security. US military generals fumed in private conversations and later in public that Google couldn't control its workforce, that it was perfectly willing to help China but wouldn't help America.

The campaign spread far beyond Whittaker. Engineers, activists, ethicists, and a broad swath of computer scientists and technologists including Stanford students were protesting against Google and Project Maven. The Pentagon was losing its attempt to claim the moral high ground and craft a patriotic message that the best of America's brains should arm the country's fighting forces with the best technology.

Wareham, an antinuclear weapons protester from the age of eleven, appealed directly to arguably the highest moral and diplomatic authority available to her cause: UN Secretary-General António Guterres. She saw a direct line from the ultimately fruitful seventy-five-year campaign to ban land mines to her hope for a preemptive ban on lethal autonomous weapons. Even if countries refused to join on the first round, a ban would become "a baton of shame," just as the 1997 landmine ban treaty shamed the likes of the United States, Russia, Israel, and China.

Her movement gained traction. In September, Guterres expressed concern at the "weaponization of artificial intelligence" at the General Assembly at the UN headquarters in New York. "The prospect of weapons that can select and attack a target on their own raises multiple alarms—and could trigger new arms races," he said. "Let's call it as it is. The prospect of machines with the discretion and power to take human life is morally repugnant."

At a November tech conference in his hometown of Lisbon, Guterres added an unscripted twist: "and should be banned by international law."

The furor over Maven had buttressed a new moral argument, drawn fresh global battle lines around the future of AI in warfare, and extinguished any dream Cukor had about enlisting the team associated with DeepMind for the battlefield.

"Holy shit, what else do we not know about what's happening out in tech companies?" General Shanahan wondered. He was right to ask. Dissent was rumbling at Clarifai too. Clarifai "is an active participant in Project Maven," Zeiler announced in a June 2018 blog as the furor mounted. The goal for his company's contribution to Project Maven was "to save the lives of soldiers and civilians alike," he said.

But Clarifai, like Google and others, would never be read in on how the algorithms might actually be used in classified operations. Maven's AI was getting involved in targeting after all. "Google didn't know because it was classified," one Mavenite told me. The workforce should not risk the survival of humans globally just so Clarifai can list on the stock market, declared a scathing three-page public letter from an employee a few months later. "Will Clarifai participate in projects that might lead to large scale warfare, mass invasions of privacy, or (perhaps a bit dramatically) genocide?"

Yet Cukor maintained a studiously sunny outlook. "It was just a lot of noise," he said. "We already had multiple backups."

Cukor orchestrated a Google apology tour. Sundar Pichai, Google's chief executive since 2015, quietly visited the Pentagon in person. Pichai explained how bad he felt. General Joseph Dunford, the chairman of the Joint Chiefs of Staff, accused Google of working with China but refusing to work with the Pentagon. Pichai was a gentleman and "took the blows," a former defense official recalled.

Besides, when one door closes another opens. As Google receded, others stepped up. Microsoft showed up loaded with algorithms, while AWS finally picked up the cloud work it had hankered after all along.

"We ended up getting a great relationship with Microsoft, which really turned the tide for us," Cukor told me. Now he had two Big Tech companies and a handful of smaller startups, with more on the way. "I could not stop the flow of people wanting to contribute. Just massive."

"Whatever you need," he remembers Microsoft chief executive Satya Nadella saying. Andy Jassy, chief executive of AWS, was the same way. "Drew started leaning on us more," a member of the AWS team recalled. "We're not going to quit on him."

Slowly, Big Tech shifted. C-suite objections to AI warfare started to fall away even as it became a reality. Years later, in February 2025, Cukor sent me a link to a news story. Google had updated its AI Principles, the ones they formed in the wake of the public revelation of Project Maven. There was no other commentary from him. No need. Gone was Google's refusal to develop AI for use in weapons.

DeepMind's independence diminished after it fused with Google Brain to become Google DeepMind in 2023. "I personally think that having autonomous weaponry is just a very bad idea," Hassabis would still say in a 2024 documentary, made by the director of *AlphaGo*. But a shift was underway: companies, governments, and organizations should work together to create AI "that supports national security," wrote Hassabis and his colleague in a February 2025 blog post. "Western democratic values are under threat," Hassabis told an Axios reporter at the same time, citing the rise of LLMs and saying the company was "duty bound" to help. He didn't address how AI targeting accorded with Western democratic values, but said benefits had to substantially outweigh the risks of harms.

A couple of months later still, Cukor sent me another message: a picture of Google's latest offering on show at a conference. "Google AI for ISR Operations," read the banner over the screen showing a map of Iran and its neighbors. "Target Area," read a box on the side. "What a difference eight years makes," he messaged me. "Google is now all in supporting

DoD." In Cukor's twenty-year horizon for delivering the future of war, he could stomach the odd eight-year snag that resolved in his favor.

During the course of Google's dissent, Project Maven kept moving forward on "Maven time"—meaning no letup—taking overnight red-eyes and going straight to work. Somalia was just the start: over the course of 2018, Maven's AI furtively found its way into more than sixty sites in multiple countries, including Jordan and the UAE.

In each site, the Project Maven team battled bad networks and reluctant users. They expanded beyond drone footage. They repurposed Maven's machine learning algorithms to analyze feeds that come from security cameras, and even to rifle through imagery taken by blimp balloons tethered above overseas bases. Their algorithms analyzed text and pictures, working their way through the phones and digital files that US forces captured from raids on jihadis. Models could detect building rooftops, walls, entry and exit points, roads, and more.

I learned that Google's war algorithms even lived on, passed down from Goliath to David. Google's leadership simply gave the work the company developed for Project Maven to the government. Xnor, the Silicon Valley startup that became one of the first four companies producing algorithms for Project Maven, subsequently maintained Google's algorithms for the project after Google's contract concluded in 2019, several people told me, a development I corroborated in a review of documents dated March 2019 that refer to the Xnor transition. And despite the moral fury of its employees and activists worldwide, the algorithms created by Google went to war in Afghanistan after all.

PART TWO

FIX

10

THE ALGORITHMS HAVE NO CLUE

"The AI is just a bag of potato chips."

BRIAN WARD SPOKE with a kind of impossible bonhomie. Everyone fighting a war, or undertaking any work on behalf of the country, or just delivering a drink to him was A Great American. No matter where we met, Ward made instant friends. He liked to keep it simple, he said. He showed up in a bar with a baseball cap emblazoned with a Venn diagram of three overlapping circles. One said "What," one said "The," and one said "Fuck." As people made out the intersection of all three in the middle, he would be greeted with immediate confusion, and not a little alarmed delight. He could disarm even the most guarded of people in one throaty laugh or hello. I watched the surliest of waitstaff start fist-bumping him within moments. His drink was a beer, or a cider, or a sour, or a double whisky. If he drank kinda hard, he ate kinda light. One meal a day, he said. The rest of him was all protein shake, as far as I could tell. He was stacked. At two hundred pounds, he said booze didn't affect him. Sixty pounds of that was working out. That's how much lighter he'd started out as a pole-vaulting teenager in Indianapolis.

"Even though it was scary, I liked to freaking run with that damn stick," he said.

Now thirty-five, he laughed as he told me his age. I didn't say so, but he was right when he announced I had assumed he was a decade older. "Everyone does," he said.

It wasn't only strangers who loved Ward. Everyone I met who knew him would do anything for him. He was warm and wild and he worked with never-ending passion.

His wild warmth came with strange twists. Violence was the only answer, he'd say. He wasn't aggressive, and there was no glint of secret malice in him. He meant—I think—that nothing ever improved except through crisis. But violence was also his métier. He'd been in America's drone wars going back fifteen years. After he took to writing songs in a home studio, producing three hundred as his marriage fell apart in 2016, he went straight back to Afghanistan with JSOC, in time for the Trump administration to drop the "Mother of All Bombs." It was the largest non-nuclear ordinance ever detonated by the US military, on a cave system in Nangarhar province.

"We are not heroes or villains, but when it comes to this work, we are more than willing," went the lyrics of one rap he wrote.

Ward bent over backward for people. When friends needed to move across the country, they called him for help.

"I knew I wasn't their first call," he told me. A sudden shot of pathos. He was the one who'd say yes.

I'd never met a two-hundred-pound people pleaser.

Maybe all people who worked in the military were like this. Ma'am, and service, patriotism and duty, consideration and generosity—and an acceptance of violence. But he said no one in the Pentagon understood him either.

He left jobs regularly. Not because people didn't appreciate him. But because he simply refused, in his telling, to deal with bureaucracy.

"He obstructed my ability to suffer," is how Ward summed up his most recent run-in with one government official. It was a phrase he said often. "I want to suffer." "My life is meant for suffering." "What I won't put up with is anyone ever obstructing my ability to suffer."

One good friend of his told me barely anyone in the Pentagon or other national security agencies had a clue what Ward meant by that, or how to deal with him. The Pentagon was a place where office workers wanted to clock off at 4 p.m. to catch the bus out of the compound, according to another member of Maven. Ward was different; he was driven by something else.

Ward's life, he told me, was about "the warfighter." The person out there risking their neck for America. And his mama. Late into the evenings, "Madre" would flash up on his phone. He was the fourth of five children. She raised them all as a single parent. And Ward looked out for her nonstop. He had sent his paycheck her way for as long as he could remember, right since he joined the Air Force in 2009. One of the first times we spoke he was helping her move to California. He drove through the night from his home in Maryland to do it.

"You're too nice," she'd tell him.

His commitment to Project Maven burned in him as a standalone, all-consuming passion. He simply committed to getting things done. He flew out to bases. He installed servers. He ran through the jungle. He didn't sleep. He generally spoke as if there were an invisible "No" dangled always just in front of him that he was constantly trying to swat away.

He was willing to deploy anytime, anyplace. He was magnetized to austerity and the singular focus that came with it. "Drop me off somewhere terrible at night alone and I'm happy." He got Maven onto nineteen different networks, two different people told me.

"I'm just a loud and obnoxious American," he'd say, quoting imagined detractors with glee.

I began to realize that Ward perhaps had worked on Maven the longest, flown the farthest, taken the bureaucratic blows and remained just as undaunted in his love for it. And just as under the radar. He liked to call himself a "garbage man." He did the dirty work, unrewarded, and that is how he said he liked it.

Some of Maven's team had started to develop a siege mentality, laboring under the impression after the Google debacle that everyone from tech companies to defense high-ups wanted to kill it off. Ward just wanted it to work. He went to Afghanistan, Iraq, Syria, Jordan, Djibouti, and the Philippines for Maven as a contractor. What people didn't understand, he'd say, was that Maven would save lives.

Enlisted into the Air Force in 2009 aged nineteen, he had become a geospatial intelligence analyst. He would digitally immerse himself in US combat zones all over the world. "You see all these different parts of the world via a freaking screen." He'd learn other peoples' lives, watching potential suspects in the mornings, with their kids, in the market, switching out their phones to make what might be furtive calls. It was all part of establishing the details of a pattern of life that could result in a legal targeting package to kill them.

In early April, 2011, Ward was working at the Terre Haute Air National Guard Base in Indiana, his eyes glued to screens. He'd done more than fifty combat missions and 480 hours by then. Live footage was regularly rolling in off drones flying thousands of miles away over Sangin Valley in southern Afghanistan. A land mine was going off every day; battles every day too. One Wednesday morning, a group of US Marines on patrol in Helmand province started taking fire. Men scattered. On the other side of the world, coming up on midnight, Ward watched the feed live on his screen from a Predator drone that slewed over the area twenty minutes after the fighting started.

He saw three men on screen laying prone with their muzzle flashes pointed west—shooting away from the other Marines, not toward them. He entered his assessment into the system chat: "3 friendlies in FOV," he typed, referring to the drone's field of view, "pers are shooting W." To a senior officer who Ward called on for a second opinion, it looked as though the men might even be facing in opposite directions, the way Marines taking fire lay to cover more ground. But elsewhere, others thought the men were shooting east, toward the Marines hunkering down in the fire. Ward couldn't positively identify them as friendlies and retracted his statement, but he couldn't be sure they were enemies either ("disregard. not friendlies. unable to discern who pers are"). *Did they have custody of their men?* As he typed, the ground team requested to prepare a Hellfire strike. After a strike is requested, geospatial analysts are not allowed to enter anything else in the main chatroom unless women and children enter the area or the Law of Armed Conflict is about to be broken. But when Ward saw a fourth person arrive two minutes later and lay down with the three others he took an extra step, sending a message known as a "whisper" on a different chat platform used to signify concern: "pers are shooting W and the convoy is to the E," Ward reiterated in his private chatroom note to a mission center in California. Ward's assessment wasn't passed on to the ground commander, the radio controller, or the drone pilot. In line with his training and following the rules of his strange faraway job looking down on war, a twenty-one-year-old Ward didn't interrupt the mission further. Communication is purposely kept to the functional minimum. Seven minutes later, a Hellfire missile unloaded from the same MQ-1B Predator that was providing Ward's video feed.

The strike landed seventeen seconds later not on the enemy but on their own: a Marine and a Navy medic. Fratricide. The first known case involving a drone strike.

Captain Colin Carroll happened to be four or five kilometers away at the time on foot in another part of Sangin Valley. He heard the boom of the missile strike, heard what happened over the radio, and paced to reach the forces as they evacuated, arriving to complete chaos, blood, and insoluble fury. "An algorithm would never have taken that shot," he told me fourteen years later: the Taliban don't fight like that, they shoot and run. AI trained to look for patterns and allied tactics would have identified three Americans fighting defensively in the prone position. Next came a moment typical of Carroll—cavalier delivery, laced with despair. "We kill the wrong people all the time," he said. "The machine can't be worse than a human."

Jerry Smith, father of Marine Sergeant Jeremy Smith who was killed, said he knew whoever was at the drone joystick was devastated. "If I could meet them, I'd hug them and tell them I don't have any ill feelings toward them. I know their daddies are just as proud of them as I am of my son." The investigation would blame "a lack of overall common situational awareness."

No one particularly got over it. Ward would instantly lose respect for protocol and resolve never to hold his tongue. It never left him. Everything else, unless it were life or death, became noise to him.

The next year Ward deployed to Afghanistan. He'd work twelve-hour shifts, listen to nine different things while watching ten things all at once. He had his whole sensory system overloaded and could still do his job. He was the ISR tactical controller—a drone strike guy. He deployed with SEAL teams and the 75th Ranger Regiment, worked with the CIA. "He's been on tons of missions that you'd hear in the news," his friend told me.

Ward told me something else: "I don't have a personal life."

His marriage had lasted barely two years, and he didn't hold out much hope for the prospect of trying again with anyone new. "I'm between divorces," he'd announce, triumphant at his own joke.

He was attracted to chaos. But then in one of our long chats, any glib ease went away. It was early evening in a bar in downtown DC, still light, but suddenly it felt as if it could be 3 a.m.

He held his hands to his face, palms flat, fingers rubbing at his eyes. I felt the sudden, fleeting weight of pain, and exhaustion, and discomfort. He disappeared.

"Sometimes I would come back from deployment . . . " he began.

His sentence was already complete in my head. He was going to say *and I was just so tired*, or *and it just hurt so much*, or *it was all too much*, something like that, I thought. Some kind of concession to defeat. But that's not what he said.

He slowed down.

"And I was just. So."

I waited.

"Disgusted."

I didn't follow his meaning as quickly as I should have done.

Say more.

He was disgusted that people back home went about their business, complaining about their daily lives.

He couldn't stomach fripperies. The only things he could allow to occupy his brain, could tolerate thinking about, were life-or-death decisions. The callouts he made all day and all night long. The forty-eight hours in a row he didn't sleep. The deaths.

And the mindlessness of people back home. All that death, just so Americans back home could live free complaining about their unhindered lives.

Cukor hadn't been sure about hiring Ward for a more extensive position: leading Maven's global operations, based at the Pentagon. Cukor wanted someone who knew military deployment, which Ward did, but also someone who was technical enough to understand AI and the systems into which it would embed, which Ward was not. "Well, sir, you're looking for a unicorn," a Maven contractor told Cukor. Ward got the job, got the plane back to Afghanistan within a week, and turned out to be something of a unicorn, nonetheless. Ward never slowed down, didn't need supervision, and Cukor came to see him as fantastic. "I needed Brians," he said.

Cukor had got an enthusiastic green light from General Tony Thomas, the commander of SOCOM, to deploy into his formation. But there were many more rings to kiss, and getting Maven into Afghanistan was harder than getting it into Somalia. The stakes were higher, the units skeptical to the point of hostile, and the bases spread all over the country in austere conditions. By the time Ward arrived in Afghanistan in the fall of 2018, Colin Carroll, along with other Mavenites and contractors, had already managed to get Maven into eight sites across the country, starting in Bagram that summer. The Soviets built Bagram in the 1950s, but, after 2001, the sand-blasted air base had expanded into an unlikely fortress for US culture—Burger King, Pizza Hut, swimming pools, movie nights, chai lattes, and war. Some at Bagram joked the team had flown Skynet in to take their jobs. As in Somalia, Maven didn't land well.

The algorithm would identify trees as people. Rocks as buildings. Buildings as cars. Distracting boxes everywhere over the screens. It was completely pointless. Dangerous even.

Some were suspicious at the arrival of a crew they didn't know tampering with equipment at their remote sites. Others asked if it was even legal. Nobody wanted to touch it.

The team was relying on the models they'd used in Somalia, none of which had been trained with data from Afghanistan. If they managed to get Maven onto computer systems, the operators would have it minimized on screen, running unseen in the background.

Another Maven contractor who was a former screener and so was all too familiar with eighteen-hour drone shifts for special operations missions was recruited to convince his former colleagues in Afghanistan to pick up the tech. But he wasn't even a believer himself, and pretty soon concluded Maven wasn't going to work out. Encouraged by Ward's zeal, the pair of unlikely missionaries set about finding vignettes to convince operators and commanders that Maven really was the new frontier of war and not a waste of their time. Ward somehow had a way of breaking through the harder people. *Honey not vinegar,* he liked to say.

Ward's job got easier once the team got Maven models working on video feeds from MQ-9 Reapers, not just smaller tactical drones. On one occasion, screeners were scanning for civilians before launching a deadly drone strike. Right as they were about to launch the strike, an analyst called out that he'd spotted a civilian. It had taken the analyst forty seconds to spot the person, a farmer with a herd of sheep. They aborted the strike. But it was far too close for anyone's comfort. When the Maven team replayed the feed with Maven overlaid, the AI detected the man "as soon as he walked onto the view, within a second, and tracked him the entire time," the Maven contractor told me. AI was instantaneous.

An email went back to the Pentagon that Maven would be useful in a scenario involving a strike. But Cukor's response was sharp. Don't say strike again, came the feedback from Cukor. By phone, Cukor made clear that no one at Maven should associate the project with striking targets, especially by email. Even though the AI worked and the use case was about saving a civilian, the timing—in the middle of 2018 as the Google debacle roiled—was too sensitive given tech workforce outrage that Maven might be using AI in targeting and strikes. But that was exactly where it might be relevant, thought the Mavenites pushing people to adopt their system. That was when it might save someone's life. But in those first two years, Maven had to fight for budget. Cukor didn't want AI blamed for anything.

He was wary that Maven could be used as a scapegoat, and that a single mistake could kill off the program. Cukor's point was that one bad call, or a strike in which civilians were killed by mistake, even if it wasn't Maven, might be used to blame Maven for all and any wrongdoing and imperil its future.

Later Cukor told me his vision was always that Maven would be about strikes. It would shorten the targeting cycle. "I won't lie. That was always in my head, it was in all of our heads."

AI would accelerate one of the hardest tasks of war at speed: to locate allied and enemy forces on a map. "It's a very complicated thing perfected over the last hundred years," he told me. "It's just a big problem and always has been. Could we just use the AI to identify everything?"

But the AI was nowhere near good enough. "The AI is just a bag of potato chips," Cukor would tell Ward and others, dismissing the algorithms. He was pissed off at the AI but he counted on the algorithms getting better one day, along with the data that fed them. His aim was to build the system in which AI could live, getting feedback from operators to improve the system as they got used to the idea and waited for the AI to get better.

In the meantime, they made tweaks. Microsoft, which was producing many of the algorithms, decided to forego detecting vehicle subclasses like tank and truck (one early goal had been to identify the specific type of tank). Soon it would deliver models that produced only two classes: people and vehicles. That got the scores up and the flickering down but frustrated users with ethics codes, commanders, and the laws of war hanging over them: they usually did not care about whether AI could distinguish a sedan from a truck, but they cared very deeply about properly distinguishing a child from a man.

Computer vision also struggled in different environments, as Brian Ward would soon find out when success levels dropped from 70 percent detection rates in Afghanistan to as low as 30 percent in the Philippines.

AI couldn't seem to spot people or vehicles. The problem dawned on the team: in Afghanistan, the cars tended to be white cars and black Toyota Hiluxes. Vehicles in the Philippines, by contrast, were often simple carts topped by umbrellas. And while outlines of people could be picked out against dusty yellow background in Afghanistan, people in the Philippines were walking along pathways masked by edge lines of thick green jungle overhead.

NOT EVERY USER IN AFGHANISTAN hated Maven. In October 2018, a solider in the 75th Ranger Regiment started working at a Joint Special Operations Command (JSOC) unit focused on eastern Afghanistan.

Newcomers were rarely trusted at first. Some of that came down to the small bits of material stitched onto either shoulder of a uniform. These patches were small woven signifiers of toughness that doubled as the war-grade equivalent of the playground pecking order. The new soldier, an intelligence officer, arrived conspicuously limping: one scroll, no tab. That signified making it through elite Ranger selection, although without doing Ranger School.

"Some people aren't very nice," a member of the JSOC unit assigned to the area told me. The new intelligence officer also deviated from usual practice, deciding to experiment with Maven's tech and poring over WAMI footage to sleuth out clues from AI detections. Identifying specific objects with WAMI was hard from so far away. The picture tended to be a morass of dots. But AI could help suck in all the data collected over several hours and rewind the history of routes taken by specific vehicles to highlight missed patterns, a process Cukor told me was known as backtracking.

The 75th Ranger Regiment in Afghanistan was focused on the Afghan wing of the ISIS group that had established and then lost an Islamist caliphate across much of Syria and some of Iraq between 2013 and 2018. The ISIS-Khorasan chapter that emerged in 2015 had grown stronger and now wanted a new caliphate of its own in Afghanistan and surrounding neighbors. US intelligence analysts were warning of a "stiff fight" between the Taliban and ISIS in parts of Afghanistan after the two groups declared war on each other in 2015. "What we see now is some pretty hard fighting going on between those groups in certain locations," Defense Secretary Mattis told me when I accompanied him on a visit to Kabul, the Afghan capital, in September 2018.

The US started trying to target ISIS with so much help from the Taliban that some Rangers quipped they were "the Taliban Air Force." Sometimes it could feel as though the Taliban were tantamount to calling in the US drone strikes themselves.

The Taliban was seeking to negotiate a role in the government and had just struck a deal with the US following secret negotiations. ("A superpower just lost a war," is how one regional official put it to me at the time.) ISIS-K was not interested in a ceasefire. They attacked a Save the Children office in Jalalabad that January, killing six and injuring twenty-seven. In late August, the group claimed responsibility for a rocket attack against the presidential palace in the capital. I landed in Kabul two days after a suicide bombing killed twenty-one people and injured ninety. The same week, US service members were killed in separate incidents in eastern Afghanistan, the ISIS stronghold.

"My team was hunting them," said the member of the JSOC task force focused on eastern Afghanistan. There could be three or four strikes a day, the person said.

Even so, US forces never really understood how the ISIS network in Afghanistan functioned. "One of my greatest regrets is we didn't get it better," said a second member of the JSOC task force, emphasizing the same group would be responsible for the August 2021 suicide bombing that killed thirteen US military personnel and 170 Afghan civilians at Abbey Gate during the hasty US retreat.

The US had determined hundreds of ISIS-K were hiding in the mountains of southern Nangarhar, and suspected they planned a lot of high-profile attacks from the region. But it was hard to tell how much mission control and planning happened in the mountains and how much in Kabul. The US intelligence operation often relied on word-of-mouth tips for its strikes. In one case, an informant paid by the US government erroneously told JSOC that ISIS-K was holding a meeting at a building in Kunar province, according to the first person recalling the events. The informant had been wrong before. Usually a mission needed multiple sources, but the rules of engagement were looser as the US was trying to amp up pressure in pursuit of a deal. The mission went ahead.

An infrared feed picked up three people leaving the building. It was dark, but watching them from above you could see their outline with the help of thermal imaging, and it showed the bodies of the three people then walking down the street. Despite having no other corroborating evidence, the US decided to strike, the person told me. Down came the Hellfire missile. The next day, local news said three civilians had been found dead. Relying on single-source intelligence from paid informants could be fatally unreliable. "It was extremely sad," the person recalled. "Everyone was pretty messed up about it."

The US had other ways to uncover information besides unreliable paid spies but the new Ranger was the first to give Maven the time of day, experimenting with Gorgon Stare, the eye-in-the-sky attached to Reaper drones and for which Google developed algorithms. "When it worked it was beautiful; you would see an entire city move," one Mavenite told me, saying the greatest value was being able to trace routes taken by specific dots of interest to reveal overlaps and meeting points.

The ability to track people made it much harder for an insurgent to hide out in an open-air village where US aerial assets hovered overhead, compared with a covered urban environment. "We can just fly and watch you all day," said the second member of the JSOC task force.

Project Maven flew in AI systems for two Gorgon Stare–equipped Reapers from Dover Air Force Base in Delaware on the C-5 Galaxy, America's largest military cargo plane. These AI systems depended on heavy computing power, which the team quickly established in Afghan outposts: the main Project Maven data center in Kandahar was the WAMI powerhouse, known as "the motorcycle," while a smaller second data center at FOB Fenty at Jalalabad in Nangarhar province was known as "the sidecar." Project Maven installed $15 million of GPUs, backup generators and a chiller system in a box the size of two tractor trailers.

But almost no one wanted to experiment with the system. It opened up a fire hose of data into their workflow, and the AI initially made that worse. "Maven was never our bread-and-butter system," the first unit member told me, adding the AI was "pretty crude." Intelligence operators could spot vehicles with their own eyes, but the AI system saved time and offered the most potential when it came to tracking a vehicle. "It definitely had bugs, but they were refining it and we understood that."

Explosives in this area were produced for specialized usage in suicide vests in high-priority ISIS operations in Kabul and against the Hazara people, said the second member of the task force familiar with US assessments. Moving homemade explosive materials of the sort that killed American forces required infrastructure and possibly communications. The new intelligence officer experimented with turning filters off or on to give a sense of how to use the AI. The newbie began to think Maven might help identify a complex ISIS network used to move explosives around, according to two people who recall the work. Aggregating AI track detections found by Maven helped reveal a decidedly low-tech network: a donkey train appeared to be moving explosives through the Hindu Kush mountain range. The donkeys were traveling not on roads but on winding mountain paths too rugged for wheeled carts. "Gorgon Stare was very capable with the donkeys. I wasn't sure it was going to work but I thought it was worth trying," said the second person.

The intelligence officer mapped donkey routes and assembly points, developing increasing confidence that even though they looked identical from above, one particular donkey train contained not pine nuts but explosives.

The newbie pitched the donkey train strike at the afternoon targeting meeting, which planned drone runs and priority strikes for the next day. The commander at first dismissed it. No one really believed anyone had actually been able to figure out a random donkey train from an eye so high in the sky, but the brief was compelling, the second person recalled to me. The commander went through the logic, asked exacting questions and began to come around.

The donkeys carrying heavy-laden pouches over their backs across the wild remote mountains of eastern Afghanistan also had a bunch of guys walking with them, that person recalled.

Deliberately striking unarmed civilians is a war crime. Make a wrong call and lives and careers are at stake. The commander would be risking legal consequences, and whatever moral authority the US had left in Afghanistan.

At that time the US was running what the second person described to me as a "highly permissive environment" for strikes in southern Nangarhar. The US was focused not on striking a deal with ISIS, but on destroying them. "The mission as it pertains to them is focused on their elimination," a spokesperson told me at the time.

Commanders on the ground didn't need additional authority to strike adult males in this area. There was a presumption that anyone the US struck in that area was ISIS, the second officer told me: "It was so routine for us to strike people in southern Nangarhar." That could go very wrong. A US drone strike in the region the next year killed thirty civilians and injured another forty. US operators thought it was an ISIS training camp; it turned out to be a group of farmers resting at the end of a long day harvesting on a pine nut farm.

The US shared mission information scrubbed of classified detail with the Afghan government representatives who sat in the office beside the unit to confirm whether they had any objection, the second person told me. The US told their Afghan counterparts they believed they were terrorists, and heard back that they had no objection to the US conducting a mission against the suspected terrorists.

The commander made the call.

"You don't shoot at unarmed people," the second person told me, stating the risk. "He was making a decision to kill people who did not have visible arms who turned out to be part of a terror network."

A Hellfire missile rained down over mountains of southern Nangarhar. As the remnants of donkey carcasses started falling down the hill, secondary explosions started nearly instantly, the second person recalled. The team took the secondary explosions as evidence the donkeys had been carrying explosive material.

When the newbie intelligence officer wrapped up in Afghanistan, a little less new, the donkey hit had been enough to win the respect of previously skeptical comrades, with or without a Ranger tab. A third person who heard about the operation remembers feeling sorry for the dead animal. But among the Special Operations Task Force; not so much. As a parting gift, the task force presented the targeter with a toy donkey.

11

HARBINGER OF DOOM

"We're going to apply AI to find submarines."

TARGETING A DONKEY carrying homemade explosives was hardly the job many had in mind for Nvidia's high-end GPUs when the Defense Department conceived of Project Maven.

Colonel Cukor always knew that the Pentagon's efforts to develop AI for war was about something else far beyond Afghanistan: preparing to fight China. I would learn Maven went deep under the sea and into the shadowy world of submarine warfare. But I could almost hear Cukor wince when I asked him about Maven's submarine work. Like anything operational, he wouldn't go there with me. Yah, that's sensitive, he said. And that was that.

The chess match of deterrence that constantly played out between the United States, China, and Russia was at its most furtive—and most existential—under the water. In 2018, former Deputy Defense Secretary Bob Work told me he thought the only area where the US war-fighting machine still maintained clear superiority over China was with its submarine fleet. (In the same interview, Work raised the prospect of preemptive strikes, arguing that the use of guided munitions in any future war would be so "widespread and profound" that it will make "a lot of sense to be the one to shoot first.") Three years later, a US think tank report would suggest China was likely to erode the US advantage in undersea warfare at some point between 2026 and 2031. Losing US advantage under the sea could potentially have profound implications for the global order.

A former Naval Special Warfare operator tried to explain the mystique—and the stakes—to me. Submarine warfare was the absolute most secret part of the US strategic military arsenal. It all came down to nuclear war and the end of the world, he let me know in cheerful matter-of-fact tones. If the US homeland was attacked by a nuclear weapon and rendered incapable of firing back, it would fall to one of America's fourteen stealthy "boomers"—ballistic missile submarines that carry more than half of the US nuclear arsenal—to lob back a nuclear weapon. That was known as "second strike."

The job of a boomer was to make a hole in the water and simply disappear for months on end. Sometimes they would pop up unannounced, in Korea or Japan or Guam or Norway, just to show they could creep up on China and Russia any time they liked. Such was the strange diplomacy of nuclear signaling. A maximum of three-quarters of the boomer fleet is likely deployed at any one time. They carried twenty-four ballistic missiles apiece, each with between eight and thirteen warheads, and there was always one boomer on constant alert on either US coast. Once a year, a missile test (without the warhead) went to show the US could hit a dollar on a football field anywhere in the world.

Most of the rest of the fleet, the country's forty-nine nuclear-powered fast-attack submarines, spend their time hunting out other countries' subs. The way this former operator saw it, keeping the location of the Navy's nuclear submarine force secret, then, was a constant matter of life and death. And, if you wanted to apply diplomatic and military pressure, and save the nation to boot, figuring out where China and Russia's nuclear-armed submarines are at all times was a top priority. If you can find it, you can attack it.

The US has a pretty good idea when "bad-guy subs" leave their docks but it is a cat-and-mouse game after that to chase where they go next, he told me. America liked to think of itself as the cat. But China had gotten good at hiding when its submarines launched into the sea, building submarine pens deep into caves and underground bunkers to conceal their movements over the past few decades. The holy grail for a Chinese or Russian sub, conventional or nuclear, would be to make it undetected to either seaboard alongside the US. An enemy sub secretly hiding out there could send out enough warheads on multiple missiles to take out every big city and military facility throughout all America with no real time for the US to react. Game over.

I was beginning to learn how closely the US guarded its submarines and their secrets. In mid-2024, I visited Guam, a US island territory in the middle of the Pacific. From the port, I could just make out the soft gray curve of a submarine's back. During my time on the island, I met a man who told me the US Navy flew him more than halfway around the world to fix one. In the end, he half-joked, all he really had to do was press the reboot button. The rules stipulated only he was allowed to do it. Submarines counted as a Fort Knox of the Navy's own creation, as Project Maven quickly found out. But there were ways to hunt them.

A SUBMARINE, MUCH LIKE A WHALE, makes a noise underwater. And just as it can be hard to learn the language of whales, so it can be hard to speak submarine. Every submarine has its very own acoustic fingerprint. That sound signature can be recorded, converted into electric outputs, transcribed as a spectrogram image, and analyzed.

For years, the Navy has maintained a prodigious database of submarines from all over the world and the signature sounds they make. Naval submarines also listen to the world with the help of hydrophones—microphones that pick up sound underwater. Subs tow hundreds or thousands of small hydrophones in the water, trailing them away from their own noisy engines. Inside a hatch at the bow of a submarine is another sound system known as a spherical array—a sonar detector that looks like a huge pitted golf ball. Both can convert acoustic signals into electric signals to create a detailed acoustic picture of the underwater environment.

The US has other ways of listening in to the underwater world besides submarines: America has surreptitious listening posts everywhere—hydrophone arrays dragged from the back of boats, fixed into place snaking across the seabed, on buoys dropped from planes. Some scientists were even experimenting with adding digital ears to the subsea fiber-optic cables that help carry the internet around the world.

"That's how we view the world," one submariner told me. "Passively."

But parsing meaning from the overwhelming volume of recorded data had become a problem. It was hard to think for all the noise. And the "bad-guy subs" weren't foolish—they would experiment with disguising their own digital sound signature. It might be better to look for anomalies than a specific script, for everything instead of one thing. But as with the drone feeds that lay on the cutting-room floor, underwater acoustic data was also going largely unexplored.

As the US warily eyed China and Russia's undersea developments, the US military machine faced economic and social challenges in staying ahead. Building new sub-hunting submarines is expensive, and relies on experts who can be hard to lure to work at remote sites. America's industrial strength was at a disadvantage compared to China. Tariffs would only get you so far. In the absence of a manufacturing leg up, AI might provide a viable shortcut. Project Maven saw its chance to bring AI to the Navy Nuclear Submarine Force and see if it could unlock the secret sounds of the sea.

Maven would soon get to know about US Navy Captain John McGunnigle, a Boston submariner with a computer science degree whose missions had already taken him under the ice sheets at the North Pole and deep into the western Pacific. By 2018, the former submarine squadron commodore was working at the Undersea Warfighting Development Center in Groton, Connecticut, and had conjured a nascent Navy project that soon got a codename of its own: Harbinger.

In July 2018, Captain McGunnigle had an idea to apply AI to submarine acoustics. With support from the Office of Naval Research, he got funding for an eight-week pilot study. Much of the work was done with help from the Woods Hole Oceanographic Institution, a scientific research center on Cape Cod, Massachusetts. Decades earlier, urgently looking for a way to detect the German U-boats that kept torpedoing allied convoys during the Second World War, researchers here figured out just how far low-frequency sounds could travel in the ocean.

Nearly seven decades later, underwater drone and robotics expert Carl Kaiser was among the small team that went to work in August 2018. They put eight hydrophones on a surfboard in Buzzards Bay near Cape Cod and flipped it over. The team drove a fiberglass motorboat around it, and then drove an underwater drone around it. And then they tried out a simple algorithm to see if AI could tell the difference between the two. It did alright—better than a coin flip at least. But the point, McGunnigle figured, was to see if there was a way get hold of data and test it.

A five-minute boat ride south of there, the team put a GoPro camera along with a hydrophone under the water at Woods Hole, right by Martha's Vineyard. Then they tested an algorithm to see if it could generate alerts to distinguish a ferry from a normal boat, and identify machinery noise on a parked vessel from the thrum of a fishing boat. "They were good enough to show that if we could do all that in eight weeks, it was obvious that a real application of effort could succeed," said Kaiser.

It wasn't the first time experiments along these lines had been attempted, but it was the first time they ever landed with the right people inside the US Navy. "Technological breakthroughs happen when society is ready," said Kaiser. "In this case, society was in no small part John McGunnigle."

Meetings with Project Maven soon followed. As soon as General Shanahan and Colonel Cukor understood Captain McGunnigle was a persuasive and effective figure who could deliver reams of passive acoustic submarine data, I'm told they asked him what he needed to make it happen. The answer, as always in the Pentagon, was money and bureaucratic support. McGunnigle got both. Maven immediately allocated millions of dollars to its new submarine effort.

He also got a team in the bargain. Mike Hunter was a former Navy intelligence officer who'd deployed to Iraq. After he left military service, he joined the Defense Intelligence Agency, and was working with the Navy SEALs as a contractor in support of JSOC when he was recruited to Project Maven through Jay Hurst, a JSOC friend who'd joined Maven a few months earlier. The interview—with Lieutenant Colonel Garry "Pink" Floyd—was customarily short.

"You speak Navy, right?"

"Kinda," Hunter offered. (He'd been a ground intelligence officer most of his career.)

Besides a couple of SEALs working Maven's relations with special operations units, Maven had no one at all from the Navy on the team. It was enough to seal the deal: Hunter joined Maven in June 2018 and became the project's lead for all naval efforts.

An enterprising young submarine lieutenant working at the Digital Warfare Office, along with another Harbinger action officer, visited the Maven team at the Pentagon to explain the project.

"We're going to apply AI to find submarines," began the lieutenant, the technical brains of the project who had taught himself how to build and train machine learning algorithms.

"Cool," said Hunter.

And that was the extent of the explanation: Hunter couldn't fathom any of it.

Even after the lieutenant went into more detail, Hunter got maybe 5 percent. He'd spend the years after trying to learn the remaining 95 percent, and eventually master it well enough to become chief executive of a commercial startup devoted to the issue by labeling acoustic data to help train the algorithms.

Now the idea was for Project Maven to repeat its computer vision process for submarines: get hold of data, label it, show it to the algorithms, then test it. The dream was to deploy a Maven algorithm on a submarine. As with the rest of Maven's pursuit of AI, the toughest job was getting hold of good data.

The team traveled to landing sites for Boeing aircraft known as P-8 Poseidons all over the world. These were America's flying submarine hunters. They carried cameras on board equipped with multiple sensor types and also dropped sonar buoys equipped with hydrophones into the ocean. But to the team's dismay, they discovered that the P-8 data was not being saved. Hard drives were mostly wiped at the end of every mission.

"The Navy has said for a long time that data is a strategic asset, foundational to how we fight wars, but the Navy treated its own data horribly," Hunter told me. "What we found out was the amount of data needed to train the algorithms didn't exist."

Worse, Hunter wasn't sure if he could get anyone to change their minds about it. It could sometimes feel as though pockets of resistance inside the Navy might doom them: China's centralized system would be able to collect data to label and train algorithms far more easily than they could.

A low point came on the penultimate Monday of October 2018. It was close to 8 p.m., and Hunter had just caused an unexpected stink.

"I am requesting an immediate halt to his efforts," wrote a systems engineer at the Space and Naval Warfare Systems Center in Newport, Rhode Island, which worked on technical support for command and control systems for the Navy and the rest of the US force. The engineer was responsible for enterprise storage, and had got word Hunter was fiddling around with IT systems down at an operations center in Florida.

Mr. Hunter, the engineer wrote in an email to eight Navy colleagues, was in Jacksonville requesting data from one of her sites. "He has gained access to the building," she wrote, adding he was asking for configuration changes. Her email went on to cite issues such as "Insider Threats," "Securing Classified Information" and "Need to Know."

For Hunter, the objection was nuts. He had got signed travel approval, passed top secret security clearances, and done an office call with both a captain and the commanding officer of the operations center ahead of his visit. Now all he wanted was for Navy Fleet aircrewmen to upload videos onto the Navy's SIPRNet, the department's shared classified network. And when for some reason that didn't work, he asked them to install Google Chrome on their desktops and do it that way. The IT team sought permission and instructions from the engineer, who flashed a red light. (Hunter, the engineer wrote, had ignored her explicit instructions to route such requests through her.)

"I don't think people truly appreciate how disruptive we were being, how much we upset the status quo, and how far the institutions were willing to go to fight back," he wrote to me seven years later, the outrages of his time at Maven still fresh for him. "This is basically what I ran into anytime I talked to the Navy; not just inertia, but outright hostility."

Hunter ultimately battled through the red tape, secured data and tape readers, and sent everything to the US Naval Submarine Base in Groton, Connecticut. The Maven team got its way with the data from the P-8s. The Undersea Warfighting Development Center chipped in with hard drives of submarine data, which would go onto an AWS Snowball—a rugged device made by Amazon that approximated cloud in a box—and from there it would be uploaded onto an AWS secure cloud.

The data was used to train up algorithms, and, one day, they got Harbinger algorithms onto a single *Virginia*-class nuclear-powered attack submarine. They were stuck using an old laptop but at least it came ready-equipped with the right security settings. The young submariner repurposed it for the job: he built a user interface that relied on one single GPU.

"When we first deployed it was something like, meh," recalled a Mavenite who worked on the maritime effort, likening the reaction to a balloon slowly deflating. They persisted: the Navy had now deployed AI on its most important and secretive geopolitical program. Hunter intended to put AI on more submarines and make waves. The US was changing how it fought wars at sea. Multiple companies started developing algorithms to rake over the data. Soon the Navy's ambition started to grow. It caught on to the broader appeal of AI, asking if algorithms could work on video and data feeds from port security cameras, ship cameras, special operations boats—and even a periscope. Harbinger algorithms also for a while went to work on specialized surface surveillance vessels known as SURTASS ships, which trail Y-shaped arrays to receive acoustic data and can also help detect submarines and underwater mines. Smaller expeditionary SURTASS components produced more recently do the same more furtively. The Navy began to explore the world of the possible too: a slide in a 2019 presentation from the Navy's Undersea Warfighting Development Center wondered where AI and machine learning might help not just with navigation and sensors, but also weapons.

Harbinger algorithms were going to work on another of America's most sensitive surveillance systems too. The Integrated Undersea Surveillance System (IUSS), run by the commander of undersea surveillance, is a shadowy undersea network established during the Cold War. Classified until 1991, the subsea system first created to spy on Russia is now squarely focused on China too.

For the past few years, the US has started slowly upgrading the network to help sonar operators and acoustic analysts detect and classify sounds coming off a cacophony of fixed sites on the ocean floor and mobile sensors. The US was separately inhaling data from some of America's clandestine underwater warfare platforms. The US collected data from wet SEAL Delivery Vehicles used for short distance insertions (meaning the operators wear wetsuits and are also submerged in water), dry mini-submarines used for slightly longer special operations missions (meaning the operators stay dry while they are inside) and the secret network of static underwater sensors known as the Fixed Distribution System. Every single military object could double as a sensor, and the likes of Maven, Harbinger, and others wanted to suck up every byte. Some data was collected, beamed across the world via Starlink's new military Starshield terminals and then into an AWS cloud space.

In 2018, Beijing was already one year into a new effort to build its own network of acoustic sensors known as the "Underwater Great Wall," and a few years later was developing AI-enabled underwater drones, surface vessels, and undersea surveillance systems to try to prevent US submarines from lurking between the first island chain and the second island chain that ring China. Some of China's newer underwater drones also doubled as listening devices.

In the three years after Russia invaded Ukraine in February 2022, ties between Moscow and Beijing grew closer, and Russia moved seven submarines into the Indo-Pacific region, I learned from a US defense official.

But there were limits to this collaboration. Russia was not sharing its nuclear quietening technology with China just yet. China's Navy was already the world's largest fleet, but it had only six ballistic missile submarines and six nuclear-powered fast-attack submarines. Even as late as 2025, Chinese subs were still noisy enough to find. (That might change in coming years: a new ballistic missile submarine known as Type 096 promises more space to try to absorb and deaden the sound of the engine and is set to join the Chinese fleet.)

If the US and allies could pool some of its voluminous data with allies, it could set an underwater submarine sound trap. In September 2021, the US announced an agreement with the UK and Australia to team up on advanced submarine tech to counter China's ambitions in the Indo-Pacific. Part of that deal, known as AUKUS, meant Australia would buy conventionally armed nuclear-powered submarines from the US rather than France, triggering an explosive diplomatic spat. But the other part, known as Pillar 2, meant Harbinger and submarine surveillance. The idea was for the allies to pool the data they collected from all their sonar buoys and run joint algorithms to create a three-country, world-spanning undersea sound dragnet.

Things went slowly, of course. Australia was already a full "frigging" eight years behind the US when it came to military AI, one Australian defense official despaired to me in 2025. Maven had "bent over backwards" to help the Australians participate in the data collection and labeling process, spurring discussion of an Australian AI defense center in 2018 modeled along the lines of Maven. But the early effort fell flat until AI remade the battlefield in Ukraine and Australian defense officials suddenly started asking in a panic to "buy AI," one official told me. They set up the Maven-like center in July 2024.

All sides struggled with long budgeting cycles and red tape. In May 2023, an Australian security official named Ninh Duong complained to me live onstage in Washington, DC, that bloated US bureaucracy was slowing down the transfer of AI and other technology. He blamed "a permafrost layer of middle management" in the US government. Wading through the US system feels, he subsequently told me, like "death by a thousand cuts." That month, the US State Department gave the go-ahead to sell a $207 million expeditionary SURTASS system made by Lockheed Martin to Australia, along with associated software and processing systems, so it could detect "enemy submarines."

After these teething troubles, the submarine acoustics effort progressed. Under the AUKUS agreement, Australia, the UK, and the US would now contribute their sonar acoustic data into a shared cloud known as the common development environment. The data in the shared cloud would be used to train the algorithms. Harbinger stayed with Maven for several years and then transitioned to the Navy, where it lives on today.

In 2025, the process was working. More than 250 users in Australia, the UK, and US were working on joint data in a shared cloud classified as "Secret" to train and test algorithms for acoustic data gathered by underwater sonar sensors, according to documents I reviewed and corroborated. In the space of three years since the spring of 2022, the allies have tested more than six hundred models for parsing data collected from deep under the sea.

China was making headway of its own: in September 2025 it showcased its new extra-large underwater drones at a Victory Day military parade in Beijing, also attended by Russian and North Korean leaders. Experts immediately framed them as intent on disrupting subsea listening posts and subsea cables. And US experts worried that the number of academic research papers put out by China on passive acoustics dwarfed the work of the US.

Drew Cukor and his team helped spin up the first efforts at AI target-hunting around the world and under the sea. But some worried they risked being too slow and too late.

12

ARMS RACE

"This is what's going to allow us to kick your ass."

IT WAS THE SEVENTY-FIFTH ANNIVERSARY of the founding of the People's Republic of China. And I was just arriving at the Chinese embassy in Washington, DC, on the last day of September 2024.

A smattering of campaigners outside were protesting against Beijing, aiming the sounds of their lonely, persistent megaphones for the gates in the US capital's leafy diplomatic district. The granite and limestone monolith fusing East and West beyond them took up as much space as four nearby embassies put together. It still conveyed what the *New York Times* at the time described as the aims behind the building's design: "the sense of importance of China and China's role in the world today." Or, as Kurt Campbell, the Biden administration's future Asia national security czar, wrote in 2008, to "let the Americans know that, while they were away fighting in Iraq and Afghanistan, China arrived."

When the embassy opened in 2009, China was enjoying low tariffs thanks to joining the World Trade Organization, and had just become the largest holder of US debt. Two years later China overtook Japan to become the second-largest economy in the world, and by some counts overtook the US too by 2020.

China's defense spending trailed far behind the US, but that too was on the rise. In the ten years leading up to 2007, Beijing more than doubled its military budget, making it the second-largest after the US. It would take another decade after Campbell's warning before any US administration made China the focus of its national defense strategy. In January 2018, the first Trump administration prioritized China as the number-one challenge facing the US, and Defense Secretary Jim Mattis raised the possibility of "major combat."

When I arrived at the Chinese embassy six years later, the fortress provided a grand spot from which to contemplate the possibilities for a putative World War III. After descending a white stone spiral staircase, the crowd dutifully filed into an auditorium hall so large that the ambassador's speech was simultaneously transmitted live to video screens around the room. One attendee walked up to a screen and toasted the ambassador's glass of red wine with his own during the live broadcast.

During President Xi Jinping's speech from Beijing to mark the anniversary, he warned China must be "more mindful of potential dangers" and prepare for worst-case scenarios. At the event I attended, officials played a video showcasing US-China friendship, flashing up images of Americans posing for grinning selfies on visits to China. In the restrained speech that followed, Beijing's ambassador scolded the US for describing China as America's "pacing challenge"—a rote new talking point from the Pentagon intended to invoke competition without conflict. The ambassador said it amounted to "fear mongering." (A second piece of Pentagon jargon described China as a "near-peer" competitor, although in private many confided there was little "near" about it.) A US State Department representative smilingly indicated he agreed with almost nothing the ambassador had just said, but had been allocated only sixty seconds in which to offer a return speech. It was enough to emphasize the distance between the two superpowers.

Then the crowd poured into a reception. I studied replica ancient urns and mountain scenes in pen-and-ink on the way to a buffet laid out with tripe and durian cake. Servers pushed wheelbarrows full of dark-broiled whole ducks across the white marble floors. Milling among the crowd was Liu Zhan, a major general in the Chinese People's Liberation

Army (PLA). The previous year, he'd taken the stage to assert Beijing's claim on Taiwan. "Taiwan is China's Taiwan," he'd said, adding the two must be "and will be" reunited. "We will not promise to renounce the use of force."

The US assessed that President Xi wanted the PLA to be capable of taking Taiwan by 2027. Washington practiced a policy of strategic ambiguity, meaning it would not say whether or not it would defend Taiwan from invasion. Even so, the island became the focus for potential conflict between the US and China, and the subject of much debate about whether AI warfare might hand an advantage to one side or another.

Many experts were cautious about the language of an AI arms race, but Colonel Cukor never held back announcing it was underway. He thought it was existential and civilizational. "If the enemy has AI and we don't, young Marines and platoons are going to get killed out on the beach and we're not even going to understand what went wrong," one person recalled Cukor saying in 2018. That year I went to see Frank Hoffman, a retired Marine Corps officer and military strategist who advised Defense Secretary Mattis in the early years of Maven. The avuncular don, whose job it was to train up the next generation of generals, described Cukor as "razor sharp," and was blunt about the US contest with China: "We've been in an arms race for ten years."

Both countries had already made explicit—and public—their eagerness to fold AI into warfare. China released its own vision for AI in the military days before the launch of Project Maven in April 2017. Lieutenant General Liu Guozhi, director of the Chinese military's agency for developing cutting-edge weapons, expected AI "will lead to a profound military revolution." A couple of months later, China released a new AI development plan, aimed at making it the global leader in AI by 2030. The PLA anticipated a shift to "smart" warfare, in which AI in military applications would underpin decision-making, weapons systems, and all military power. In November 2017, China held its first Target Recognition and Artificial Intelligence Summit Forum at an automation research institute. Intelligent machines, it said, could become "primary warfighters."

The embassy's defense attaché General Liu told me he knew about Project Maven, of course. The Americans, in his view, were far, far ahead when it came to bringing AI to the battlefield.

But what about "intelligentized" warfare under development in China, I asked, referring to the unwieldy translation some US researchers gave to Beijing's own emerging theories of AI warfare. Some translations suggested "smart warfare" might be more appropriate; others suggested "mind control." What did it really mean?

"It is very difficult to know what's in the mind of your enemy," he answered with cryptic chutzpah.

Publicly available information about the PLA fielding specific military AI systems remained sparse. Occasional reports would trickle out: a 2023 obituary for a Harvard University graduate who returned to China in 2014 related his development of a "superbrain" for military decision-making. He simulated man-machine and machine-machine contests in chess, and led military AI programs with names like "War Skull I" and "War Skull II." The PLA used these to bring together intelligent decision-making for multiple combat platforms in air, land, sea, and space, according to one report.

The embassy officials I met were set on convincing me that what China really wanted was safe AI. They made their argument through the lexicon of the silver screen. The US public watched movies like *The Terminator* that fantasized about a deadly robot onslaught set on destroying humankind at the hands of Skynet, another Chinese diplomat told me. The Chinese public was instead shaped to view AI as heroic, the diplomat went on. US officials publicly fretted about incorporating AI into command and control systems for nuclear war. *The Wandering Earth 2*, a 2023 sci-fi epic blockbuster from a Chinese director inspired by watching *Terminator 2* as a kid, sees China save the planet from solar assault with the help of a superintelligent computer.

Elsa Kania, a US think tank researcher, was among those who were unconvinced. She translated and analyzed Chinese military writings focused on how Beijing planned to incorporate AI into concepts of war. She laid them out in a November 2017 report and subsequently briefed

Pentagon seniors on her findings as Maven raced to deliver the first algorithm to Somalia. She wondered whether China would face "fewer or different ethical or legal constraints" over using AI in combat than the US.

Still, if there were an AI military arms race, Kania's perusal of the documents suggested America might have started it. Deep into her report was a footnote on page 16: "To date, the PLA appears to have primarily adapted such concepts from U.S. writings," she wrote. PLA researchers also "closely tracked" Bob Work's Third Offset Strategy that gave birth to Project Maven and the pursuit of human-machine teaming, she added elsewhere. China, she wrote, had few clear definitions for the ideas it was espousing.

The US was more worried that the Chinese military might be quicker on the uptake. Mike Griffin, the Pentagon's undersecretary for research and engineering, despaired to me in 2018 that it took the US an average sixteen years to field a new operational capability, versus seven years for China.

The Google debacle at Project Maven only deepened the contrast between two vastly different procurement systems in the AI age. China could count on industry to push game-changing AI to the military; US efforts were stalling in the face of domestic opposition from civilian engineers.

The fallout from the Google dissent triggered multiple overlapping efforts to get Big Tech and the public to switch sides in the great debate about AI and war. The most favored lever, deployed in Congress and the Pentagon, was to ram home the threat from China. The effort to reshape the US public in support of AI warfare came in three waves.

First, a few weeks after the first public outcry over Google's work on Maven, Congress set up the National Security Commission on Artificial Intelligence. The independent commission counted as a who's who of AI defense hawks and technologists. Eric Schmidt, the former Google chairman who supported the Pentagon and Maven, joined as chair. Maven's original defense backer, Bob Work, joined as vice chair. Leaders of Microsoft, Google, Oracle, and AWS came on board as members, along with the CIA's investment arm In-Q-Tel. The Defense Department was attempting to align the interests of money and modern munitions with a big-hug embrace of Big Tech.

Nearly two hundred briefings and three years later, the commission would publicly argue AI would be central to the future of warfare. It raised alarms about a rising AI military threat from China. It urged the US to develop, field, and adopt AI weapons of war. The group acknowledged AI weaponry could lead to conflict escalation, weapons proliferation, and instability. But America had no choice, went the drumbeat. Battlefield success may come down to AI, and the commercial experts who could pull it off needed to recognize as much.

Second, in June 2018, the Pentagon set up a new office to accelerate the adoption of AI throughout the DoD. Lieutenant General Jack Shanahan left Project Maven to lead the new AI center, the Joint Artificial Intelligence Center (JAIC), for which Cukor and team had written the proposal as early as August 2017, according to drafts I reviewed. Going public with this pivot and refocusing it on ethics, Shanahan told me, was absolutely part of the Pentagon's response to the Google "episode." Alongside trying to build out AI for defense logistics, maintenance, administration, healthcare, and operations would now be heavy public discussion of what the Pentagon started calling "Responsible AI" intended to mollify critics. Shanahan went on a year-plus listening tour.

Third, US military leaders went loudly on the attack in Congress. Project Maven's pursuit of AI for the military was "not about doing something that's unethical, illegal, or immoral," said General Joseph Dunford, who as chairman of the Joint Chiefs of staff was the top uniformed officer in the country (and a Marine Corps officer). He had "a hard time" with companies that looked for work in China but didn't want to work for the US military, he said in November 2018. "We are the good guys and they should know that," he said. AI could be existential if it came to a fight: "I do think that whoever masters AI is going to have a decided competitive advantage."

And all the while Project Maven fought, successfully, to expand its budget. Cukor was typically aggressive: "We don't have a choice not to do it," he would tell Congress in briefings, arguing the US was in an AI arms race with China. "We have to do it or we will fall prey to the Chinese who are going to do it better than us and beat us."

Congress heard his plea: Project Maven's budget jumped more than ten times, from $16 million before the Google fiasco in 2018 to $93 million after it in 2019. It would rise for the rest of Cukor's tenure, totaling $1 billion. And from now on, whenever the US needed to justify the pursuit of AI for war, the US would turn to the specter of losing World War III.

Despite the high stakes, I wondered if there might be something potentially hollow, or perhaps manufactured, about the grand claims made for and about military AI so early on in its development. Eight years on, I asked retired Air Force three-star general Jack Shanahan if he thought Project Maven was an information operation—propaganda aimed at intimidating and inhibiting China.

"Not intentionally," he told me. "But of course it was."

Shanahan was in charge of the US military's influence operations in the early 2010s, supervising overseas messaging and outright propaganda intended to deceive foreign adversaries. The US military legally wasn't allowed to lie to its own people, but it was allowed to lie abroad.

As overall director for Project Maven, Shanahan saw the benefit of a message about AI carefully crafted for consumption in Beijing and the PLA's Eastern Theater Command, the military command that would likely be responsible for any invasion of Taiwan.

It wasn't a straightforward line to tread, though: he didn't want to oversell a nascent capability, especially if it might induce Beijing to go faster at developing its own military applications for AI. But he also wanted to let China think the American military's AI was strong and successful enough a high-end capability that it wasn't worth taking on the US in a fight.

"We had a message—this is what's going to allow us to kick your ass."

Shanahan was briefed that China devoured whatever the Defense Department put out about Maven. It was reported, translated, and absorbed back in China. "Everything we said," he told me. Shanahan wished Cukor would give interviews, but the colonel always refused: "I thought it was a missed opportunity."

One morning, though, Shanahan got an unwelcome though cordial 6:30 a.m. call from Patrick Shanahan (no relation), the deputy defense secretary who succeeded Bob Work.

"The boss ain't happy," one Shanahan told another.

Details of an interview General Shanahan had given to a reporter had been included in the Pentagon's "early bird" press roundup that goes out to the defense secretary (then Jim Mattis) each morning.

"I got hammered," Shanahan told me. "He thought I was revealing an operational capability to our adversaries."

Cukor, by contrast, never gave a single interview. Besides one early conference appearance in mid-2017, he said barely anything in public about Maven at all.

As Cukor considered the Google fallout, he took steps to shield Maven from publicity and media altogether. Even though Maven had successfully sought to declassify data for commercial vendors, in December 2018 Maven got a national security exemption so it didn't have to answer any of the many media requests that came in under the Freedom of Information Act, including one from nine months earlier.

"I still don't know how he got that," Shanahan told me. "People were scratching their head at how he pulled it off."

After the Google walkout, Project Maven simply stopped talking. "Everything went quiet," said another Maven team member. "We were on an unofficial gag order," recalled Lieutenant Colonel Garry "Pink" Floyd. No one was allowed to so much as post where they worked on LinkedIn. The aim was to advance the capability and avoid more drama. The Defense Department wanted the American public to support algorithmic warfare, but it no longer wanted to share with them what or where it was, or how it was going.

Cukor told me the Google debacle had nothing to do with the decision to exempt Project Maven from FOIA requests. Information associated with Project Maven pertained, according to the December 2018 determination, not only to the "capabilities" but also the "limitations" of critical defense applications making use of Project Maven's AI. That would enable an adversary to identify capabilities "and vulnerabilities" about the Defense Department's approach to AI development and implementation, it went on. I thought back to the algorithmic systems that creatively navigated software safety rules, confused tanks for rocks, and couldn't tell the difference between a man and a child.

Cukor argued Project Maven should have been exempted from FOIA requests from the start because it pertained to critical infrastructure security information. He attributed the post-Google timing of the exemption to an oversight; someone on the team (whom he did not identify) dropped the ball in the chaotic early days, he told me.

I was unable to confirm this reason, and late one evening during my research I pored over an internal Pentagon-commissioned review of the project from 2024 for answers. It upheld my theory: following the Google uproar, it said, Project Maven became "less transparent and more insular."

THE US MIGHT HAVE DECIDED to keep quiet about its new war tech in public, but allies were soon clamoring for more information. At Dome Plate, a twice-yearly conference for the United States and its closest "Five Eyes" intelligence partners—Australia, Canada, New Zealand, and the United Kingdom—Lieutenant General Shanahan briefed Project Maven's pursuit of algorithmic warfare to the heads of defense intelligence for each country.

One senior Australian defense official was bowled over. "We really need to get on board with this," he goaded his team back home. "You could see they were onto something."

But the British were fed up. Air Marshal Phil Osborn, head of UK Defence Intelligence, was never shy to offer an opinion. Project Maven was already far out ahead of the British. Don't do this to us, he appealed. Don't run so far ahead on AI and then turn around and wonder where your partners are. We need to stick together. Osby's comment stung. "He was right," said Shanahan.

The Americans had done this to the Brits before, they complained, citing a gnarly initiative to create a single IT environment for intelligence agencies. From now on, Shanahan tried to read in the Five Eyes early, and a UK national security official described Maven's efforts to coordinate from then on to me as "sincere and generous."

In 2020, Cukor went to Herefordshire in the UK, the home of Britain's elite Special Air Service. I learned from others present that he briefed the SAS about Maven, and that they took it up enthusiastically. The UK national security official briefed on Maven's evolving efforts told me the British saw Maven as means to compensate for limited industrial capacity. The UK might not have a sufficient arsenal of guns and munitions, but Maven could help allies decide where best to aim their limited firepower for greatest effect. Even so, by 2025, Cukor told me that if you want to see the cutting edge of AI, you go to Shanghai. "AI is almost free in China: we should be focused on adoption, adoption, adoption," he told me. "We are in an existential fight with China."

The day after I attended the Chinese embassy event, General Liu emailed me politely. "As I mentioned, the Chinese government has been an active member of the international community on governance of AI technology," he wrote, sending me four "important" policy documents. Together they suggested the pursuit of global cooperation along with conservative use of AI.

By way of return, I asked General Liu about reports from US think tanks that detailed China's development of AI for military applications over the past five years, seeking his view. One suggested China could rush to deploy untested and unreliable AI weapons systems and another that China was using AI to war-game the military takeover of Taiwan. Beijing, said a third, wanted to "speed up the development of unmanned, intelligence combat capabilities."

General Liu didn't write back.

But I remembered something else he'd told me the evening before. The US, he said, had been experimenting with AI directly in battlefields. He told me if I really wanted to understand what AI could mean for the future of warfare, I should look at a single US company: "Palantir."

13

DADDY KARP

"Safe means that the other person is scared."

I HAD MADE IT to the outside of Palantir's boardroom. The only thing visible above the uppermost edge of frosted glass that ran across the length of the room was a shock of white and gray-streaked hair bolt upright. It belonged to Alex Karp, the twenty-first-century AI arms dealer.

The rest of him—much like the future of warfare—was cast in a shadowy outline. All I could perceive, struggling to see through the opaque wall of the large office room in early December 2024, was that he was standing up.

The parallels with theoretical physicist Albert Einstein went further than Karp's hair. In an op-ed for the *New York Times* the year before, Karp approvingly cited Einstein's 1939 letter to President Franklin Roosevelt entreating him to "speed up" efforts to build the atomic bomb. Six years and $2 billion later, the US detonated atomic bombs named "Little Boy" and "Fat Man" three days apart on Japan's Hiroshima and Nagasaki, killing more than 100,000 people instantly. Attempting to corral and limit rivals' development and use of the same weapons would shape the next eighty years of international diplomacy.

The capabilities of the newest AI technologies are "equally significant," argued Karp, calling for swift action to breathe a new arms race into life all over again. Karp omitted to mention what Einstein said fifteen years later: signing that letter to Roosevelt was the "one great mistake" in his life.

Now the martial impresario was on home turf, gesticulating in his boardroom. He was in his customary wild mood, and he had money top of mind. A couple of days earlier, Palantir had overtaken Raytheon to become the top-valued defense contractor in the world.

Karp was self-appointed promoter-in-chief for AI warfare. As CEO of Palantir Technologies Inc., he had spent the past few years hawking AI weaponry around the world. Ukraine. Israel. America. Europe. More than a dozen countries. A master of hyperbolic bluster, Karp fashioned his own hymn sheet—patriotic, madcap, fast, cerebral, and warlike—and sang from it at every opportunity. The company benefited from its reputation for allegedly having helped find Osama bin Laden (suggested in *The Finish* by Mark Bowden), that I could not corroborate, but it earned spy detractors too. The eavesdropping National Security Agency mostly stayed away from Palantir, and the Army leadership briefed against it for years until Palantir sued and won a contract.

The day after I visited, Karp would roll out his vision for weaponized AI at a national security conference. The only way for Americans to be safe, he argued, was to make sure America's adversaries live in fear of the country. "Safe means that the other person is scared," he said. America's adversaries "need to wake up scared and go to bed scared."

Karp described peace activists as "an infection inside of our society," wanted to rename the Department of Defense the "Department of Offense" (the second Trump administration would go one further and change it to the Department of War), and hung the future of America's next nine hundred years on speeding up the sale of AI weapons to the world.

The second Trump administration was about to burst back onto the stage, with a fulsome mandate it could take all the way to the bank. In his outgoing speech, Biden would warn of the dangers of the rise of a tech-industrial complex five days before chief executives of Amazon, Google, and Microsoft would stand alongside Elon Musk at Trump's inauguration ceremony. (Karp, a Democrat, would donate $1 million for the celebrations.)

Even before Trump's November 2024 reelection, Palantir enjoyed a sudden bull run: in September 2024, the company made it into the S&P 500, and in October its value crested $100 billion. Palantir, a mere infant twenty years earlier when the CIA first invested in the startup, overtook Boeing, a ninety-year-old defense juggernaut.

What was going to happen next in the defense-tech world, Cukor told me in December 2024 with the foresight to which I was now accustomed, was the arrival of new business unions between software makers and hardware producers. "Everyone is getting ready for the next administration where it's widely understood there's going to be a big push on AI," he said. "Everyone is arming on the contract side. You are seeing rapid alignments."

He was right: in quick succession, Palantir expanded a tie-up sharing battlefield data with Anduril Industries, a defense-tech startup that briefly participated in Maven, valued at $14 billion. Palantir also struck a deal with Booz Allen. The starkest move came from OpenAI, the company at the heart of the global generative AI explosion. Up until early 2024, it banned users from working on weapons development and warfare. No more: the week I visited Palantir, OpenAI announced a partnership with Anduril, bringing its AI directly onto the battlefield. These new AI tie-ups would make defense-tech platforms stronger, all the more ready to duke it out with the five big traditional prime defense contractors under the Trump administration and reshape the landscape, Cukor told me. Money would flow into new AI defense companies, from investors and buyers alike, as never before.

Now, on the afternoon of December 5, Karp was signing one of those deals to extend his company's AI-enabled platform into autonomous warplanes on the very front lines of war. And I was ambling around Palantir's New York office in Chelsea solo after the company's press handlers cut me loose.

For a company dedicated to offensive weaponry, it had its dissonant cutesy elements. One row of meeting rooms was named after bagel toppings, and I spent close to an hour chatting through the niceties of military contracting with Palantir's chief technology officer Shyam Sankar in a toasty little room named "Poppyseed." Slices of rare steak were piled high on an inviting food counter at one end of a large gathering spot filled with high black tables and chairs more reminiscent of a jazz bar than a martial business venture. Perched up on one of the chairs, I heard talk of "a World War III thing." And all the while Palantir's stock was rising, reaching a record high the day of my visit, putting the company at $164 billion. That was four times higher than last year. War talk seemed to pay.

Sankar hankered after the days when Americans who bought cars, cameras, and cereal were also subsidizing arms manufacturers. He rued the decision from one nineteenth-century industrial giant, named Ball Corporation, to stop making spacecraft to concentrate on making beer cans. He despaired that the US military ecosystem had no chance of producing weapons at scale without producing for the civilian population at the same time. He wrote as much in a heretical September treatise he said he metaphorically nailed to the metro entrance outside the Pentagon, styled after Martin Luther's cry for religious reformation back in 1517.

The era felt about right to him: the Pentagon always seemed to want to build a faster horse, rather than invent an altogether new and better means of transport, he complained. Sankar recalled a senior government official who was stuck to his old guns, telling him: "Look, if I ask for a three-wheel Jeep, give me a three-wheel Jeep." "There's no more offensive statement to me than that," Sankar told me. Six months after our meeting, Sankar would commission as a lieutenant colonel in the Army Reserve, leapfrogging ranks into a new Executive Innovation Corps alongside others from OpenAI and Meta. The move drew praise from some quarters and consternation from others over a potential conflict of interest.

Sankar wanted to compete with China on manufacturing, fashioning a new era of dual-use companies dedicated to contributing to the country's lethal power. He found his instructions for the future of the war in the past. Like others I met, he seemed to revel in the era of total war and World War III always seemed just around the corner. After we met that day at Palantir, he emailed me a quote later that evening.

"The coming war is going to be a technology war," it read. "We believe the US is woefully unprepared and ill-equipped for a war driven by advanced technology. The only solution is to outsource advanced weapons systems development outside of the traditional services and government ecosystems and hand the development to civilians." It came from Vannevar Bush, cofounder of Raytheon who oversaw the Manhattan Project to build the atomic bomb, in 1940. "It could have been said today!" Sankar messaged me, triumphant.

KARP WAS IN THE ROOM to sign a deal with Shield AI, a defense-tech startup that had spent the past nine years developing a way to fly aircraft autonomously with the help of artificial intelligence. The company had started with small quadcopters and soon graduated to producing drones the size of washing machines. Its AI system could fly warplanes in the absence of GPS and satellite communications, including purportedly in Ukraine.

In May 2024, the company was so confident in its autonomous systems that it managed to convince the seventy-five-year-old secretary of the Air Force, Frank Kendall, to take off in a F-16 fighter jet modified to be flown by AI. Soaring above the desert over Edwards Air Force Base in California, Kendall swung around in the front seat as AI flew him at 550 miles per hour. Facing off against another F-16 flown by a human, the two went twisting and looping through the air in a dogfight. The human just edged out his AI rival, and some experts I spoke to dismissed it as a stunt. It was still early days for autonomous air combat training. But after his near nose-to-nose experience, Kendall was convinced. "They don't get tired, they don't get scared," the retired army lieutenant colonel said after landing.

Shield AI had raised $1 billion thanks to a constellation of venture capitalists, and was now valued at close to $3 billion following its latest, sixth round of fundraising. Now the startup was about to land an even bigger catch: Palantir. The new strategic partnership between the pair aimed to combine Shield AI's autonomous flying platforms with Palantir's data analytic platforms. That would merit a military breakthrough: it would bring AI-powered intelligence and operational control to autonomous flight. It would be the future of war, today.

"You're going to see these systems start to do things on the battlefield that you previously haven't seen," Brandon Tseng, the president and co-founder of Shield AI, had told me the day before. Now Brandon and his older brother Ryan, who together started the company in 2015, had their eyes on supplying hundreds of autonomous drones to Taiwan, to help the island fend off a military invasion from China. The two sons of a Taiwanese electrical engineer had made repeated trips to Taipei in recent months, seeking a billion-dollar sale. Beijing had noticed, and just that morning imposed sanctions on Shield AI and a dozen other companies, to prevent Chinese components from reaching them. Brandon took it as a medal of sorts: to him it meant the Chinese were afraid of his company's tech. Before long he would accuse China of copying it.

Standing outside the boardroom, scanning the shadows beyond the frosted glass wall, I realized everyone in the room was standing up. It struck me as curious. Perhaps the meeting was wrapping up, and they were all about to leave.

Just then the door opened and Karp burst through toward me. Gabriel, one of Karp's two Austrian executive assistants who had been hovering alongside me outside the door, made a quick and kindly intervention on my behalf. (Karp jokes that his two assistants control his life, Gabriel told me, but the reality is Karp controls his own movements, and no one can keep up with him.)

Karp no sooner spotted me outside than he abruptly turned, me in tow, ushering me in after him, and suddenly I was in the room where it happened.

Fanning out around Karp and a long boardroom table were ten more men, an assortment of high-ups from both companies. They seemed as surprised as I was that I was there. They were all standing in a circle around the table, hips thrust forward in hoc to Karp's frenetic energy, and somehow I now had them cornered. Karp cornered me back. As I made my cheery opening gambit, my back to the door, my British accent rang out. Karp promptly announced he could send me home soon.

"You probably can," I only half-laughed, understanding he aspired to deport me to the UK.

"Pretty soon, pretty soon," he jousted, more ominous than I was expecting. Then he moved on to the meaning of today's tie-up between two companies focused on AI weaponry and autonomous flight.

"I've always had the fantasy of having a drone company that could take fentanyl-laced urine and spray it on short sellers," Karp announced to me and the instantly horrified room. "And now I found the right company." With that, Karp was gone.

It wasn't the first time Karp had laid into short sellers, who had for years taken bets against Palantir, driving down its stock price. But it was his choicest language by far and the first time he'd invoked a death threat.

"We were all taken aback by the opening exchange," one of the company officials assembled in the room told me later.

Karp had also humiliated the handful of Shield AI leaders in the room. The strategic partnership was meant to be just that: a partnership among two champion companies to deliver the weapons platforms of the future. Palantir was a whale to Shield AI's minnow, but it wasn't buying Shield AI; the agreement was for an exchange of technology—I learned each company was paying the other exactly the same amount of money, measured in low millions of dollars, for access to each other's precious tech. But Karp wasn't about to mollify me—or his new partners. I chased him out the door, asking what he meant. He motioned for me to go back into the room and talk to everyone else.

"They're very sensible, I'm not very sensible," came Karp's parting shot.

Karp had been fine-tuning his obsession with short sellers for a few years, goading Wall Street in the past for seeing him as "batshit crazy" and a "freakshow leader." In March, Karp had claimed short sellers "just love pulling down great American companies so that they can pay for their coke." Despite the company's strong valuation throughout 2021, when its worth regularly hovered between $50 billion and $70 billion, Palantir's share price spent the next two years in the doldrums. Company revenues were down, it was missing forecasts, and the market was less than impressed. In December 2022, the company's value fell to a two-year low of $12 billion.

Midway through 2023, a short-seller report described Palantir as "an overhyped data consultant." "The Bear Cave believes Palantir is an AI imposter," wrote Edwin Dorsey, a twenty-five-year-old Stanford economics graduate who had made a name for himself for his twice-monthly Substack newsletter read by 50,000 subscribers. "The retail investor hype around Palantir is hard to miss," he added, saying later that he felt the company was misleading its own investors.

Palantir's share price was often supported by what the company's friends and competitors called Karp's fanboys. A group of 60,000 retail investors congregated in one online subreddit feed alone, referring to him as "Daddy Karp." Another appreciative social media account was named "KARPGOD." Palantir fans flocked to the company's defense, but Dorsey stuck to his guns.

"I think now would be an opportune time to bet against them if that's something you're looking to do," he told the host of Palantir Vision, a YouTube channel devoted to the company, a week after his newsletter ignited the furor. Karp defended the company as being a profitable software company, pointing in vague terms to the effectiveness of its work on the battlefield in Ukraine.

"My real answer to the short people is: Ask the Russians," Karp told Bloomberg TV when asked about the short-selling report. "Ask people who are on the battlefield," he said.

Early in 2024, the company pushed back up through $50 billion in value. Revenues started rising, and it was now riding record highs. But the doubt was obviously hard to shake off.

Even on the day I visited, as Palantir's value soared to a new all-time record, Wall Street generally doubted the company as a good long-term bet. More than twice as many analysts recommended selling Palantir shares as buying them.

BACK IN THE NOW-KARPLESS ROOM, the remaining ten men left standing offered me the whites of their eyes, and suggested I come back in an hour. But somehow I stayed rooted to my spot, gazing at this strange scene.

"Why are you all standing up?" I ventured from my corner by the door, to nervous titters in response. For the length of the corporate meeting, including signing the deal twenty minutes earlier, I began to understand the men had spent the entire time standing.

"Well, we were just having a conversation," one ventured tentatively, as if that counted for an explanation.

"First of all, Karp doesn't sit," came the more convincing response from Shyam Sankar, Palantir's chief technology officer who had been with the company for eighteen years. With Karp gone from the room, they no longer needed to mirror his macho energy. "I will start a trend here," Sankar proposed, sinking into his seat and offering one to me. Slowly everyone sat at the long table, and we morphed into tamer creatures.

I asked everyone what I needed to drink to keep up with the teetotal Karp, suggesting whiskey.

"Stay away from the drone fluid," came the only half-cogent response.

However dismissive Karp might have been, Brandon Tseng, the co-founder of Shield AI, was going to have to forgive him. If Shield AI didn't raise another round of financing, or list itself on the stock market, it would eventually run out of money. Selling the company to someone like Karp might be exactly what his company needed. Besides, Tseng had another reason. He felt he owed his life to Palantir.

Brandon and Ryan Tseng grew up highly competitive. Raised in Texas and then California, they dueled on rival paper routes as the middle two siblings of four children. Brandon played water polo and took mechanical engineering; Ryan pursued electrical engineering like their father. In early 2008, Brandon joined the US Navy and was stationed on an aircraft carrier as an engineer. But that was never enough for him.

Since watching action thriller movies *Under Siege* and *The Rock* when he was ten years old, he had his heart set on joining Navy SEALs, the special operations US commando force first created in 1962. His parents expected him to change his mind, but he never did. The first time he tried, he didn't make selection. He saw that disappointment—eventually—as a massive wakeup call.

"After I removed every excuse from my head, it's just simply because I was not the top candidate," he told me. "And so that's what shaped me. I'm like, if I'm ever gonna do anything ever again, I'm gonna go after it one hundred and ten percent. I never want to experience a setback in my life again." At the end of 2009, he got there. At the Navy SEALs training base in California, hopefuls must survive seven intensive weeks of "basic conditioning." Midway through comes Hell Week. "The only way out is through," says a slogan I saw daubed in white paint around the training ground when I visited in 2018. The facility, in Coronado, San Diego, is a spot on the beach, beautiful and simple, Brandon recalls. "But that place has a ton of pain and suffering associated with it. It's amazing how awful a place can be just with cold water, sand, and some boats."

During my visit, I saw a big brass bell hanging up on the training ground, not far from the beach. Anyone could ring it at any time. Most people did. Three times and it meant they were giving up on trying to become a SEAL.

Brandon never rang the bell, and he went on to join Navy SEAL Team 7, stationed on the West Coast. It was one of the two newest teams in the elite group, created a few months after the attacks of September 11, 2001. "Forged in War, for War," goes SEAL Team 7's mantra. In 2013, Brandon deployed on his second combat deployment to Afghanistan. This time, he was a squad commander.

Stationed in the south of the country, Brandon's platoon regularly went out on raids and sometimes just to meet elders in the villages. They were housed at a forward operating base in Oruzgan province, doing operations further south into Kandahar, Zabul, Ghazni, and Paktika provinces. The Taliban remained expert at planting makeshift bombs known as improvised explosive devices (IEDs), digging them into roadsides and anywhere else that might target US troops.

IEDs were in houses, component parts strapped to motorcycles: everywhere. The year before Brandon deployed to Afghanistan, 237 US forces were killed in action. Nearly 3,000 more were wounded. Several US Navy SEALs had also recently been killed, including by IEDs during combat patrols in Zabul province and elsewhere.

During Brandon's seven-month deployment as a ground force commander, they went on more than seventy missions, some lasting three days. Their task was to kill or capture Taliban and al-Qaeda leaders. Firefights were a constant.

"Most times we would go out, there would be some form of kinetic activity," recalled Lieutenant Andrew Cornelius, an elite US Navy bomb disposal officer who was attached to Brandon's platoon, and who joined them on operations throughout May to November that year. He answered to the nickname "Corny," and it fell to him to plan the operations, trying to dodge known hazards. The problem was, as Cukor had rued since 2001, almost nothing was known. For the US forces who arrived each rotation to poor handovers and little information, it had often felt like they were fighting a new war each six-month deployment. Different US forces still fell prey to the exact same traps year after year.

Corny was wary of almost anything that might create a chokepoint for the US forces—roads, buildings, even land features, he recalled to me. But before he flew in, a friend had told him to make sure he got access to a Palantir account. He had never used one before, but it changed everything for him. Individual military services like the Army and Navy all submitted their own reports noting down hazards encountered on each operation, but accessing them all and overlaying the data at an individual planning level and trying to draw out patterns was cumbersome, and risked gaps.

Palantir systems aggregated all the detail, plotting past explosions and other threats on a map used for planning. One time, ahead of a raid on a village against Taliban threats, Corny turned to the system and it directed him to a report written by another bomb disposal tech the year before that he would never otherwise have turned up. Corny knew him, so he got directly in touch to talk through the details, safeguarding the mission. When they did spot an IED, Corny would carefully crouch down and disarm it. Other times, he would blow them up, along with the motorcycles they were strapped to.

Toward the end of their deployment they were supposed to do an operation in another village. It was a turnover operation—meaning the men comprised personnel from the next incoming platoon as a way to break them in—but these handover missions were historically very dangerous precisely because no one had the same style of operating and had never worked before as a team.

"When I did the search, it looked like the map had chickenpox," Corny said of suddenly coming face-to-face with a record number of IED warnings popping up on the satellite imagery. He showed Brandon and they agreed the area was simply too hot with IEDs. "We had never seen something like this before," said Brandon, who partway through his tour had taken over as platoon commander. Worse, it was meant to be a nighttime raid. It would be too dark.

"Our principal way of avoiding IEDs is with our eyeballs," Brandon deadpanned. "And even then, like, we're not super great at avoiding IEDs with our eyeballs."

To him, it was a suicide mission. And it was bullshit. "The village had no strategic significance whatsoever," he said. But bringing any of this up with the Special Operations Task Force commander in charge, who was pushing the raid but was not a Navy SEAL and was new in the region, was almost scarier than doing the raid. The Navy SEALs execute orders, they don't question them. The next four days, as Brandon made his case to swap out the raid for a different one, were one of the most stressful things he'd ever experienced in his life. He recalls losing weight.

"I'll do any other mission," he told them in the face of heavy pushback from the dogmatic commander. "A lot of American taxpayer dollars goes into building, developing, manufacturing Navy SEALs, and you don't want to lose them on these operations." The Palantir data counted as ammunition to convince them to pull the operation. Eventually, they agreed to switch the target to a village farther south. For Brandon, the impact of that and other missions was clear. "Palantir saved my life," he told me. "The intelligence that they provided my SEAL platoon in 2013? Like, holy smokes." It was "very, very visceral" for him, Brandon told me, describing himself as "a forever Palantir fan." "It prevented us from getting blown up. Multiple people on my platoon have the same reaction."

At the time, in 2013, Lieutenant Cornelius could only check the Palantir systems on a laptop back at the base. But by the time he deployed to Iraq to do a similar job in 2015, there was a mobile version on an Android phone, and he took it with him everywhere he went on raids, checking the map as he went. That same year, Brandon got out of the military after his seven-year stint, and suggested to his brother Ryan they start a defense-tech company so disposable robots could be first to enter buildings—instead of precious US forces.

Ryan, a tech entrepreneur who had sold his own Wi-Fi charging company, initially dismissed the idea. But even after he warmed to it, the brothers failed to convince different investors. Slowly, seed money came and they built the company up from scratch, expanding beyond tiny flying robots to midsize drones and fighter jets, and they have now installed their AI pilot on seven other companies' aircrafts. In 2021, they snagged Doug Philippone, a former Army Ranger who was Palantir's long-running global defense lead, who put money into the company and joined their board in 2022. From the start, Philippone foresaw an eventual tie-up between the two companies.

It was everything that advocates of AI warfare had spent years salivating over. It was not only the development of AI, but also its employment on the battlefield. The idea was that AI could now be everywhere: not only helping commanders find targets in real time but also directly linking back and forth to powerful weapon systems and their sensors on the frontlines.

Shield AI's autonomous drones were already deployed in Ukraine, India, Latin America, and the Pacific, although it wasn't clear to me how well or how autonomously they worked. Now they would instantly start sharing data back and forth with Palantir systems. One day soon those drones would fly as a swarm of warbirds. And they would surely carry munitions too.

"I just saw this vision," Philippone told me, sitting opposite me in the boardroom the day they signed the deal. "America is only gonna win if we pair the very best companies," he said, describing Shield AI hyperbolically as the very best autonomy company in the world, and Palantir as knowing how to move data around like no one else. "I want America to win, period. Like, if we lose the next war, money doesn't matter," he said. But Karp, a philosophy-student-turned-billionaire, had made it clear just how much money was indeed part of the game.

Holding court at Palantir's offices later that day under a sign reading "Fortune Favors the Fast," Karp started talking off the cuff about Drew Cukor. He was the only person who had seen the value and implications of large language models for warfare, years before they started arriving on the scene in reality, he said. Karp referred to him approvingly as "crazy Cukor." He almost seemed to worship him. Two weeks earlier, at a by-invitation-only national security event I attended in Washington, DC, Karp referred to Cukor as the "founding father of AI targeting."

Now, Karp argued, as AI remade the world, large language models would function like uranium. Chatbots would power a new form of warfare that could threaten the world and would remake diplomacy. It was a dark and explosive vision for AI, one that Karp said began on the battlefield with Project Maven. "It's really changed the history of the world," he said of Project Maven. But, Karp went on, he has never been allowed to tell that story.

14

PALANTIR, PALANTIR, PALANTIR

*"When I started Maven most people thought
Palantir was exiting the Pentagon."*

A FEW MONTHS LATER, I wandered near a canal in the rarified Georgetown area of Washington, DC, to reach another Palantir enclave. This one comes with a seven-foot-high model of *Star Wars'* Chewbacca and a wall poster dedicated to the imaginary hip-hop hits of co-founder and chief executive Alex Karp. One notional track, named "Public At Last," commemorates the company's 2020 listing on the New York Stock Exchange.

I was more interested in a different part of the office. It was April 2025, and I wanted to find out more about the Project Maven story that Karp said he'd never been permitted to tell. The one he claimed had changed the history of the world. It would certainly help change the history of the company: in August 2025, Palantir bagged a $10 billion agreement with the US Army for multiple licenses, and the company's value soared to a record $411 billion, ranking it among the top twenty-five companies in the world.

"It's hilarious to watch Palantir skyrocket," Brian Ward told me as the company's value reached record highs. "I don't know how anyone can argue it wasn't because of Drew Cukor."

If that sounded like hyperbole, Akash "Aki" Jain, Palantir's CTO and the president of the company's US government-focused subsidiary, wasn't far off in his own estimation.

"A lot of our success in defense absolutely is tied to Maven; that's very obvious," he told me. Nearly 60 percent of the company's revenue comes from government customers. But besides specific contracts, Jain thinks the first thing Project Maven and Cukor did for the formerly AI-skeptic company was enable it to embrace artificial intelligence.

"Drew forced me to visit my priors on AI and accept there might actually be value there," said Jain. "He was insistent you've got to figure out AI." Describing himself to me as an AI skeptic in the 2017 timeframe, Jain told a Special Operations Forces conference in May 2025 that he now eats, sleeps, and dreams AI. Today, Palantir has a devoted AI platform.

Besides suggesting the company should ready itself for "technology built for a post-AI world," Cukor also counseled the company on how to reconstitute itself inside the Department of Defense. The outcome could have been very different. When Maven started, Palantir was a source of opprobrium not only for some in the Army, but also for several other parts of the defense and intelligence world that was slowly pushing them out, spanning the National Security Agency to Special Operations Command. "We brought Palantir back to life via Maven," another member of the team told me. "They were completely done." (Some Palantir officials sought to temper this view, citing several smaller defense contracts at the time.)

I was here to see where the turnaround, and the future of Maven, started: the lab.

First comes the virtual reality room, a small space with a bare round cork table. Under the guise of augmented reality goggles, the small round table becomes a proxy for a bright green island, missiles swinging around from behind. The room starts spinning even on solid ground; the brain races to catch up. It's a weird physical and visual bonanza that can leave a person dazed as if emerging from days of helicopter rides. The cork island could be Hawaii, or Taiwan, and there comes the sudden sickening feeling of being physically subsumed by war. Only it's a war as trite, bright, easy and fleeting as a video game.

Beyond the table is another door. Trapped inside is a jumble of wires and headsets on shelves around white walls. A whiteboard covered in teacher colors of green, red, and blue sketches out boats and plans. A warm-hearted golden retriever with thick caramel fur emerges, prone to jumping up on visitors in pursuit of stretches and food. (In the past, the lab has had a lab too.) Human-canine teaming is just the start, though. Human-machine teaming is the higher agenda. Tucked in the far end of the lab, with a large US flag hanging at his back, Robert Imig works at his standup desk.

His colleagues at Palantir say they think of him as a unicorn—a humble and intelligent co-worker who gets both the business and tech side of things. When I meet him that spring, he is six years into his journey as Palantir's tech lead for Maven. That pretty much makes him the man who birthed America's first bespoke AI-enabled warfare platform.

When Imig joined Palantir, it was his first job out of school. Back then, in 2012, he was a young computer engineer living in the Bay Area and interning for Cisco. One night he was out with his sister at a bar, and remembers some guy going off about jumping out of planes with servers. He was agog. Where do you work, he asked. He'd never heard of the company the guy answered with, but looked it up, applied and got a job.

For Imig, Palantir's national security associations mostly counted as hazy mystique rather than a driving factor or reason for concern. He had no direct experience of any of it. In 2018, he was working on the commercial side of the company, producing applications for business clients. There wasn't a parachute in sight.

But one day in June 2018 a military task came calling nonetheless, when he was approached by Anand Gupta, a fellow Palantirian he'd not met before. Gupta liked to talk fast. He struck Imig as a hyper crazy guy. Gupta was Silicon Valley, through and through. He liked to think he grew up on a bean bag under his mom's desk in Menlo Park. She'd dropped out of a PhD program in AI at Vanderbilt University in the late '80s to join an AI startup. Following in her footsteps, he (temporarily) dropped out of Harvard to join an unorthodox fellowship in 2012, set up by venture capital billionaire and Palantir co-founder Peter Thiel. Gupta began what he described as "a terrible startup" focused on biotech of his own at age twenty before returning to Harvard.

In 2015, he joined Palantir as an intern with big ideas. And three years later he was an excitable blur of drones and video and computer vision algorithms, and had a new task to do for the Defense Department. Gupta was talking about Project Maven. Imig thought it sounded awesome. "Where do I sign up?" he asked.

BY 2018, the Maven team was doing its thing across multiple regions of the world. Besides Somalia and Afghanistan, they were in Jordan, the UAE, and more. They had made an algorithm; now they needed a permanent home for an algorithm to live. They moved on from the ad hoc efforts of the first launch in Somalia when they simply shoehorned AI into the existing system. The team decided to commission a custom-made Maven appliance (known as a MAPP, or mini-MAPP for the smaller version) to carry the GPUs necessary to run all the AI models and deliver inferences—meaning the trained AI's conclusions about new data that showed up as a detection—for up to four drone video feeds at a time. The AI outputs would then plug into platforms already widely in use for combat operations so the object detections could show up on a screen.

The first time Colin Carroll tried to get the new Maven server made, he disregarded the design advice of ECS contractor Katya Volkovska. Eight massive overheating GPUs literally melted the $2 million boxes they put them in.

"You were right," he told her, calling to interrupt her holiday. "They're on fire."

Volkovska thought Carroll was smart, passionate, dumb, and stubborn all at the same time. It's the only time he's conceded she was right about anything.

They married in September 2022.

ECS, the prime contractor, got the Maven appliance that carried the AI models remade in a bigger box, with the help of circuit boards from

Supermicro and GPU chips from Nvidia. Then the team hooked the AI outputs from the Maven appliance into a number of different image processing systems that the US military already relied on for operations. The idea was the AI detections would simply show up over the drone feed viewing systems they already used. One of these systems was Minotaur, a government-owned software suite made by Johns Hopkins University Applied Physics Laboratory with software help from defense contractor Alion Science and Technology. Minotaur fused data from sensor feeds, including drone videos on a common platform, and then AI detections would show up as boundary boxes or dots over the drone footage in video players used by intelligence analysts and "screeners." Another of these image processing systems was Novetta's software suite, the Automated Information Discovery Environment (AIDE), which also came with various tools already widely used by the Air Force and others throughout the world. These included popular ways to play drone video feeds with frame-by-frame precision, map overlays to help show and locate the environment, and establish and track patterns of life by analyzing intelligence data.

Cukor wanted to go a step further though: he wanted everything including the AI inferences to show up all in a single everything platform, and he wanted the platform to do much more than just display AI over video feeds. It would be the vehicle for his dream to bring intelligence into operations. A slick platform for special operators would entice the people who conduct combat missions—rather than the intelligence analysts they rely on—to lean in to AI too. If he could embed AI-enhanced video players directly into operational mapping, planning, and intelligence tools, then they might really get to a system where users would right-click a white dot, get a coordinate, and then go shoot it. That meant a push to get AI into the targeting workflow.

"This is where we start opening up an entire can of worms," Cukor told me.

Maven's money was earmarked for intelligence, not operations; plenty of overlapping software options already existed for military operators, and bringing AI into intelligence, let alone operations, was already deeply controversial. Besides that, they had to consider what company would actually make the thing.

America's national security apparatus already had options for a kind of all-consuming "common intelligence picture." The National Reconnaissance Office (NRO), which sent up spy satellites and doubled as both an intelligence agency and a combat support agency, ran a platform produced by an intelligence program called FADE: Fusion Analysis and Development Effort. FADE started in 2007 and rather than relying on commercial tech was produced in-house. It aimed to detect the changes in time and space that can characterize "patterns of life." It sucked in more than two hundred different intelligence data feeds, so analysts could investigate and display them on a screen with a set of tools made by defense-intelligence contractor CACI. FADE was used for mission planning and battlefield forensics rather than operations itself. One of FADE's mapping tools known as MIST (the Multi-Intelligence Spatial-Temporal tool suite) "was the only show in town," one Mavenite recalled to me. MIST aggregated clues from millions of data points, such as shifting phone pings in a specific location that might indicate an operation was being planned.

Cukor reached out to NRO, but he decided they were developing in a way that was incompatible with Maven's approach. At NRO, the government owned the software, whereas Cukor thought the commercial sector should own, operate, and update the software. NRO's remit was intelligence alone; Cukor was looking to build something for combat. Then there were logistical and bureaucratic headaches like platform compatibility and permissions to put Maven on the NRO platform, which would have taken a year at least. Cukor said the discussions were "cordial" but others on the team felt fed up and it would prove a source of continuing tension. (NRO did not respond to requests for comment.)

"Every step along the way we were dealing with defense-industrial bullshit," a different Mavenite recalled to me.

Waiting a year was a nonstarter for Cukor and "Maven time." But creating a new rival platform risked a fundamental duplication: NRO data feeds would be at the heart of anything Maven could muster up for AI to analyze.

Cukor had wanted Google Earth to create an operational platform that could fold in targeting but as it became clear that Google was going to drop out in catastrophic fashion for his project, he decided to call up his old friends at Palantir. "Hey, it's been a minute," he said on a call to Aki Jain around April 2018. "Could we meet?"

Today, Alex Karp is all in on AI. But back then, he was a skeptic. Palantir had watched IBM's $4 billion investment in Watson, an AI-charged super project, crash and burn with not a little schadenfreude. One Palantir official half-wondered if Google dropped out of Maven because AI turned out to be a lot harder than they'd expected. There is no magic AI button, went Palantir's mantra. But some at Palantir were also growing concerned about the threat posed by its AI-curious commercial rivals, including Novetta and its AIDE platform. Palantir risked falling behind on new tech.

The company was also falling behind on contracts. Despite winning its 2016 court case against the Army, which helped it expand from the intelligence community into the military, it still faced a number of problems at the Pentagon. Suing the Army created a lot of bad blood. "There were a lot of scars, raw nerves," says Cukor. That came on top of what Cukor described as the company's hard-sell techniques that sparked outrage in a genteel Pentagon. "They were their own worst enemy."

Jain was all too aware of the company's polarizing reputation for aggression and gall. But nobody could tether Karp. "Even five years after the lawsuit there was still a lot of overhang around this," Jain told me. "The lawsuit upset a lot of people and sullied our name in a lot of quarters in the Pentagon and in the Army in particular."

Jock Padgett, a marine who led IT innovation at JSOC in North Carolina's sprawling army installation at Fort Bragg, typified some people's attitude toward Palantir. He listed his worries at the time to me in an assured drawl: back in 2018 the company had a reputation for being expensive, and he worried he might not be able to get the military's own data out from Palantir systems that voraciously collected and analyzed their own data. Besides that, Palantir could be uncomfortably arrogant. "It was an attitude problem," Padgett recalls, citing meetings with Palantir officials in 2016 and 2017.

Still, Karp embraced national security work where other tech companies rejected it. The former doctoral student in German philosophy billed himself as defending the values of the West. But he would also trot out throwaway lines about death with little mention of the impact on victims and combatants or how mistakes in ethics or the rules of engagement could lead to tragedy. Padgett found him "chauvinistic." For some in the special operations world, he was too brazen, too trite, too much in the spotlight. A civilian glamorizer of war slaying.

"Candidly you're kind of a sideline," Padgett said of Karp, as if talking to him. Karp would never go to war himself; he would never have to risk his own neck or the necks of fellow combatants and civilians, Padgett explained. "We kind of have to. It was a slap in the face."

Few had plans to pay for Palantir once the wars in Afghanistan and Iraq drew to a close. By 2018 Palantir had no contracts with the Navy, its contract with the Marines had ended and it had only a slice of work with the Air Force. Even their final contracts with SOCOM were expected to end as the wars came to a close. "When I started Maven most people thought Palantir was exiting the Pentagon," Cukor told me.

Cukor's outreach to Aki Jain couldn't have come at a better time for Palantir. As the public outrage over Google's secret contract with Maven was building, Cukor flew to Palo Alto in late April 2018 and briefed a small group of Palantir officials.

"Here's the issue," Cukor began. In a long and intense meeting, Palantir officials could barely get a word in or demo a single thing. Over the next three hours, Cukor sketched out his vision for Maven for the coming ten years. The data sources, the transportation of the AI, the ways the workflows of combat would change forever, and the arrival of tech companies into the heart of war and government. He wanted AI on every single piece of intelligence, all around the world, all the time. This was how the US would run the wars of the future.

Key to Cukor's vision was making the AI relevant and easy to use. Maven had got the AI roughly working, but the team was finding that users didn't want to go near this thing because they couldn't engage with it. The hardware was clunky, the way the AI presented itself was distracting and often wrong. Cukor always thought AI itself would get better; he knew a whole new era of data-hungry transformer models was coming. These were deep learning models based on neural networks that were first described in a 2017 paper by scientists working at Google and promised to deliver much more in terms of image classification than the convolutional neural networks most Maven models relied on. These transformer models would eventually put the "T" into ChatGPT when it launched in November 2022 and unlocked the promise of generative AI.

Cukor never wanted AI for the sake of AI. He wanted an interface that military operators were willing to use. AI detections should show up over video, on maps and with precise locations that could immediately link to geographic coordinates and cross-reference with other data sources. A smart system. He didn't want Palantir just to repurpose their old software, Gotham. He wanted them to reimagine a user interface for AI from the start. To build something bespoke. Send me a white paper, he commanded.

The Cukor blizzard flew out the door and the Palantir officials were left smarting: they were data management platform experts, not producers of shiny things like interfaces. The white paper they sent at the start of May made a pitch for both, according to people familiar with it. Maven gave them a small contract for a prototype, but otherwise Cukor got them to build the platform essentially for free, another Mavenite told me.

Palantir wasn't the only company in the race. Cukor also tapped Anduril to produce a prototype interface, along with Novetta (which produced AIDE) and Johns Hopkins University Applied Physics Laboratory (which produced Minotaur). Everyone had six months. Cukor talked with them all, but perhaps never as extensively as he did with Palantir.

"It was a natural Darwinian process," Cukor claimed. He had to be looked at as a neutral player but he had a hunch who the winner would be. From the start he had given his special attention to briefing Palantir, and he was proud of what emerged: "You don't hear about AIDE or Minotaur except in small corners of the Navy anymore. It's just Palantir, Palantir, Palantir."

Anduril quickly fell by the wayside. At the time they were more focused on hardware than software. Before long, leaders at Palantir told me they would spend almost half their time furiously trying to build the system Cukor wanted.

Cukor was alive to how much animus the company elicited. He counseled Aki Jain on undertaking a high-end apology tour. That listening tour, as Jain preferred to frame it, helped soften some of the relationships the company held with some of its most important national security customers. Cukor help "mentor" him through, encouraging him to switch to a much more collaborative approach and to apologize "for things that were probably more aggressive" than Jain said they needed to be.

"'Not everybody loves you,'" Jain remembers Cukor warning him about the company. "We knew that, but it was important for him to remind us. Nobody had ever said to us: 'To be on the team you need to work on this. We love you, you're brilliant; but stop being assholes. This is a no-fail for our country; I expect you to conduct yourself in a particular way.'"

Jain moved to Washington, DC, to work on Maven, spent about half his time doing so, and felt as though he were working for a startup all over again. Cukor initially played down the company's contribution, figuring that associating Maven with Palantir inside the Pentagon would only backfire. His own team expressed dissent about Karp's company. Cukor asked company officials to take the Palantir company name off presentations about their work on Maven. Palantir teams even switched out the company's own font from briefing materials. "Cukor kept it very much under wraps; he didn't want anyone to know he was working with Palantir," one Palantir official recalled to me.

But Palantir simply couldn't help itself: branding would creep back into presentations. Sometimes a tiny logo would appear in a corner. Briefers might try to slip their own name into the title of Cukor's new "smart system" when they pitched other branches of the military: "Maven Palantir," they'd say, or "Palantir Maven." Cukor would glower at the Palantir team in the middle of a briefing, exasperated the company didn't see their indiscretion might torpedo the whole outreach.

Palantir never dropped its combative spirit. "We're expensive compared to what?" the usually soft-spoken and contemplative Jain bristled to me one day. "I'm so tired of hearing this complaint. Our shit works."

Special operations forces already relied on Palantir platforms to run tactical operational missions—Gaia for mapping, and Gotham for rooting out the meaning behind intelligence data. But they had never integrated computer vision detections into their platforms before, and, unlike their rivals Minotaur and Novetta, they had no video player at all.

The company's pilot with Maven was set to run for six months, meaning that's how long Robert Imig had to build a video player from scratch. From his prior Palantir experience he knew how to file group patents for complex tasks such as how to parse computer data to figure out an individual user's location. But this was new territory. "How the hell do people do this?" he asked himself.

With little idea how to proceed, Imig turned to one of the biggest video platforms in history: Netflix. Palantir set about creating war tools using the same sort of technology that underpinned sites like YouTube, Netflix, and Hulu. Imig researched streaming protocols used by all three. "We were like, hey, these big guys can do it; this is the way we should do it."

Several Mavenites strongly disagreed with the approach: streaming platforms might be good for playing video forward, but that had very little to do with the close frame-by-frame rewinding work, bookmarking, and annotating of individual frames required by intelligence analysts peering in for excruciating details on drone feeds and kills. They also had a mission-critical need to display feeds in real time rather than play movies on demand at leisure.

As Imig was building out an initial prototype, staring at a screen for endless hours, he got a feeling of an eyelash in his left eye. It didn't count even as a scratch. But he couldn't shake it.

The cancer diagnosis came quickly. He felt the acute irony of trying to make a video player with the help of only one good eye. He had surgery in September. Today he speaks with two perfectly light blue eyes; one of them a false replacement. He shipped his video player, and in October it faced off against its rivals.

A trial event took place in a room at Johns Hopkins University Applied Physics Laboratory (JHU APL) in what became known as the Maven Integration Lab. On one side of a workbench was Minotaur, the system JHU APL was using to show up AI detections. On the other was the video player from Imig's team at Palantir. The servers went head-to-head.

The room quickly soaked up more than fifty people, many of them "on blast" at the top of their voices. Cukor started barking at people. *Why can't the Microsoft algorithm detect this vehicle? Why is your feed a second late?* The pressure-cooker setup was "one of Cukor's secrets," Imig recalled. Bit by bit, things improved.

Palantir's approach focused on direct contact with users. That elicited sometimes brutal feedback, but also could buy trust. The company fielded engineers and field service representatives all over the world, who sat next to users and improved the software by the day. That marked a huge change for a military used to ten-year horizons for new tech. For Maven, Palantir's breakthrough came with help from a community it already knew well: special operations.

As the integration lead, Colin Carroll arranged much of Maven's spread around the world but he was also among those who were down on Palantir. He thought going with Palantir meant Maven would end up overpaying for a fancy duplicative platform the US didn't need and distracted from the whole point of the project: artificial intelligence. So rather than send Carroll to see the Navy SEALs, Cukor sent a Navy SEAL reservist who worked under Carroll to smooth the way instead.

The SEALs were a bit like a sorority, one SEAL told me with a laugh. Anyone who wore the Trident—the gold badge that signified membership—was part of the club. One Maven SEAL reservist had spent nine years on active duty taking him to eight deployments around the world. Although he was of two minds about whether Palantir was a good idea, he knew several SEALs in high command and helped get Maven Smart System hardware into various locations. It wasn't straightforward: sometimes the SEAL would show up with hardware headed for a unit's server room and the unit would try to throw him out just because it was Palantir. My boss outranks your boss, the SEAL would say, and get the equipment in.

The biggest breakthrough came when Maven got into SEAL Team 6. The team, officially known as Naval Special Warfare Development Group, or DEVGRU, took on some of the fiercest combat operations around the world that Maven hoped could serve as testing grounds for AI, and they came with their own command and IT infrastructure. The SEALs were also among the most experimental of America's fighting forces. Whatever tech they took on and developed often led to technology and techniques spreading through the wider force.

In October 2018, the Maven team installed the Palantir server for Maven on the server rack at Dam Neck, the location by Virginia Beach where Navy SEAL Team 6 and other military groups are based.

The team was wary, though. Maven had to get this right. If SEAL Team 6 tried Palantir and got fed up with the AI it could be game over. Palantir also worked to earn goodwill: they would stay up all night, pushing out updates and improving the software to convince DEVGRU's AI skeptics and Palantir haters.

That still wasn't enough to convince the SEALs. One Navy SEAL tried to eject Palantir and their hardware. Adding to the pressure, some users at Dam Neck used historic data to see if AI could perform well on something known as slant count. This signifies perhaps the most consequential and sensitive task that went into planning any kill or capture mission. The slant count logged how many men, women, and children were confirmed at a location. The count was central to US collateral damage assessments and whether to undertake a strike or raid against a high-priority target, based on an assessment of how many civilians might be harmed or killed. A single mistake from AI misidentifying who was on the ground could mean a dead civilian, the end of a SEAL career, and maybe a crime.

It would also imperil the future of Maven and AI. "Giving AI literally the most sensitive task was probably the world's worst idea," one person trying to field Maven told me. "If it goes wrong no one is ever going to trust AI. It was a terrible use case." They would never, they added, stake their career on an AI-generated slant count. "I don't want to go to jail."

The users switched to a lower-stakes scenario that made more sense: tracking mopeds. Counting mopeds could sometimes suck up a person's entire life. Analysts followed the mastermind of the USS *Cole* bombing for months on screens before the raid to capture him was launched. AI could save time by maintaining custody of a potential target, a former senior member of JSOC familiar with Maven told me. Maven's tracking AI could also help reacquire a target if it briefly went out of frame and then back in, the former member told me, adding that it could also potentially function as a check on human bias, the person added.

There were limits to what Maven could achieve on the Virginia coast. The team wanted to test out AI on real raids with real data in real time. I learned a senior SEAL commander took a risk: he gave permission to send Maven into live operations. The first, primitive version of Maven Smart System was now bound for Djibouti.

I'd visited Djibouti in 2015, when the US military was undertaking "the biggest active military construction project in the entire world," and I was trying to establish whether China was planning to build its first overseas military base there in pursuit of a new world order. A spy hub at a choke point for global maritime trade, Djibouti was nestled among Somaliland, Eritrea, and Yemen between the Red Sea and the Gulf of Aden. The US ambassador there told me Djibouti was "at the forefront" of US national security policy. And soon enough I found evidence that Beijing was indeed building its first overseas military base there. When I visited, the jihadi threat was also growing—a suicide bombing at a restaurant popular with western military personnel the year before shook the populace and the authoritarian government in charge. One member of the US military who served there called it "a hot hell box in the armpit of Africa." I'd be drawn to a more upbeat description from a friend: "Somalia meets Las Vegas."

Camp Lemonnier, the US military base in Djibouti, ran alongside the airport and was visible from the air. US forces ate at Combat Cafe, watched films at the Oasis Movie Theater, played ping-pong, poker, and Xbox 360 in the gaming room, went to the gym and joked to each other to "have a Djiboutiful day." But the base still managed to have "a secret side" on the other side of a fence that most of the five thousand US forces would never get near, never mind enter.

There, at Camp Titan, special operations task forces ran operations across the continent. They were targeting al-Shabaab and the local al-Qaeda affiliate along with several others including ISIS. As US troops withdrew from Afghanistan, Djibouti became the active center for what US soldiers at the camp refer to as "g-wot": the global war on terror. JSOC operators there started using the prototype Maven Smart System and AI on drone operations across the region, spanning Libya to Somalia. The feedback was blunt: they thought it sucked. That spring of 2019, Palantir made sixty changes to the platform. Even so, users complained the company was always behind on a deluge of requested upgrades and many turned it off and went back to the old intelligence tool suite provided by NRO. Maven wasn't mature enough at the time to use its detections as confirmation for a commander's decision to strike, the former senior member of JSOC told me. But what would become Maven Smart System soon started to improve. Before long analysts on live operations realized the system was useful for tracking target vehicles, especially mopeds, meaning analysts could look away. It also reduced the time to prepare a strike. By the middle of 2019, Maven Smart System was providing an operations picture in every war zone where the US was fighting.

"We had the entire JSOC community," said one person.

15

PALANTIR SPLITS THE TEAM

"The program as we built it will never recover once we hand the keys to Palantir."

WHEN BRIGADIER GENERAL Christopher Donahue got to Bagram in 2019, US forces likened his arrival to a hostile corporate takeover. The Army one-star was the incoming commander for US and NATO special operations missions in Afghanistan. He knew the US presence in Afghanistan was winding down and he saw it as a chance to develop hi-tech solutions for everything. He was prepared to give AI a shot.

Colonel Cukor visited Bagram in June 2019, and hit Donahue with his usual pitch: "You have a dumb system," Cukor would say, clicking through his deck of a hundred or more slides. "I want to give you a smart system."

His ambition for what would become Maven Smart System spoke to the twin issues that would soon tear Maven apart: Palantir and personalities.

General Donahue told me he was all-in on Maven Smart System from the start, though. His team on the ground was screaming for it, he said. But he would also give it a tough ride, arguing Maven needed to fail and frustrate people in order to get better. He knew it had taken the US military nearly twenty years just to get night vision right, and in his view it was now obsolete. He was the one pushing for augmented reality headsets.

Before Cukor returned to the US, he visited Bagram's Joint Operations Center. Three live-action screens were up on the wall.

"I want our AI on all those screens," Cukor announced.

Matt Barchick, Maven's newly arrived liaison officer in Afghanistan, gulped at the command. "Oh shoot," he thought to himself.

Getting Maven Smart System up on the screens would become the first major test for Project Maven, but going all-in on Palantir would also break the team. Detractors were dumbfounded by what seemed like duplicative cost: the new Palantir server for Maven sat beside the old Palantir servers for Gotham and Gaia that Palantir had installed two years earlier. They needed totally separate accounts, and, under the Palantir license, no live data was allowed to travel between the two. Cukor had told his team members he wanted total separation so there was no risk an experimental AI system could mess up live operations. Plus, back then delivering software updates to Maven required server downtime that live operators wouldn't accept, and they integrated ten years' of historic operations data. But special operators rolled their eyes at the Defense Department for shelling out twice on expensive Palantir boxes sitting right beside each other. A Palantir official argued to me that the Pentagon was simply buying services like a cake, in layers. (It occurred to me later that cakes are usually sold whole or in slices.)

Others figured Cukor was backing the Palantir option because he didn't want users complaining about how bad the AI was to the Hill, which would risk Maven's fast-growing pot of money. They saw the Palantir platform essentially as a fancy COP—a common operating platform—but not AI. Palantir's new platform overlaid video feeds with augmented reality, painting brightly colored mission planning lines over footage to mark out routes as safe or dangerous. Concentric circles over the map known as range rings would radiate out from a potential attack site to indicate the likely area where casualties might be affected by collateral damage. The overlay also numbered specific buildings, making communication between disparate teams much easier. Units all the way up to commanders loved it.

But the decision to back Palantir made life harder for founding members of the team like Joe Larson, Cukor's right-hand man. Larson had worked at Palantir for nearly five years prior to joining Project Maven, and had to be shielded from any defense work with Palantir to avoid conflicts of interest. Palantir was a subcontractor under the prime contractor ECS Federal. But as Palantir and ECS butted up against each other, the former would soon be made a prime contractor on Maven too, equal to ECS. And as Palantir's role in the project grew, it prevented Larson from doing more and more of the Maven work he loved. To little avail, some on the team jokingly started referring to Palantir by a coded Pig Latin variant, Tirpay, in an attempt to limit Larson's exposure to office talk about the company.

For Cukor, Larson's struggles couldn't be helped. He'd always argued the users needed an appealing way to access AI, but Colin Carroll thought Palantir's interface had little to do with AI and wasn't working particularly well. Another person confided that in 2019 it was "totally unusable."

The display could be particularly irritating. Every second of video was made up of at least fifteen individual still frames. The AI might behave differently on each frame, which could result in constantly flickering AI detections that made it too distracting for humans to use. And sometimes objects the AI didn't recognize all got lumped together as a new amorphous category: "Object Unknown." Users could offer a thumbs-up or thumbs-down on errant detections, but despite what some thought that feedback system wasn't live; someone would have to bundle up the results and send them back to the algorithm vendor to input the results into a frustratingly manual training process.

Carroll thought the answer to delivering effective AI was to sidestep the users altogether. He didn't want a smart system, he wanted a smart sensor. He wanted to get AI directly onto sensor platforms of the weapons themselves, right up on the MQ-9 Reaper drone in the sky. That way users wouldn't need to stare at the screen, and it would save huge amounts of processing bandwidth. Cukor had given him go-ahead to start experimenting with that, and Maven developed its own line of effort known as "Edge," which was intended to put AI directly onto flying surveillance platforms and weapons rather than into the systems that processed their footage back on the ground. But none of it was sufficient to sate Carroll.

On May 21, 2019, Carroll had landed a little after 2:30 a.m. and sent Cukor a private email. "I am still sitting here on the plane typing this waiting to deplane, and just exhausted," he wrote to his boss. He'd be moving off his orders as a reservist come July, but still expected to stay on at the Maven office as a subcontractor, assuming Cukor would still have him, or perhaps move to the Pentagon's new AI office, the Joint Artificial Intelligence Center (JAIC), he wrote. He did apologize for the email, but he was "struggling" with the question of whether to stay or go.

He'd really enjoyed his two years on Maven as a reservist, saw the potential for machine learning and computer vision for intelligence work, and thought the team had taken the Defense Department as a whole a few years into the future. But at some point in the past six months he thought the program had "lost its way." The vision had changed, and it's "not the vision I think DoD needs for AI." Maven had sold out to "propaganda," he emailed his boss, and started focusing too much "on the elixir of Palantir."

Carroll thought Palantir had ruined its Defense Department business around 2012 "with just terrible business practices," and he thought committing to Palantir now would come back to haunt them. He explained how he saw it: it just wasn't in the company's DNA to integrate systems, he said. Plus Palantir was at the time focused on data fusion software, not AI, and operated as a business development machine. "The program as we built it will never recover once we hand the keys to Palantir," he warned.

Cukor would make no concession to Carroll's view and, at least on the surface, seemed unfazed. He encouraged freewheeling feedback and nothing much bothered him. There were purists, he'd tell me, but Maven wasn't about AI for the sake of AI. And the Defense Department should bring on commercial platforms instead of building things in-house. Acquisition rules said as much. He wanted to have a material impact on daily workflows, and he wanted practice to make perfect.

Cukor always tended to respond to any question of mine, however uncomfortable, in the same way. It was as if he were a medium-altitude, long-endurance drone himself. Neither too close nor too far. Context first, slowly routing back to answer the question, the heat taken out of it. He loved Colin, he'd tell me sometimes. He could live in a world where people have different opinions. But every so often he'd step toward a view of his own. "Colin gets in his own way sometimes," he'd start to say. "He's so aggressive he doesn't have the world view—sometimes you have to slow down for other people."

Eventually, as AI improved, Cukor argued a single platform would help fuse together multiple different layers of AI, not only computer vision, but "reasoning models" too, large language models that can break down complex problems into smaller steps, to deliver a single picture. Not just "there's a tank," but AI might help detect what it's doing there, what it's going to do next and how much fuel it has. Cukor believed so much in this future that, long before the arrival of ChatGPT-3, Maven also put IBM, Northrop Grumman, and others on pilot contracts to develop reasoning models to analyze the output of dots, images, and other data.

But that was tomorrow. This was today. And it became quite clear to Matt Barchick, the Maven rep scrambling to get Palantir's Maven Smart System up on those screens at Bagram JOC, that users on the ground thought Maven and AI were cloud-cuckoo land.

"They hated it initially," Barchick despaired.

Like most Mavenites, Barchick was a Marine Corps reservist. But unlike the rest he'd never been deployed before and didn't like traveling. He simply did his dues: training with his Marine Corps unit for a weekend a month and two weeks a year. He knew nothing about AI. He knew nothing about special operations. He knew nothing about Afghanistan.

"I was just so green," he recalls. "Like, me and AI doesn't make sense."

Barchick had joined the Maven team after a perfunctory fifteen-minute interview with Carroll. On his second day in the Pentagon in mid-April 2019, Carroll informed him he would be going to Afghanistan. Barchick thought Carroll might just be messing with him because he was the new guy. But within a few weeks, there he was. "I was like, what?"

AI detections initially seemed even more cumbersome with the new

Palantir system. Detections showed up with dots so big they covered objects, making it impossible to distinguish between adults and children. In a crowded market scene, it would be impossible to tell from the obscured video whether a person of interest was buying food or stashing weapons.

"This is crap, this does nothing for me," one analyst at a smaller site complained to Barchick.

Barchick would try coaxing users to stick with it. He knew it wasn't great yet. "But like it isn't about you," he would cajole. "This is about great power competition." If users didn't start telling the team how to make it suck less now, it might be too hard to catch up later. The US could be on the back foot in any conflict with China: "We don't have the manpower."

That argument seemed less compelling than a specific inducement Maven relied on: twenty $6,000 brand-new gaming-style laptops they handed out at Bagram JOC to run Palantir feeds.

Palantir was also feuding with its project rival, Novetta. Both companies had reps at Bagram working on Maven tech, but gave exasperated users contradictory instructions not to use each other's video players. "What the hell am I supposed to be using dude?" users would ask, according to another Maven rep. "Whatever makes your job easier," came the response, aimed at conciliation.

Palantir's platform was good but people complained the video player was still crummy, plagued with delays that might be fine if you were watching a sports game but not if you were planning a real-time strike. One time it froze up so badly the video player was half a day behind, relaying drone footage of a bright sunny day as operators crowded inside the JOC at 2 a.m. in the pitch black. Cukor would send down the message to Palantir: you need to fix this now.

Imig would regularly get calls from overseas reps relaying the feedback at 2 or 3 a.m. "I got beat up for probably six months. Just yelled at, constantly."

Cukor preferred to make his point in person. Every Friday evening, he'd leave the Pentagon and arrive at the Palantir offices in Georgetown. It would be 8 p.m. when the meeting would convene and last till past 10 p.m. It was the only time that worked for him. "Cukor would beat me up about like two seconds of the live feed," Imig recalled.

On August 12, the nascent Maven Smart System reached a new milestone. Joe Kernan, the undersecretary of defense for intelligence who was ultimately responsible for Maven, was on a visit to General Donahue and the team. Barchick had finally delivered: Maven Smart System was up on screen in the JOC, and it was being used for real operations. The day Kernan visited, AI detections popped up over footage displaying a live mission. The resulting strike against the target was judged a success. It was the first time Pentagon leadership could see what really might be possible.

Maven's reputational success was fleeting. Toward the end of September 2019, special operations forces were preparing to launch a mission to kill or capture a high-value al-Qaeda target so sensitive that multiple people from the US were remoting into the system to follow the mission. The network crashed. No internet connection, the radio chat went down, video too. "Everything just freezes," one person recalled with horror. It was a Sunday night, and other US personnel on base were just getting settled in to watch the football game playing back in the States. An army captain responsible for command and control stormed into the TV room. Everybody turn off Maven, came the command. It was imperiling the high-priority mission, he claimed.

It was never quite clear what was responsible for the network outage, and accusations against Maven, AI, Palantir, and other culprits flew about. "We kind of crashed it," said one observer. Barchick insisted the crash had nothing to do with AI or Maven itself, that it had been the number of people who remoted in. But after the incident, many still believed Maven had caused the outage.

"Over the twenty-year global war on terror we've been blamed many times for many things," Aki Jain told me when I asked him about this episode. "Each time it's almost always watching Monday night or Sunday night football or there's a spike in usage."

He did concede that georectification of full-motion video—the effort to orientate the video to the map—also uses a lot more bandwidth. He said the company "did a lot of work" to optimize that in different scenarios, but argued the problem essentially showed how US forces on dangerous missions were reliant on old, decrepit network tech.

Another time, Maven Smart System helped out in a mission that went wrong. It was the start of a night raid to kill or capture targets inside a compound in Helmand. The courtyard was empty: the raid was on. More than half a dozen Marine special operators had filed out and were bunched up against the outside wall of the compound, ready to breach it. But then came an explosion from behind the wall. Booby trap. Ambush. The wall collapsed, dust kicked up. And suddenly the Marines were taking immediate fire from a bunker too. Under a hail of bullets, emerging from the cloud of dust, no one could pick out where the attack was coming from. Then came the radioed distress call: troops in contract, requesting close air support. Maven Smart System wasn't an approved method at the time, but intelligence analysts looking at the screens of drone feeds above them could already see AI detections marking out US forces. Boundary boxes picked up Marines running out from under the smoke. Counting out each one with the help of the boxes, the analysts confirmed none was trapped under the collapsed wall. Then they used the augmented reality layer to take the grids off the video within seconds to make sure the incoming strike hit the bunker and not their own men, a process that would usually take minutes to do manually. The A-10 attack aircraft came in. It "eliminated" the bunkers. After that extremely close shave, the Marines wouldn't stop using Maven. Soon an Air Force colonel who ran MQ-9 drone missions was convinced too. She wanted Maven mandatory on all feeds.

But whatever success might be coming around the corner for Maven Smart System, it was too late for Colin Carroll to see or believe. It wasn't just his objections to Palantir and the future of AI at war. He was still working nonstop for Maven but yearned for the early scrappy days of the project.

"The big family started to fight," said Brian Ward.

"Egos were bumping into each other," said another witness.

Major Carroll seemed increasingly irked by another Marine on the team, Sy Poggemeyer, one of the younger members and still only a captain, who Cukor considered "just amazing." He was hardworking and impassioned about trying to solve the project's systemic data challenges, but he could also flare up. Some team members felt insulted by him, as did external contractors at ECS, according to multiple people I spoke with. Tensions built.

"I have had it with Sy and how he treats people on our own team," Carroll emailed Cukor, copying in two others, on May 24. It wasn't just him and Maven colleague Gene Whipps, Carroll emailed, but the ECS team the week before, and the Johns Hopkins University team (that Carroll was about to join) the week before that. Carroll refused to work with him. "I will not deal with him any more as the data lead," he wrote. "It's come down to this—its [sic] me, or its [sic] him."

Carroll fumed that rather than fix Maven's data woes, Poggemeyer—who was now running the data team—seemed to think he knew more than a PhD and four vendors "who literally do this for a living."

Whipps was the engineer with the PhD. For his part, the US Army Research Laboratory researcher told me he was struck by how "fast-paced" Maven was, as it moved to streamline computer vision models that, at the time, he said were state of the art. By email, he told me through a representative that he found Maven "overall" a positive and rewarding experience and enjoyed it so much that he extended for a second, final year. Carroll by then had also started to date Katya Volkovska, who worked at ECS Federal, Maven's prime contractor, so, in addition, he wanted to move to a job outside government to avoid any possible conflict of interest.

One night a little after midnight on July 4, 2019, as Carroll prepared to move off the team, his fury boiled over. In an email with the subject line "Some Program Thoughts," Carroll wrote to Cukor and one other person accusing Maven—Cukor's baby—of being a "failed startup."

He'd worked his butt off for Maven and now the team was "rapidly falling apart," he wrote, according to the documents I reviewed. Carroll seemed to be half-pleading that Maven was his baby too. They were still on the same team really. The legacy of the Maven program "will be my legacy as well," he wrote.

But Maven was wayward. It had expanded into too many different things and erroneously brought on Palantir "on a false promise" that its platforms were ready for licensing and would be operational quickly, Carroll continued in his email. This decision handcuffed the potential success of the program to sole-source delivery from a company he wrote was "motivated by money alone." Palantir didn't view itself as Maven, didn't sell its capability as Maven, didn't even mention Maven at a Smart System pitch to an Air Force intelligence unit on behalf of Maven, he wrote. They weren't even doing any of the hard work associated with actually building AI, they were just the user interface.

He missed the old days. He wasn't the only one. Maven's engineering lead Krishnan Aiyer had just that week told Carroll it had become "a lot less 'Maven-y' around here."

It may be that he was wrong, Carroll quivered toward the end of his long, six-point-bulleted email—that these issues were in the past and his points would be irrelevant.

But a clutch of people was already leaving and he expected still more to leave soon too. He reached his finale: "I am sure there are others waiting for the dominoes to start falling before they make their moves, as no one wants to feel like they are on a sinking ship."

He pressed send.

16

A STRIKING OPERATION

"Someone just made their company a lot of money."

DONALD TRUMP WOULD SAY the next morning that the mission to kill or capture Abu Bakr al-Baghdadi, "the world's No. 1 terrorist leader," had been the top national security priority of his administration. The "impeccable" two-hour nighttime raid on his hideout in northwest Syria in October 2019 involved a clutch of Special Operations Forces on the ground, eyes in the sky, massive airpower, ships in support, a shootout, and the target at the end of a concealed tunnel detonating his own explosive vest. "He died like a dog," said Trump.

The president, talking to reporters for nearly an hour, himself seemed staggered. The mission was "an unbelievable success," he said, praising the US personnel as "incredible." They accomplished their mission "in grand style," he said, praising a large crew of brilliant fighters.

Before the end of the operation, Baghdadi would flee into a dead-end tunnel, along with two children, and explode himself. The task force dug him out of the collapsed tunnel. A dog was injured. A DNA test on the spot confirmed it was Baghdadi. "Jackpot."

Trump didn't want to say how he was able to watch the operation live, but he said it was like watching a movie: "The technology there alone is really great."

At least a fraction of that technology, I later learned, meant Project Maven.

Colonel Cukor wouldn't address my questions about the Baghdadi or any other operation, but he had previously told me Special Operations Forces, especially the more sensitive units, "used this thing extensively, all day long, working aggressively," he said. "They really were our biggest champion."

The operation was the culmination of three years of US efforts. Baghdadi was the founder and leader of ISIS, the vicious and violent group that announced its arrival in June 2014. The so-called Islamic State went on to rule over millions of people across Syria and Iraq. Baghdadi had declared the existence of a new caliphate just over ten years after he'd been held by US forces at detention centers Abu Ghraib and Camp Bucca in Iraq. He killed his victims in public beheadings, live on camera. His jihadis killed prisoners of war. Militants in Burkina Faso, Mali, Mozambique, Somalia, Afghanistan, the Philippines, and elsewhere pledged their allegiance to ISIS. His group had massacred Yazidis, a Kurdish religious minority, in Iraq in 2014, and kept up the killing and kidnapping for the next three years.

After the 2017 fall of ISIS strongholds in Mosul (Iraq's second-largest city) and the group's headquarters in Syria's Raqqa, the US and others set about destroying what remained of the caliphate. By the spring of 2019 there was barely anything left, but Baghdadi had eluded Obama as well as Trump, who said he regularly asked after him.

"We would kill terrorist leaders, but they were names I never heard of," Trump said. "They were names that weren't recognizable and they weren't the big names. Some good ones, some important ones, but they weren't the big names. I kept saying, 'Where's al-Baghdadi?' And a couple of weeks ago, they were able to scope him out."

Earlier in 2019, the US had tracked Baghdadi to Idlib province and six months later nailed his exact whereabouts. In Erbil, at a headquarters unit for a Special Operations Task Force in Iraq, a planning team set up maps and readied video footage in advance. The US grew "completely confident" there were no other ISIS forces nearby. But they expected other armed groups.

On the eve of the operation, the weather looked good. The moon shone only the slightest of slivers. President Trump gave them the green light. What weighed on the president the most?

"Well, just death."

Baghdadi was holed up four miles from the Turkish border. The compound on the western outskirts of a village named Barisha was enclosed by a wall, and underneath were a series of mostly dead-end tunnels. The team knew there would be children at the compound.

"This was a very, very dangerous mission," Trump said.

After dark that Saturday, eight armed helicopters flew low and fast for over an hour. Most people had already turned in for the night. To make sure, helicopter gunships let out two volleys to frighten civilians into staying indoors.

The eyes in the sky were mostly trained on the compound. But drones, satellites, and manned aircraft were also on the lookout for anyone who might try to come to Baghdadi's rescue. "Basically any bad guys that were gonna advance on our guys," recalls one person who watched the raid unfold live.

The cameras mostly relied on infrared and kept scanning the area: video—and AI—can only identify threats if the camera is already looking directly at it.

Two of nearly 35,000 air strikes against ISIS were delivered during the raid, when US airships rained down strikes against local gunfire coming from a multistory building near a T-junction a little way east of the compound. Grainy camera footage shows the sudden obliteration of a group of people on foot.

"That gunfire was immediately terminated. These people are amazing," Trump said.

Friendly forces set up a roadblock a few hundred yards to the east of the compound on the way into the heart of Barisha village. After the helicopters landed, a string of at least ten Delta Force commandos approached Baghdadi's compound from the southwest on two sides by foot, snaking through a sparsely dotted olive grove.

The special operators blasted through the wall, avoiding a booby-trapped door. "By the time those things went off, they had a beautiful, big hole, and they ran in and they got everybody by surprise," the president recalled the next day. "Then all hell broke loose."

An explosion of noise and confusion. They called out in Arabic for those inside the compound to surrender. Out came eleven children and two men, who were detained. But not everybody came. Four women and a man failed to surrender. The US forces killed them. Baghdadi was still in there.

The screens at the operations center for the task force headquarters in Erbil were a mass of information. A request to get Maven Smart System at Erbil had already gone in toward the end of May, and now, five months later, it was put to work. The team received a report that a vehicle was on the move and sent a drone to search it out.

The drone camera slew to the location. As soon as it did, Maven's AI suddenly pinged. Up popped a telltale detection on the video.

One of the guys in the command row at the screen clocked it at once. *Oh look at that dot*, he said.

It was pinpointing a vehicle on the video, and the vehicle was on the move. Not even the screeners, whose job it is to stay glued to the multiple feeds, had spotted it yet. AI had just beaten a human on a high-priority mission.

The AI detection showed up on the map too. The dot was moving south through Barisha, the spread-out village to the east of Baghdadi's compound. At a T-junction, the vehicle turned to the west, heading down the road toward the site of the raid a thousand or so feet away. The task force commander put the information together fast: Those guys are driving toward the task force, toward our guys at the building.

The US didn't think they would be ISIS, but they knew there were other militant groups in the area and the vehicle displayed what General McKenzie would later describe as "hostile intent."

The task force commander got on the radio and told the ground force commander in charge of the raid about the vehicle. The ground force commander called in a hit: *Take it out*.

Thirty seconds later an attack helicopter swooped in. Gunfire rained down.

A man ran out of the van to the north side of a roofless building on the edge of the road. The helicopter fired a strafing run, blowing out the north and west walls of the building.

Five minutes later, the airships launched a rocket too.

"And the van is gone," the person familiar with the raid recalled.

The shot-up carcass of the white-panel van, blackened with smoke and pockmarks, remained on the roadside beside a neat line of olive trees.

It was the first time that the commander had ever used AI to help find a target. It couldn't have happened on a more consequential raid characterized by many fast-moving, dangerous parts.

The speed at which AI highlighted the vehicle left an impression with JSOC.

"Someone just made their company a lot of money," the commander remarked.

AI appeared to be able to cut through the fog of war. The task force commander and others started leaning wholesale into Project Maven's technology.

What, I asked the person familiar with the feed, if there'd been civilians inside the vehicle spotted by the AI? Or near it?

"Those were definitely bad guys that the AI ended up picking up," the person told me.

Barakat Ahmad Barakat would go on to tell a different story. He told NPR he'd been driving home to Hatan, a village several miles to the west of Baghdadi's compound, that night after finishing his day's work at an olive press in the north of Barisha with two friends.

"Suddenly I felt something hit us," the thirty-year-old told NPR.

The van came to a sudden halt and the men fled. His friends—two cousins, Khaled Mustafa Qurmo and Khaled Abdel Majid Qurmo—died in the aerial attack that followed. Barakat cradled one of them in his lap on the roadside, just as an airstrike arrived. It blew off his right hand. He later lost the whole of his arm. His left hand doesn't work well.

"Am I Baghdadi? How is this my fault? I'm just a civilian. I didn't have any weapons," Barakat told NPR in another interview. "We're farmers.

I make less than a dollar a day. Now I'm handicapped, and my two friends are in their graves."

The US has acknowledged 1,437 civilians were unintentionally killed in five years of operations against ISIS in Iraq and Syria to the end of October 2019. According to Barakat and NPR, that number should be at least two higher. (The monitoring group Airwars has logged the total number of civilians killed to be higher possibly than 8,000.)

The US military has investigated the claims twice: the first report did not uphold the claims. But after NPR sued to see a redacted version and questioned a number of its conclusions, the US did a second investigation. The results of the second report have not been made public.

THE DEPARTMENT OF DEFENSE says it protects civilians not only because the Law of Armed Conflict requires it, but also "because it is the moral and ethical thing to do." What role AI will play in this remains an open question. The US faces accusations going back years that it was taking insufficient steps to prevent, acknowledge, or investigate civilian harm. In 2018, Congress required the Defense Department to produce an annual report on civilian casualties resulting from US military action. But even after that, several in Congress would complain the department continued to "significantly undercount" civilian deaths, in part because the department reportedly only counted a civilian casualty if the military itself determined it were "more likely than not" that civilian casualties occurred. The US military wouldn't accept outside reports, but it also often didn't check what happened on the ground. A 2024 Government Accountability Office report cited media reporting that the DoD was mistaking civilians for combatants.

When I asked the person who watched the feed about the investigation, they insisted the commander's decision to strike was the right one, even if later on it might come out that the van was carrying civilians: "In the heat of a mission, all the task force commander and ground force commander sees is anything moving is a potential threat," the person said. "In the moment, that was a good strike."

The AI, the person went on, wasn't the only source on the existence or whereabouts of the van, and had no role in the strike: "That's a human decision; not an AI decision."

AI had functioned as intended: it correctly and quickly identified the van. That had created more time for reaction, including any assessment and warning, than if the US operators had noticed and then pinpointed it later on.

Even so, the fear of AI getting it wrong inhibited usage. "A lot of folks in the Pentagon were still very reticent to use AI because they were worried about indemnity, especially when it comes to strikes," another Mavenite told me.

General Jim McConville, a four-star Army general who retired as Army chief of staff in 2023, knew how important taking the time to pause and check could be. He remembered being dispatched in his helicopter as a much younger man to shoot out a white van that had been accused of firing at a base for US forces in Iraq, he told reporter Jen Judson in 2023. His weapons were aimed at the van but he hesitated, and instead brought his helicopter around directly in front of the vehicle, which promptly came to an abrupt halt on the road. The van and hovering helicopter eyed each other warily. And then a family got out. Had he fired, he would have killed civilians. Years later he worried AI had no morals and wouldn't manage decisions like that. "It doesn't really think, it doesn't feel," he told Judson. "My experience has been sometimes AI can go horribly wrong because you start getting into targeting and depending how things can happen, who is responsible when AI does the targeting, they do the shooting, and you put a missile into a van with a family? Who owns that?"

Successful identification was part of a broader problem with AI. "AI only sees what it sees; it doesn't see good guy, bad guy," the person who watched the raid went on. "Not even man, woman, child."

AI researchers often warn of machine bias. A ping in the fray of a high-adrenaline raid could incentivize operators to action. AI announcing "I see a van" might make it more likely for someone to react, triggering responses that might end up as a death warrant.

The person argued that the US always confirmed a target through a separate source and that AI added to awareness rather than taking away from it. "Most of the strikes that we saw where mistakes happened, AI was not involved at all. But there's a lot of people who were trying to throw AI under the bus at the time," said one Maven representative. The Maven team would react quickly, asking AI's in-house accusers: Was our system involved? Was it turned on? What did it say? What were the outputs? "Most of the time when it was turned on, we got more data than they would've gotten otherwise."

Lieutenant General Jack Shanahan, speaking several weeks before the Baghdadi strike, said Special Operations Command and JSOC were almost "pounding the table" asking for AI capabilities. He argued AI would contribute "a massive increase in situational awareness." "It allows things to go faster, it helps mitigate the chances of human mistakes," he said.

That would come with limits though: "I'm never going to say that AI is going to eliminate friction, chaos, the fog of war. Never. I'd be very careful about that," he said. "We know it's not perfect yet, we're here to wring it out and tell you how to make it better."

Identifying the physical presence of a van is a narrow act. Information that might make a van a true target—besides its existence, speed, and direction of travel—lies in all sorts of other places: who is in it, why are they there, what are their intentions, are they armed, where are they going, where have they just been. Seeing, and more to the point, understanding the big picture rests on far more than AI detecting objects alone.

Wes Bryant, a former special warfare operator who led strike cells and was steeped in the practice of targeting, told me he thought AI could help address this problem too. Until spring 2025, he spent nine months at the Defense Department's new Civilian Protection Center of Excellence. As branch chief of civilian harm assessments, he spent hours visiting US Central Command and Special Operations Command.

In his research, Bryant found it was rare for the US to decide to drop munitions on a target where the anticipated collateral damage—meaning harm to civilians—was high. Deliberate misdemeanors were also rare, he said. More often, mistakes came down to a faulty assessment about quite how many civilians were at a targeted location, or the target itself was misidentified. Bryant said the US was also "horrible at response" and failed to sufficiently investigate and acknowledge errors. In December 2021, the *New York Times* would detail far greater numbers of civilian casualties throughout America's wars in Syria and Afghanistan than the US military had admitted, building on years of prior reports.

Precision drops were not so much the problem. Since Desert Storm in 1991, when nearly 90 percent of all the mostly unguided bombs dropped by the US Air Force missed their mark, according to post-action review, US guided weapons now came with pinpoint accuracy.

Modern errors were instead more likely to occur as a result of faulty context, ignorance, and cognitive bias that could underpin a mistaken decision to strike.

Some officially acknowledged blunders stood out in America's targeting history, highlighting a mismatch between tremendous US firepower on one side, and inadequate insight, out-of-date databases, and miscommunication on the other. Digital platforms such as Maven Smart System might help address such deficits. In February 1991, a US bombing raid killed more than four hundred civilians sheltering in a Baghdad bunker: the Iraqi military command post had long been converted to an air raid shelter, but the US said it hadn't known. In May 1999, the US would blame "a series of errors and omissions" after a B-2 stealth bomber released five two-ton GPS-guided precision bombs, each individually programmed with precise coordinates for their targets, on the Chinese embassy in Belgrade, rather than the Serb weapons supply office two hundred yards down the road that the US insisted was the intended target. The CIA mission had relied on maps and processes that failed to update US military and intelligence databases the embassy had moved three years prior, and "missed phone calls" meant those who tried to raise the alarm went unheard. In 2015, the US conducted a bombing raid against a hospital in Afghanistan as workers there frantically tried to raise US military officials and diplomats in the middle of the night. Médecins Sans Frontières said forty-two were killed. The US had not directly targeted the hospital, as many assumed, but relied on faulty coordinates that put the target in "the middle of this field," and then guesswork to find a likely target building nearby. The US would later say the faulty strike was the result of "a combination of human errors, compounded by process and equipment failures," and took action against twelve US military personnel. The US would also acknowledge on separate occasions that it bombed civilian wedding parties and harvest farmers in error.

Bryant saw room for AI to help correct and fill in for human flaws. He argued AI might be able to help understand and interpret patterns of life. After the August 2021 US withdrawal from Afghanistan, the US would admit "a tragic mistake" when it launched a strike against a white car it suspected of harboring ISIS jihadis moving explosives after tracking it for eight hours. The US strike instead killed a humanitarian aid worker moving water canisters in a compound along with seven children and two others. None had any link to ISIS.

Maven was not involved in this strike, but Bryant argued AI might give vital context: not just the presence of a car but an alert flashing up with a potentially useful fact. Reminding a commander that the overwhelming percentage of cars in Afghanistan are white Toyota Corollas might help trigger a life-saving thought: "This one might not be the right one."

The best possible function of AI in warfare, Bryant argued, could be not to introduce certainty—but doubt.

17

DATA HELL

"He poisoned the entire pipeline."

THERE WAS ALSO such a thing as too much doubt, which often arrived in new and unexpected ways.

One day when Brian Ward was in the Maven dungeon, people were already chortling. It was pretty easy to make out the word.

One of the vendors had just called the team to complain about a set of images they'd been sent. These images were the vital feedstock used to build the war algorithms. When the project launched, Cukor said publicly Maven's "immediate focus" was developing algorithms for thirty-eight classes of object. The Maven team would commission labeled images for each of those categories—taken from a variety of angles, distances, weather conditions, and drone sensors—and then send them to AI vendors.

Brian Ward looked at the screen. Instead of tidy boundary boxes, he saw a freehand scrawl over the entire picture.

"F U C K."

And over another frame plucked from drone video footage there was no mistaking the crude hand-drawn doodle.

"It was a dick pic," Ward told me.

Once the team stopped laughing, it dawned on them how unfunny it was.

"He poisoned the entire pipeline," said Ward.

The American military's cutting-edge AI brain was being trained to recognize errant appendages. The aggrieved joker who had drawn the graffiti was feeding bad data to the algorithm. Labeling errors required more processing and could muddle painstaking and expensive efforts to teach an algorithm to recognize the objects, meaning it would be less reliable and less fast at identifying objects and underperform or even mislead in war. Learning incorrect patterns could make the AI more prone to hallucinations, faulty outputs, bias, and drift—leading to decreased accuracy and algorithmic failures over time. Humans were getting in the way of Project Maven's artificial intelligence.

"He poisoned the entire pipeline," said Ward.

Rather than pay for an expensive data-labeling company, Cukor initially decided to keep costs down and use military labor instead. Members of the armed services were encouraged to open a software labeling tool from Figure Eight (formerly known as CrowdFlower), and click out boundary boxes around objects on drone screen grabs.

"I was convinced we could have millions do this," Cukor said of the Defense Department's two-million-strong workforce. "Man, I tried that so hard."

Rear Admiral Tim Szymanski—the commander of Naval Special Warfare who first opened up the Navy SEAL data pipes to Project Maven—took pride in supporting the labeling work. Szymanski's intelligence chief undertook a "monumental effort" to mobilize Naval Special Warfare reserve intelligence folks to help categorize and label, he told me. Here's a vehicle. That's a building. There's a person. Under each class, such as vehicle, there might also be a subclass, such as car, and sometimes something more specific than that, such as a Toyota Corolla.

Maven gamified the task, and invented a scoring system to dole out points for top labelers. Sometimes Cukor would hand out a letter of thanks to a particularly productive service member along with a "Maven coin"—a round military token with cheery robot cartoon on one face designed by one of his daughters. He tasked the whole Maven team to label images at some point or another so they would know how much it sucked. To Ward, the work was monotony itself. "It's like your brain melting through your eyes and every orifice that brain matter could come out," Ward recalled. Cukor would encourage them, preaching that the US distinguished itself from China's centralized data labeling, which might deliver volumes but, he'd say, relied on forced labor.

But to some, labeling duty clearly felt like punishment, or after-school detention. That brought a greater likelihood of mistakes, and carried risks: training a single algorithm could require 10,000 images, and every single wrong label would make it perform worse.

America's recalcitrant volunteer labeling force was suddenly not looking quite so voluntary, or so competent.

"We probably underestimated in early Maven days just how bad data quality could get," Jane Pinelis, who was director for test and evaluation on Project Maven during 2019 and into 2020, told me in a 2023 interview. She said they had trouble getting great performance out of the algorithms, and that they were hampered by "dirty data." At the time, I hadn't understood quite how literally that term might apply. "Let's just say we learned the hard way the impact of data quality on the quality of your models."

Project Maven "coins" created to acknowledge and encourage the team and supporters.

Another person on the team put it more pointedly: "We fucked up data so bad."

Google complained from the outset about "poor" quality data labeling, according to customer engagement updates I reviewed from October and November 2018. Smaller batches of high-quality labeled data would be better than low-quality high-volume labeled data, the Google Cloud AI team advised. CrowdFlower started delivering better quality labeled data to Google by the middle of November, but there were still problems with the department's "ground-truth data."

Efforts to contract out the toil to gig workers were among the first steps to professionalize the task, but they didn't go much better, several Mavenites recounted to me. In public, the Defense Department was meanwhile attempting to assure an alarmed public and dubious tech employees that military AI could be tamed. After Lieutenant General Jack Shanahan's year-plus listening tour, the DoD announced in February 2020 that combat AI technologies would be "responsible, equitable, traceable, reliable and governable." It was unclear how the department would extract such qualities out of dirty data combined with the black box of AI, and how it could achieve such standards with the help of ethical principles that amounted to little more than aspirations.

Shanahan also conceded that while he was intent on hiring an ethicist at the JAIC, ethical questions at Maven "really did not rise to the surface" every day because, he said, no weapons were involved. Carroll scorned the effort to push AI ethics: *Those who can, do AI. Those who can't, do AI safety,* I was told he'd intimate.

Matt Zeiler from Clarifai most definitely could do AI. But he was so unimpressed with the data quality that he came up with his own solution: he automated a way to clean up the labels provided to them, to better train his algorithms. The process was still messy. Sometimes they'd be twiddling their thumbs for weeks while another vendor changed the data format or an interface. Or they would discover that someone had taken shortcuts.

Each second of full-motion video from a drone could consist of thirty static frames. Professional data annotators might label, say, every tenth frame showing a frozen moment of a moving object and submit that for training. But some at Maven were furious to learn the annotators were also labeling all nine frames in between, not by drawing around the object each time, but by taking the average and interpolating a straight line of travel between the two points. If a person or vehicle swerved out of the interpolated boundary box, then Maven was paying millions of dollars to train algorithms to recognize the ground.

A frustrated Zeiler wanted to use Clarifai's own data-labeling platform. But the government wouldn't let him: Cukor wanted a single data-labeling solution that would supply labeled data to all the algorithm vendors, not just to one vendor. He was worried about relying on one vendor, or worse, losing control of the data trove.

Clarifai took on advisers, including former congressional staffers, defense acquisition officials, and a three-star Army general, but nothing helped. They would all counsel him to avoid using the world "platform." The Defense Department was clearly allergic to the idea of putting all its eggs into one basket, and "platform" sounded like an all-consuming total solution. One technical expert who worked on the project told me that a platform could also narrow the scope for cutting-edge research: it might be easier to manage and retrain an algorithm on new data inside a platform, but much harder to try out an entirely new way to create the model in the first place.

In the early days, Cukor tried to generate synthetic data to help, particularly in cases where only a handful of precious images might exist of a US adversary's sensitive military equipment. The idea was that generating multiple artificial images of the same thing would help fill in sufficient gaps to create a full data training pipeline, and one much easier, quicker, and cheaper to make too. He had flown to meet Microsoft executives in 2017, appealing for exactly this help, and been paired with a crack team led by Dimitrios Lymberopoulos. Try as he might, Lymberopoulos couldn't generate synthetic data to work well enough for such specific objects (the AI would just "see through it," Cukor told me), but the team's failure came with an unlikely success: the models they made revealing the weaknesses of synthetic data were so impressive that Cukor asked

Microsoft to join the project as algorithm vendors. For Microsoft this counted as a huge business opportunity: the contract might not be much, but Maven was the accrediting sponsor for the Azure cloud infrastructure: AI warfare was Microsoft's way into the DoD's big cloud contracts.

Several Mavenites told me that Maven's best algorithm producer was Xnor, which took over Google's efforts on drones that produce wide-area motion imagery (WAMI), as well as keeping up its own exemplary research on full-motion video (FMV). Xnor's models managed a good balance of accuracy and speed, they said. So when Apple bought them in early 2020, it was another huge blow for Maven. Once again, the US military hit a wall in Silicon Valley: the best simply refused to work with the Department of Defense. Apple refused the Maven work, and the Xnor algorithms wasted away.

Microsoft soon emerged as one of Maven's best remaining model producers, pumping out models for WAMI and FMV and more than doubling its team to fifteen people. But, by the end of 2019, even their performance reached a plateau. "Garbage in, garbage out," Cukor would say. And it was worse in real life: algorithms that might test okay in the lab—a sanitized environment in which models were examined on data clips that were very similar to the data clips they were trained on—would fare far worse out in the field. Cukor called this stage "data hell."

IN EARLY 2020, a Maven team was visiting the Aberdeen Proving Ground, a strange place of World War II bunkers and bizarre experimental military tests on the eastern coast in Maryland. I'd visited Aberdeen in 2018, where I watched a robot attempt to walk through a vat of couscous to approximate the drag of water without electrocuting itself in the process. Elsewhere on the grounds, I stepped into an exoskeleton that could hold a weapon and assess fatigue levels. I also experienced the eerie sensation of entering their bomb vault. When the door closed it turned the room into a dark prison, its thick studded walls pockmarked by years of invented substances exploding inside.

The Maven team was at Aberdeen to see if they could figure out how to connect algorithmic systems for the big blimps that US forces tether over bases to keep watch. It was a rare offsite for the team. They had been going since before 6 a.m., but were nursing a certain sense of freedom without the boss about: Cukor was long gone on a plane to the other side of the world. Or so they thought.

They were just leaving Aberdeen when Cukor called them, impatient.

"Where are you?" he demanded, according to one person's recollection that I corroborated.

"Hey boss, I thought you were in Korea."

"Get back to the Pentagon. We gotta redo things." Cukor was fed up that results showed the algorithms were no longer getting better.

The effort would come to be known as "The Reboot." Within days, more than fifty people were convened for nonstop brainstorming sessions at the ECS Maven Integration Lab in Fairfax, Virginia. Algorithm vendors who flew in came with more than a hundred suggestions to improve performance. Cukor was always constructing high-stakes bake-offs for the best AI brains he could find, but they begged him to ease up.

Their main request was for the thing Cukor never liked to dispense: more time. He had to slow down, the vendors pleaded. "American taxpayers don't pay you to rest," he'd tell them.

But they couldn't get a new model out the door and fielded onto hardware in combat zones all over the world as fast as he wanted. They were spending brain power on logistics instead of research; the teams were all exhausted and increasingly demoralized. Slowing down was not in Cukor's nature. He yielded just a little: they'd get more time. The second major request was for better data. One analysis demonstrated how each labeling error pulled down model performance. It was time for Cukor to acknowledge that Maven couldn't rely on cheap labeling by grunts. But fixing it would cost time and money. The vendors wanted Maven to relabel the entire military-created training set of more than a million images. "We ended up not using any of them," Cukor told me.

The third change was a request for clearer goalposts. The earliest tests were judged by eye—comparing detections that popped up from one algorithm with another. This subjective approach carried high stakes: several startups were axed on the basis of unclear metrics.

Maven needed "a very defensible way" to accurately pick winning models, Pinelis had told me. The team had run their horse race behind in secrecy, using code names for rivals and hiding evaluation datasets and test scores. They didn't want competing vendors to develop ways of winning the test rather than winning the war. "It's really hard to completely and thoroughly evaluate these AI-enabled systems. And the reason that it's hard is because they frequently arrive to the Defense Department as a black box, so the systems are not extremely well understood," Pinelis told me.

Lymberopoulos, Microsoft's principal research manager who built and led the company's Maven team, started to feel Maven's evaluators were underscoring his team's models—they'd test well on Microsoft's own internal metrics but kept showing little progress on Maven tests.

The computer engineer was competitive about everything. Just getting to see data from drone videos enchanted him. Not for their gore, but for their rareness: he was excited by new angles, distances, objects. "I couldn't see anything like this on the open internet."

Cukor would try to coax—or shame—his way to success. "He would play your ego," Lymberopoulos recalls. "You guys are from one of the best companies out there, you're spending three times more than what the other guys are spending: why are you behind?"

Months after Microsoft first complained, Maven agreed to change the way it did evaluations and Microsoft test scores moved up. The resulting mark was known as the F1 score. This gave a form of average known as the "harmonic mean" between precision (of all the things that you find, how many are things you want to find) and recall (of all the things you want to find, how many did you find).

At a test event in 2021, ECS Federal wanted to keep the competitive fervor going but lacked Cukor's subtlety. They awarded Lymberopoulos and his team a prize for coming first even though there was little to choose between the top-rated algorithms.

The vendors cringed a little. It was the sort of golden trophy that a school sports team might win: angel wings and hologram stickers galore. Only this one stood two feet high and had red tape with black marker on it that read "MaVeN!"

"Please don't," Lymberopoulos conveyed to ECS, embarrassed to receive his award sitting beside Daniel Marasco, his now-smirking rival from Clarifai. "It's unnecessary." (He left the hulking trinket at the ECS office.)

A more relevant divide between Clarifai and Microsoft came down to speed and precision. Microsoft models were better than most at recognizing small objects—such as people—that might be represented on screen by only a few pixels. But they were slow: instead of running the model on each of the thirty frames that make up a second of video, it would run it against far fewer. (It also did away with all subcategories, identifying only "person" and "vehicle.") Clarifai, by contrast, tried to look at four times as many frames per second as Microsoft did. As a result, Clarifai models could detect objects faster than most human reaction times. That meant smooth displays that were better for keeping track of a moving object. These minor differences might sound dull and esoteric, except that they might one day become the basis for shooting at someone.

AWS tried a third tack altogether. At one test event, Clarifai and Microsoft submitted a handful of models as usual, but four people told me AWS submitted more than sixty. Their rivals were astounded and some accused the company of trying to game the system. Cukor told me he didn't see it that way: they were creating models specially adapted to multiple different circumstances, such as when fading light can make it particularly difficult to detect objects. But he said that the "orchestration" technology to switch between bespoke models simply wasn't there yet and so, heeding the cries of the company's rivals, he imposed a limit of a maximum of five models per vendor for future sprints.

By the time Maven got a few months past The Reboot, things were looking up. Cukor had met with a twenty-three-year-old dropout named Alexandr Wang. He ran a four-year-old data-labeling startup named Scale AI that brought all sorts of "tricks" for how to improve and automate labeling. Maven also brought on Palantir's data management platform—as the company had always hoped—which could better organize the labeled data and reveal whether certain objects and scenarios were relying too much on images taken from only one angle, one distance or one time of day, among other concerns.

In September 2020, Maven brought both Scale AI and Palantir on as new prime vendors alongside ECS Federal. The models got "substantially" better, Cukor told me. Labeling was becoming big business: Wang, who was roommates with OpenAI's Sam Altman during Covid, would become the world's youngest self-made billionaire the next year. He would soon be preaching to Congress about the dangers of Chinese facial recognition systems and the need for America to label data of its own at scale (and at Scale).

In June 2025 Wang would become an employee of Meta after the company bought a 49 percent stake in Scale AI for $15 billion. But the company's fortunes with Maven would almost immediately take a dive: in September 2025 Scale AI lost the largest data labeling contract on record yet—a much-delayed bid for a seven-year government AI labeling contract worth up to $708 million for Maven and other intelligence programs. The company still had other government contracts, but in this one it was dislodged by a minnow named Enabled Intelligence, a startup I visited in late 2022 as it was still building out its secret and top secret business. Peter Kant, the company's founder, put his faith in quality over quantity and turned to an often-overlooked labor force: people on the autism spectrum. Research showed neurodiverse workers were often better suited to the repetitive, concentrated work of pattern recognition that had so irked Maven's early haphazard, all-volunteer and error-prone labeling force. The company's technical performance was assessed as "outstanding," Kant told me, and after winning the Maven contract he quadrupled his small workforce of thirty employees almost overnight while Scale AI was left disappointed.

Even before then, Daniel Marasco, the machine learning research scientist who worked at Clarifai, said Maven ultimately created a very impressive, serious labeling program with a huge amount of data and compute underpinning it. Maven would have four hundred professional data labelers running eight-hour shifts all year round, producing better data than other government AI efforts. "Every time we get labeling data that's not from Maven it would be really bad," he told me. Lymberopoulos, who in 2022 left Microsoft for Palantir and soon hired Marasco to work for him too, agreed: "I believe Maven has curated a huge dataset that has delivered the best general CV [computer vision] models out there." And if data really was the new oil, Maven was now getting rich: it would soon have a precious store of at least 100 million professionally labeled images.

A SPAT OVER ACCESS to this precious data would lead to fracture between Cukor and the boss who had provided top cover for his crusade—and provoke major upheaval in the DoD's AI efforts.

When Lieutenant General Jack Shanahan left Project Maven to start the Joint Artificial Intelligence Center in 2018, he had expected Maven would eventually come under the JAIC's control as well. Maven, as he put it, was supposed to be an "artificial intelligence, machine-learning pathfinder project" in the department; the minnow to his new whale. The JAIC would soon staff up to hundreds of people, with hundreds of millions of dollars to spend, aimed at removing the roadblocks that prevent department-wide AI. Maven was there to shine a light but the JAIC was always meant to be the way to scale AI across the whole department. Shanahan made that much clear in August 2019 when he announced that the JAIC's biggest effort for the next year would be "warfighting-focused" on AI for maneuver and fires.

There was a lot of "bravado" about how Maven would report to the the JAIC, Cukor told me, but he argued it could never happen: Maven was funded by Military Intelligence Program (MIP) dollars and could never be nestled under "non-MIP" money focused on combat operations. I sensed the return of Cukor the bureaucratic ninja. Shanahan told me it would have been "doable" but that initially he hadn't wanted to hold Cukor back by folding Maven into the JAIC. "The only reason Maven is what it is, is Cukor," he told me. "You cannot constrain someone like that and get results." To some the future of Maven was about congressional money mandates, but to others it was clear what was going on: a turf war.

When Colin Carroll left Maven in July 2019, disenchanted over Maven's embrace of Palantir and because he had started dating ECS contractor Katya Volkovska, he ended up at the JAIC along with several of Maven's old team too. Shanahan would email Carroll the next year that he was "a force of nature" filled with "passion, dedication, technical expertise, and ability to just get things done," saying he "would have been screwed" without him. Carroll wanted to develop the idea he pursued at Maven. In December 2018, Microsoft proposed an idea to fix some of the problems they were encountering with WAMI feeds when the patchwork of imagery was always off by a handful of pixels. Instead of bringing WAMI data feeds down from MQ-9 Reaper drones to an AI processing system on the ground, why not put the algorithm up on the drone platform itself to process data from the sensor up there. That way they could cut out the data link, the transport, and the intelligence analyst altogether. Microsoft proposed calling it Smart Sensor. Cukor hadn't been interested: the plan had too much operational complexity, he told others, and bled into operations rather than intelligence alone, which could mean transgressing Maven's funding lines. But Carroll was captivated by it, pursued it as an initiative under Maven and then took it with him to the JAIC. In 2023, the autonomous sensor was tried out in the Indo-Pacific, and the Marines started using the system in late 2025.

"In the not-too-distant future, the Department will fight autonomously with human-machine and machine-machine teaming," said an internal document describing the requirements for Smart Sensor. The effort briefly went by the name Project Atropos, Colin Carroll's choice named for the third of the Three Fates in Greek mythology, who could cut the thread of life: he wanted to cut the human pilot out of the loop. (Carroll bought the domain name in 2012, and 13 years later used that for his autonomous drone startup Atropos Group.)

To put an AI brain on a drone, Carroll and the JAIC would need a store of labeled images taken by MQ-9s with the necessary variety of angles, distances, times, backdrops. Maven had the perfect feedstock: it had helped build much of the MQ-9 dataset he now wanted to access.

But Cukor wasn't about to give it up. He might have lost some of his workforce to the JAIC but he wasn't about to lose his data too. Although Maven had gone to great lengths to get much of the data declassified, Cukor argued that they had gotten the declassification on condition they safeguard the data. This effectively turned down not only his former underling Carroll, but also the JAIC leadership including his former boss Jack Shanahan.

"We have to liberate data across the DoD," Shanahan said in August 2019 remarks he made to reporters that could just as well have been crafted for Cukor's ears. Nand Mulchandani, the chief technology officer at the JAIC who acknowledged rising tensions with Maven to me, said the Pentagon was paying so much to be an early adopter of tech through Maven that it should reap the broader benefits too.

Cukor insisted to me he shared multiple aspects of Maven with the JAIC and the Defense Department, but he stopped short of sharing the data because of the risk it could leak. "Once you open datasets they go everywhere; they get legs," he said. "At the end of the day the data is more valuable than the algorithm, frankly."

By then, Cukor described Maven as having some of the most exquisite labeled datasets on the most sensitive things in the world. He didn't even give Maven vendors the data—they could only access it within Maven's training environment in the cloud, SUNet. But he wouldn't let the JAIC into that either, not at first. He wouldn't let them have the MQ-9 models Maven had developed either. "Things got ugly," Cukor told me.

"Cukor became enamored of the fiefdom he built and wouldn't share," another member of the JAIC during the 2020 dispute told me. A showdown was coming.

In a series of some of the longest and most contentious meetings of some defense officials' careers, JAIC officials tried to reason with Cukor to get access to the data. Cukor would stay polite, dogged, and unyielding. Sometimes Cukor would counter with national security concerns. Sometimes he'd suggest that so much data focused on US wars in the

Middle East that it would expose the Defense Department to criticism that it was spying on Muslims, and leaked graphic combat footage would reignite a furor worse than the one before at Google that could deepen the dent in fragile relations between the Pentagon and commercial tech workers and in civil-military relations writ large. It would be a matter of national embarrassment.

In one spring 2020 meeting to try to resolve the issue, Cukor suggested only he and his team could keep the material safe. And for the first time in all his years spent supporting and protecting Cukor, Shanahan ran out of patience. The genial three-star general turned red with anger.

"What do you think we do for a living?!" he exploded. "You don't think we can fucking protect this, Drew?"

No one had ever seen Shanahan like this. Shanahan had assumed Cukor would be loyal to him both personally and professionally, since he gone to bat for the colonel time after time. He was disappointed in Cukor, as if he were a parent whose kid had let him down, and shocked Cukor and co had showed up ready for battle. After Shanahan's outburst Greg Christ looked to some as though he might cry. But that's not how Cukor operated. He remained implacable. He was all about his mission.

"You can imagine two organizations bickering about data," Cukor related to me years later. "Ultimately we prevailed, but at great cost and there were a lot of angry people."

Shanahan stopped talking to Cukor altogether.

"My biggest regret is not solving the Maven problem," Shanahan wrote in a May 2020 email to Carroll, meaning he wished in retrospect that he'd brought Maven and the JAIC together at the outset. He had set some things in motion that might result in improvements, but both of them "know Drew too well," he continued. "Not sure if anything substantial will happen until he departs the seat."

I asked Cukor if he had become exactly the sort of office bureaucrat protecting his own interests that he had always railed against.

"I know that's how it looked," Cukor started to answer me slowly. "The irony is not lost on me."

This was by no means a concession. That data was the country's most precious national security asset, he insisted. He couldn't risk it getting out. He had his mind on China: if Beijing got access to the dataset it could learn from it, or, worse, poison it.

The other member of the JAIC at the time told me that although there is always a risk data could leak, there were ways to protect and track every download, and that neither Carroll nor the JAIC writ large would have got that wrong.

A Defense Department audit of the JAIC's and others' abilities to protect AI data and technology didn't seem so sure. The JAIC hadn't developed a security classification guide to protect AI data yet, according to a June 2020 report. It also flagged six cases of lax security controls among military components and contractors and warned that malicious actors could steal information in a way that "could threaten the safety of the warfighter" and disadvantage the US against its adversaries.

The JAIC took its complaint above Cukor's head. The issue worked its way up to the top expert on data in the department, the chief data officer.

David Spirk.

The man who had carried Cukor's computer for him nineteen years earlier in Kandahar. The man whom Cukor had counseled to leave the Marines and ascend the ranks of the DoD as a civilian. The man whose thigh was once tactically one with Cukor's. And the man whose job it now was to weigh in on data-sharing disputes inside the department.

Spirk considered the issue carefully and broke the standoff—mostly but not entirely in Maven's favor. He thought it was a silly flash point dividing the two organizations and told me he made his decision independent of any friendships. I learned he put out a nonpublic memo that codified Maven's agreement that the JAIC should have access to most of Maven's data, but not the precious and expensive Maven testing and evaluation datasets. These could neither be shared nor accessed on the grounds of data sensitivity, he ruled. Besides, the JAIC hadn't offered him a plan for how to keep the data safe. China wasn't top of Spirk's mind so much as preventing vendors from cheating by getting hold of exam answers in advance.

The decision pitted Carroll against Cukor once more. It threatened to impede Carroll's drone-based AI sensor project, requiring him—and the government and the taxpayer Cukor cared so much about—to double up on data collection, labeling, model creation, and time. Carroll would later express his displeasure in public. The failure to share data across the Department of Defense counted as "Maven's biggest failure," he would tell a podcast in 2022.

"The whole point of Maven was to become a pathfinder for AI but instead the data stayed locked up inside Maven and didn't spread across the department," the member of the JAIC during the 2020 dispute told me.

It wasn't just the data: the staffs of the two AI offices weren't cooperating either. If Maven was a pathfinder on AI for the Defense Department, the path appeared to wind around an island of its own making.

When Shanahan retired that summer, he was still raw and angry with Cukor. And now Maven and the JAIC were two rival AI houses inside the Pentagon.

PART THREE

FINISH

18

WE'LL FIND IT AND WE'LL STRIKE IT

"The whole nine yards."

BUTTED UP AGAINST the breakfast counter in the Washington, DC, hotel was a trolley. It was 7 a.m. on the last Friday of August 2024, and I was looking for a coffee pot. Slowly I realized I was looking not at little jugs of milk, but rather at tiny battle figurines.

Painted in bright hues, swords at angles cutting against the air, cartoon furies and agonies were molded into their minute faces. Soon I saw the whole floor was dotted with trolleys wheeling about these one-inch wonders, pushed along by hunched men in baggy gray T-shirts. I'd somehow entered a toy soldier convention. In front of me, poring over the breakfast menu, was a decidedly real soldier.

Thirty years in the military. Special operations. Iraq. "It's pretty weird we've landed at a fantasy battle convention," my conversation partner smiled at me before deciding on his breakfast order. Avocado toast. "And here I am, the man everyone accuses of inventing Skynet."

I was breakfasting with Joseph O'Callaghan, an Army colonel who had over the past four years become the poster boy for the US military's pursuit of AI warfare. In Colonel O'Callaghan, Colonel Drew Cukor had found his front man.

Colonel O'Callaghan's specialty was artillery, lobbing howitzers and high-mobility rockets across the deserts of the Middle East. He'd been a battalion commander within the 18th Airborne Corps' Field Artillery Brigade, spent a year planning operations in Erbil back in 2015, and fourteen months running the targeting process against ISIS from Iraq and Kuwait as fire support officer, responsible for coordinating artillery fire, air strikes, and naval bombardments. He was responsible for the very targeting process Cukor had for years told his team to avoid mentioning, even in internal Pentagon correspondence.

In October 2019, O'Callaghan arrived at the 18th Airborne Corps in Fort Bragg, North Carolina. The 18th is known as America's rapid reaction force. They were often first to respond to crises abroad, and counted deployments to Panama, Bosnia, Iraq, and Syria. "We'll probably be gone out the door soon," O'Callaghan told me between sips of coffee. "It is a rare year that we are home." He told me he always kept his bag packed and quoted a line from *The Princess Bride* to convey the cycle: "Good work. Sleep well. I'll most likely kill you in the morning." "That's what living in 18th Airborne's like."

O'Callaghan's boss was Michael "Erik" Kurilla, who became the 18th's three-star commanding general the same month O'Callaghan arrived. Pretty soon there was a picture of Kurilla, previously commander of the 75th Ranger Regiment and a former director of operations for JSOC, up on Maven's office walls. Some in the office chortled among themselves at what they saw as Cukor's obsession with the hard men of the military, treated as pinups by the wannabe algorithmic warriors of the future.

Cukor said it was simply about remembering the customer. Their customers didn't live within the walls of the world's then-largest office building. They weren't ladder-climbing bureaucrats. They didn't get to go home on time. They were the ones out on the front lines risking their necks—and the necks of their forces—again and again. Maven wasn't just another anonymous technology effort doomed to fail; he wanted everyone to confront the people who faced the horrors of combat. "What are we doing for Kurilla today?" Cukor would ask his team, tapping at his picture.

Any time someone would balk at the scale of the task Cukor expected from them, or dared say it couldn't be done without a more realistic timeline, Cukor would trot out his reply. "Do you see this guy right now? He's looking at you and he disagrees."

Kurilla was a West Point graduate who started winning best lieutenant competitions in the 1980s. He spent every single year from 2004 until 2014 leading conventional and special operations forces abroad, including six years in a row in Afghanistan. He came up with the risky idea of fighting during the day rather than under cover of darkness at night in Afghanistan—it meant they could go "from killing five guys to killing fifty guys," a Ranger told Sean Naylor for his 2015 book *Relentless Strike*. Admiral William McRaven, former JSOC commander, called Kurilla "the finest soldier I've ever known."

Kurilla was also embittered. "I'm often reminded that our American society places the wrong people on a pedestal," he said in a 2015 speech, calling out America's worship of sports stars, actors, and singers. "Those aren't heroes."

Instead, the man known to some as the Big Eagle and to others as the Gorilla, valorized the men who signed up to the military—and they returned the favor.

"Colonel Kurilla is like my dad. He would die for me," one soldier (under his command at a forward operating base named Marez in Iraq) told blogger Michael Yon in May 2005.

A few months later in 2005, Kurilla continued to fire his weapon after he took a commando roll, a bullet in each leg and one in his arm during a frightening street shoot-out in the northern Iraqi city of Mosul. Later that day, he downplayed the near-death experience to his wife in a phone call from a combat support hospital.

"Honey, there has been a little shooting here. I got hit and there was some minor soft tissue damage," he told her. Yon, the blogger who embedded with Kurilla, said the X-ray showed his femur was nearly snapped in half. "I'll be fine. Just some minor stuff."

Maven made a "huge play" for the conventional military, as former JSOC leaders like Donahue and Kurilla were promoted up and out into the wider force. Making it into "the Big Army" could mean more take-up, more money, and a bigger impact on the way America fights its wars. But it also came with greater challenges. Conventional forces tend to be less interested in trying out new tech than special operators, and were bound to programs already funded by Congress. But Kurilla came from the world of special operations, and he was set on bringing new tech to the sprawling base in North Carolina of 90,000 troops.

"He takes command and he's like, 'Bring me that Maven,'" recalls Nick Villarruel, a member of the Maven team who worked on the effort to bring AI to the 18th.

Kurilla wanted the 18th to be the Army's first AI-enabled corps. It was far from clear what that meant. AI was a black box, but Kurilla wanted to understand how he could use the system. What could he actually trust it to do? Could he plug AI into everything? Could he plug it into targeting? Video footage would not be so relevant for the Pentagon's future wars against China or Russia. The drones that collect video were likely to be shot down or jammed. Satellites and still imagery might have more luck.

Kurilla was open-minded about where AI might prove most useful—he wanted intelligence, operations, and networking to try it—but O'Callaghan said he and his team responsible for running weapons fires ended up finding "the most traction."

"We were introduced to them in about February of 2020 coming out of the strike on Soleimani," O'Callaghan recalled to me during a 2023 visit to his office, where I first met him.

I learned later from others that, just as with the Baghdadi raid, Maven had also been up on the screen in Erbil during the January 2, 2020, strike that Donald Trump ordered against Qasem Soleimani, commander of Iranian Revolutionary Guard Corps' elite expeditionary Quds Force.

The top secret JSOC mission was kept so under wraps that Maven members would only learn weeks afterward that the algorithms developed under their own program were used to track the vehicle carrying Soleimani soon after he'd landed at Baghdad airport that evening, right up until after the moment the US fired at his car. One person who watched the mission feed told me AI made it simpler to follow the vehicle. Maven

Smart System's augmented reality layer highlighted the track of the car, helping determine metrics like its speed and direction.

But O'Callaghan wanted to do more than find something with the help of AI. He wanted to shoot it too. "We want to be able to push an AI detection all the way to the gun line and then shoot this, like live fire off of it," he told Maven's vendors. O'Callaghan's aim was to shorten the kill chain—the process by which a target is identified, located, tracked, and then "prosecuted." Find. Fix. Finish.

For Cukor, who had spent the past three years avoiding saying the word "strike" out loud, O'Callaghan's explicit focus on using AI in support of weapons fires released him from his constraint. "When we were primarily an intelligence function or capability I exercised extreme caution so that we would not bleed over into the operations side. The intel people cannot overreach," Cukor said. "I never said fires or strike capability or single click."

Cukor could let O'Callaghan represent the ops side, but it had to be phased carefully. O'Callaghan, Cukor told me, "knows this dance as well."

Within a few years, O'Callaghan had put himself in the middle of the dance floor. He would give himself a new title, backdating it on his LinkedIn profile to his 2019 arrival at the 18th, stretching it out across the next six years: "Director of Algorithmic Warfare." Before long, O'Callaghan's war bunker below his office in Fort Bragg was dedicated to all things Maven. Along the hallway to a test room filled with computers, fat marker pens daubed the walls with long hieroglyphics, sketching out the principles of data and AI-assisted military strikes.

The idea was to use AI on imagery from America's classified spy satellites and unclassified commercial companies, such as Maxar Technologies, with satellites in low earth orbit. Maven's Nick Villarruel had already spent months puzzling away on how to get algorithms working on these types of platforms. Along with his deputy on the effort, Jaim Coddington, the pair tried out some of Maven's existing computer vision algorithms on still images of China and Russia from high-altitude U-2 spy planes, RQ-4 spy drones, and spy satellites.

The team faced challenges: the images, often taken from 60,000 feet up and higher, are classified. That made it harder to work with AI vendors and allies who didn't have shared security clearances. Second, the angles were tougher, since US spy planes and spy drones don't directly overfly China and Russia. Instead, they tilt as best they can to see the ground from airspace above international waters. Third, the sensor data is less plentiful and harder to parse. Besides taking photographs, known as electro-optical imagery, the sensor data also provides images in two other varieties. Synthetic aperture radar is useful because it can "see through" cloud cover, dust, and rain. Multispectral imaging delivers a high-contrast visual interpretation of a scene based on light the eye can't detect. That can help reveal hulks of metal hidden in a forest, for example, or chemicals mixed in with earth. Little commercial equivalent of this exists, and the resolution is so low it often ends up looking like blobs.

Microsoft found the work taxing: a highly sought-after object on a multispectral image might occupy only ten pixels—not much for their computer vision models to go on. A human might constitute between fifty and eighty pixels, often too few to differentiate between a man and a woman. A weapon on a shoulder might constitute three pixels. While an experienced analyst could parse an image thanks to the context, Lymberopoulos had no clue: "I couldn't tell it was a gun, because I wasn't trained."

Maven was also still navigating what Coddington described to me as a "tense, unfruitful" relationship with Air Combat Command, which provides US Air Force combat airpower to the rest of America's fighting forces. On the last Wednesday of August 2019, Maven folks were down in the southern tip of Virginia to brief commanders and staff at the ACC headquarters at the Joint Base Langley-Eustis beside Chesapeake Bay. They expected pushback, and were warned that a particular Air Force major was preparing to brief heavily against Maven. The week before, ACC Commander General Mike Holmes had taken the unexpected step of shaming Maven in public. He saw potential for algorithms to understand battle, but didn't think the US was ready to turn that over to the machines quite yet: he suggested the project's multi-million-dollar capability was no better than the skills of his aide-de-camp's three-year-old.

"He picked out all the green things. You know, green, green, green, not green. That's what we're doing with Maven. It's car, car, car, not car," he told reporters. Maven, he said, still needed "mom or dad" looking over its shoulder.

The Maven team knew their models barely worked—but they wanted to develop a use case with a willing customer. Not for the first time, the Air Force was uninterested. "There's so much nuance it's hard to have these conversations," said Coddington. "They weren't into it."

Knowing the major was preparing to shoot him down, Colonel Cukor simply filibustered his way through the briefing. He ran through three hundred slides. Some thought he was relying on a technique the Marine Corps Deception manual calls "dazzling"—overloading sensors with information. The generals left before the dissenting major ever got his turn to speak.

Even so, the Air Force stayed on the sidelines, and the 18th pressed ahead. On the night of August 27, 2020, the 18th carried out what O'Callaghan calls "the first AI-derived target" in the entire Department of Defense: an old tank hull that had been wheeled onto a firing range.

It wasn't pretty. The systems involved barely spoke to each other and the satellite image downloaded at the pace of molasses. It was close to 10 p.m. when the AI finally detected the tank, more than twelve hours after the system had started looking for it. The system asked its human minders to confirm the AI selection. Next the system sent a message to an M142 HIMARS—the wheeled rocket launcher that is a mainstay of America's artillery forces—instructing it to fire. A rocket whistled through the air and found its mark, destroying the tank. O'Callaghan keeps a picture of the moment on the wall of his office conference room, flames licking at the edge of the frame. "It showed us the art of the possible," he says.

They tried again later that year, initiating a series of quarterly exercises called Scarlet Dragon to work on AI-enabled targeting. Just as before, they would place objects in the field, run models and then try to detect them as targets. One Palantir rep embedded with the 18th was so quick and clever that he soon became known to some as "Rainman": he would go outside every time there was a problem, Slack someone back at HQ, and get the system updated remotely within a couple of hours. US Army officers used to obsolete equipment that could take years to replace were stunned by the responsiveness.

The systems often failed, however. After a satellite passed overhead, sometimes no AI boundary box would appear on screen displaying a tank detection on the resulting satellite image, even though the tank was sitting in the open field staring everyone in the face. One high-profile test event led to controversy. In the demonstration intended for Distinguished Visitors, Maven was "showcasing that we had detected these objects," said one observer. "We actually didn't detect them," alleged another, saying "Nerds . . . just drew them on."

It was hard to unravel the veracity of this claim. Two people familiar with the allegation, which circulated widely in the Maven office and worried several people on the team, told me no AI detection was ever faked. One said that the satellite was taking so long that a named area of interest was hand-drawn onto the screen, but said that wasn't the same as drawing on an AI detection, stating that never happened. Another person told me that so little about the system was working well at this early stage of its development that the team tended to set sensitivity levels "super high" so that AI identified absolutely everything and distinguished almost nothing. When I asked an official at the 18th Airborne Corps about these allegations, I was told they were the result of a "misunderstanding." Whatever the particulars in this specific case, the rumors fueled Maven's detractors and the 18th kept experimenting to improve the system.

One of the more charitable members of the team said Cukor liked to say things to will them into existence. In briefings, Maven tended to present the good as excellent and let the bad fall away, this person said. Good stories from the field earned Maven more funding. Each year, Cukor was getting budget bumps for Maven, briefing Congress on success stories and, he says, taking care to explain its limitations. The biggest bump came in 2019, when Maven's budget rose to $189.53 million; the following year, to $232.96 million. Cukor would rally the team talking about the future, to a time people would write books about what they were doing. Some felt proud and were convinced the arc of progress would deliver more reliable AI. Others saw subterfuge in such an approach.

O'Callaghan, a former competitive cyclist who owns a bike shop, was firmly in the progress camp. He liked to compare improving the military's algorithmic capabilities to the dogged way a bike racer improves their speed—each training session needs sometimes infinitesimal improvement in any of various categories from posture to equipment. He even invited in cyclists and Formula 1 racers to talk to the 18th.

One Maven representative who supported the work of Scarlet Dragon insisted the exercises showed him that AI, however halting it might be at first, really was going to change warfare. The AI could find thousands of targets within moments—something no human poring over satellite images could ever hope to do. "It was groundbreaking stuff. I could read the writing on the wall," the Mavenite said. AI didn't just have speed and scale; it also had far greater sensitivity than any human: it would pick up on changes among just a few pixels that the human eye couldn't even perceive. "It wasn't much of a mental leap to image how powerful this would be once it was refined."

It wasn't Oppenheimer, but some Mavenites began to discuss that, in terms of the future impact on human capabilities, weaponizing AI didn't feel far off. Some felt giddy standing on the precipice of a future they themselves were fashioning.

The 18th grew more ambitious. Instead of one rusty tank hull, they asked algorithms to identify twelve objects of interest. They fused data from multiple sensors, snooping and gathering radio and electromagnetic signals pulsing from radar, missile guidance systems, and aircraft. Maven Smart System displayed the corresponding detections on a map. A human team would vet the targets to check if they were valid or decoys. Actionable targets were sent through to the Advanced Field Artillery Tactical Data System (AFATDS), the software on which the Army relies to coordinate and execute weapons fire. O'Callaghan gave this AI-enabled space-based capability a new name that finally made explicit the link to targeting, hoping that would confer on it the aura of respectability and carry the authority of a new discipline: Broad Area Search for Targeting, or BAS-T.

AI was producing so many targets that the humans couldn't keep up. In a July 2021 white paper, O'Callaghan wrote that AI's ability "to digest thousands of square kilometers at once, enables the Corps to hold enemy forces at risk constantly."

THESE AMBITIONS HIT A ROADBLOCK when O'Callaghan showed the system to the soldiers who would actually have to use it. Joey Temple joined the 18th as a fires support officer the same month O'Callaghan's white paper came out.

Temple was a veteran of five combat tours in Iraq and twenty-five years in the Army. He had worked with O'Callaghan from twenty years before. The pair was just catching up, beginning to reminisce in his office, when O'Callaghan brought up AI. "He starts vomiting all this stuff. And I'm like: 'What in the world are you talking about AI?'"

Temple had no clue. "So I'm trying to write all this crap down, all these abbreviations, all this other stuff that I did not know at that time and make sense of it so that I could actually be value added to his team."

Temple oozes no-nonsense competence. He has eight children. He keeps a model skull wearing the unit's maroon beret and clenching a dagger between its teeth on his windowsill. And he has more than a decade and a half of targeting experience.

He wasn't about to put his trust in an algorithm. "This shit don't work," he recalls thinking. "I wasn't a fan."

A Palantir representative at the 18th told Temple he'd love to demonstrate the system to him. But Temple told him to go away and cussed him out. "I was like, 'I don't need another freaking targeting thing. I don't care. I have enough crap to try to manage.'"

A month later he started to change his mind. He was assisting with the ignominious withdrawal from Afghanistan after twenty years of war. US forces and their allies rushed to evacuate people from Kabul as the

Taliban took over. Ultimately, more than 120,000 people would be airlifted out of Kabul in dangerous conditions.

People were sleeping in the office. Temple was given access to Maven to help make sense of the situation. On a single screen, he could combine data feeds that tracked aircraft movements, monitored logistics, watched for threats, and showed the locations of key personnel. "I could see General Donahue walking around," he says of the commander of the 82nd Airborne Division at the time, who would become the last US soldier to leave Afghanistan. So could hundreds of people logged onto the same system from Kabul, the Pentagon and, according to Temple, the White House. Maven was running in the Pentagon's National Military Command Center (NMCC) and both Defense Secretary Lloyd Austin and General Mark Milley, chairman of the Joint Chiefs of Staff, were briefed using Maven Smart System.

Temple had never seen so many people use the same system all around the world at once. He was used to wannabe systems that would crash with only forty people on them. (And Maven had been accused of crashing an entire network one Sunday night in Afghanistan only two years earlier.) "That's when I became a believer," Temple says. He apologized to the Palantir representative and asked him to teach him how to use the system.

As the effort to bring in supplies and evacuate people turned deadly, the team used Maven to try to estimate how many people were on the airfield and trying to get out. "A lot of people didn't believe how many people had gotten on there," another person familiar with the operation told me. A human analyst would have to pause the feed and try to count heads by eye. The team tweaked a few parameters on their best models to reorientate it for crowds, and then AI did the counting instantaneously. Leaders in the Pentagon began to comprehend just how much they'd lost control of the situation that was developing in Kabul.

On August 26, 2021, as thousands scrambled to get into Kabul's Hamid Karzai International Airport, an ISIS-K suicide bomber at the airport's Abbey Gate killed 169 Afghan citizens and thirteen US service members, including eleven Marines. Two days later, US president

Joe Biden warned another imminent ISIS attack was "highly likely." On August 29, after following a white Toyota Corolla sedan for eight hours, US drones watched it enter a compound in the capital and targeted it with a Hellfire missile. General Milley called it a "righteous" strike, telling reporters the target was an ISIS facilitator.

But that's not what the people who knew the family said. All ten people killed were civilians, including seven children. The man targeted was an aid official who had been trying to get his family out of the country. An Air Force inspector general would later blame "confirmation bias" for, the *New York Times* would later report, "warping operators' interpretation of what they were seeing."

I was told Maven was not switched on for this strike. An aide on the Joint Chiefs of Staff asked Brian Ward to see if, had the AI models been used during the operation, they would have been able to detect the children. Ward sat down and ran multiple AI models over the dusty, blurry recordings of the MQ-9 drone video feeds. Not even Ward, an experienced intelligence analyst, could see any children by eye the first few times he rewatched the infrared and other video feeds. "It was witching hour and as sun goes down sensors have a hard time—there just wasn't enough differentiation between the ground and the kids."

But when he watched it over he began to detect there were a few shadows moving after all. These were the "few partially obscured forms" a US commander would later say were briefly visible moving in the compound. If a team was fatigued, at the end of a twelve-hour shift, Ward thought they'd be very unlikely to catch those moving pixels either, no matter how heightened their senses might be ahead of a potential strike and given the fear that ISIS was planning another attack. "You don't want emotions to get involved," Ward said.

When Ward ran several AI models across the feed, at varying levels of sensitivity, Maven's AI couldn't catch the pixels either: "It picked out the adult male but it didn't pick out the children."

The incident raised the question of why America was launching strikes against targets if its sensors were not good enough to deliver a clear picture in the first place, no matter whether AI or humans were watching. Abiding by the Law of Armed Conflict rides partly on the intent behind any strike, but if you can't see likely collateral damage it is hard to assess intent. America's ability to summon firepower had raced ahead of its technical abilities.

Ward saw hope for something else: the fact he knew there was something showing on screen, no matter how hard it might be for the human eye to detect, meant there was a chance an algorithm could detect them in future. He took the example to the AI vendors: "This is why what you're doing is so important," he told them.

Another person who worked on Maven was aghast: if Google hadn't refused to continue working on the project, they figured the algorithm work would have been two years further on. "Those ten Afghan civilians would have been alive if the effort led by Project Maven to deploy AI on military drone camera feed had not been delayed," the person posted on LinkedIn. A Maven official told him to remove the post the next morning, due to a prohibition against commenting about Maven on social media.

The vendors who'd stayed on the project were surprised their algorithms hadn't detected the movement. Ward thought more data, more training would help. One algorithm vendor wondered if an alert would have pinged if the algorithm had been running at a different sensitivity setting, but Ward told me he'd tried every setting going. The algorithm vendors asked to get the video so they could practice with it, but it was never provided to them, one of the algorithm makers told me. "I was disappointed," the person told me. But however much better algorithms and the data that trained them might get, a central problem would remain: AI would always miss things.

A model can perform very accurately 99 percent of the time, but then in 1 percent of cases the model and user can fail at the same time, the algorithm vendor told me. "Does it mean this thing sucks? Probably not," the person continued. But, they told me, the AI could go either way. "It's a probabilistic system," the person told me. "It will never be perfect."

The 18th would soon try Maven with moving targets too—hunting out enemy vehicles through aircraft or space sensors. They would simulate how to do the same from submarines, and to classify a truck as an enemy vehicle and track it through the Carolinas. And they started dropping weapons from B-52s, F-35s, B-1s, F-15s, F-16s, and Gray Eagle drones on AI-detected targets.

"We'll find it and we'll strike it," O'Callaghan said. "The whole nine yards."

I was given a demonstration of Maven when I visited the 18th. It was drawn from a real-life operation in early 2023, when a US Navy ship evacuated American citizens from Sudan, which was in the midst of a civil war. The system's central map was generated, in my case, from unclassified satellite imagery. On the surface, yellow boundary boxes marked where algorithms had identified ships in the Gulf of Aden. Blue areas signified places that would be included on a no-strike list, such as hospitals and schools. On the left side of the screen, icons offered separate data streams, such as vessel tracking feeds, that could be overlaid on the map. There was also a function for a "tactical data link": a method for transmitting directly between machines a commander's decision to fire a weapon.

O'Callaghan was so taken by AI he was reworking his entire targeting workflow. He analyzed the targeting cycle at the 18th, which he said relied on six key steps: a human would decide when and how to shoot at a target, assess the operational approach, assess the data collected, decide to act, communicate the decision, execute fire, and communicate what happened. With the arrival of Maven's AI, he reduced the human role "in the loop" (meaning a human had to make the decision) to only two places: the decision to act and the action itself. A human would also be "on the loop" (meaning a human would supervise a machine making the decision) during an automated collection process, but the assessments throughout would all be AI-enabled.

It still wasn't as reliable as a human: AI could correctly identify objects with 60 percent accuracy, compared with an 84 percent baseline for humans at the 18th, but it could do it farther and quicker than humans. Just as AI was beginning to earn more followers, the humans were reaching breaking point. The Palantir worker who Temple had cussed out (and had since made up with) was toiling so hard that his blood pressure shot up to 190/140—an emergency hypertensive crisis. Even as he checked himself into hospital, thinking he was having a heart attack, he was still on the phone to Cukor about work.

And in the meantime, O'Callaghan insisted it was safe to reduce the number of fallible humans involved. "It's not *Terminator*. The machines aren't making the decisions, they're not going to arise and take over the world," he said. But it was also much more real than a fantasy toy soldier convention. "It's up to humans to decide whether to pull the trigger."

Perfect or not, AI was now part of the kill chain.

19

NOBODY KNOWS TARGETING BETTER THAN TREY

"I will either be famous or live in infamy."

BY THE TIME Rear Admiral Frank "Trey" Whitworth got to hear about Maven, its most ardent advocates already feared the worst: that he would kill it off.

Whitworth was head of intelligence for the Joint Staff, a role known as the J2, and sat at the pinnacle of the intelligence function that Cukor wanted to overhaul.

The two men were born within a year of each other and had equivalent rank, a Marine colonel and a Navy captain, until Whitworth started climbing the ladder and Cukor got stuck on it.

By 2021, Cukor's promotion chances had been shot down for two years running, despite glowing statements in his file from Lieutenant General Jack Shanahan (before they fell out). Cukor shook off the implied double insult and told himself he could get more done as an unfettered colonel. He had grown up with little and now, as he approached mandatory retirement, he felt he had nothing to lose. His lack of promotion counted to him, on a good day, as evidence he was insufficiently bland and safe.

Whitworth might be a serious, deliberative person, but there was little bland or safe about his past. He was a former SEAL Team 6 intelligence director. After a career spent working in Afghanistan and Iraq, deploying as a target developer right after 9/11, he was now enjoying newfound seniority as a "warrior-admiral" among the president's top military advisers.

With his white hair, narrow eyes, and deliberate, jargon-laced speech, Whitworth came across as clinical and exacting. He also came with particular expertise in targeting, later framing himself to me as "kind of a purist" on the original targeting cycle. He had already sat on the military targeting committee for nearly two decades. Now he was the top uniformed intelligence official in charge of the joint targeting cycle. That was the process by which the US military selects not only what to attack, but also decides how to prioritize targets and match the best firepower to them. "Nobody knows targeting better than Trey," one person told me. "He's the consummate expert on targeting. There is nobody on this earth better than him."

Whitworth was among US military leaders who believed with pride—and not a few blinders on—that the modern US military necessarily held itself to extremely high standards, overwhelmingly met them, and that there was accountability in cases where it didn't.

The joint targeting cycle relies on six phases and is fundamentally about building a legal action. "Joint" meant every service—Army, Air Force, Navy, Marines, and Space Force came together for the process. For decades beforehand, the joint targeting process had become highly criticized by operators as "distorted" by imperfect intelligence, "excessive focus on weapons selection" at the expense of strategy, severe time constraints, and communication problems.

Civilian populations "may not be intentionally targeted," and acts of violence solely intended to kill, maim, or spread fear among the civilian population are prohibited under all circumstances, it reads. Particularly sensitive targets, known as STAR targets, especially if they might incur collateral damage and kill civilians, require presidential sign-off. For the Air Force such sign-off could take seventy-two hours.

Whitworth considered "distinction"—distinguishing between combatants and noncombatants—the hardest part of targeting. He spent a great deal of time thinking about it and embedding it into the military's process. When in May 2022 the Defense Department would rewrite one of the relevant documents, *Intelligence Support to Targeting*, Whitworth would insist on the removal of the "demeaning" notion of support when it came to target developers and intelligence professionals: they weren't just support personnel; they were "main effort." Mistakes meant dead civilians and possible transgressions of the law. Mistakes could mean presidential apologies, humiliation in front of Congress, the end of a career, and undermining America's own war aims.

Most specific to Whitworth, the document stipulates that the J2 prioritizes intelligence collection efforts, analysis, validation, and assessment for all joint operations and provides "major input" to the people running operations. The target development phase also includes vetting the underpinning intelligence, validating whether prosecuting the target complies with the laws of war, and determining if a target should be added to a no-strike list.

The dry language of doctrine describes the six phases like this: the commander issues targeting guidance; a target development and prioritization process rakes through the options; during capabilities analysis the best options to go after the targets are evaluated; the commander assigns available forces, sensors, and weapons systems; forces plan and execute the mission; then combat assessment determines how effective they've been and they cycle back through the phases, sometimes concurrently.

But as the work at the 18th demonstrated, Maven was giving fallible AI the chance to take shortcuts through the decision-making loop, plucking humans out from their traditional role in the cycle in four of six places. "Trey was guardian of the targeting process," Cukor told me. "Now there was me just blowing it up."

That was the context when, in fall 2021, Whitworth visited the Maven office for one of Cukor's infamous briefings. Cukor liked to say he offered "a very gentle message." His listeners tended to think nothing of the sort: Cukor blitzed listeners with slides, money, success. A billion dollars in funding. Broad area surveillance. AI targeting. Lots of information and little room for questions.

According to one person, Whitworth's reaction was immediate: he blew Cukor up.

Were they skipping steps, moving too fast, bending rules, Whitworth jabbed. To some in the room, Whitworth's pointed volleys insinuated that Maven was circumventing the vaunted joint targeting process.

Whitworth questioned the idea that Maven could potentially let an operator click a dot and tell the weapon system to shoot. "Tell me about the recordkeeping?" he drilled.

Project Maven was also pretty expensive. Was it really worth a billion dollars? asked Whitworth. He didn't seem to think so. Cukor seemed like a relentless salesman for his project. He was just this "driving force," Whitworth recalled to me, characterizing his description as a compliment. "He believed in what he was doing and was selling it all the time."

Whitworth, who told me he didn't recall the exact specifics of the meeting but did remember "a little bit of a tense conversation," told me later it would have been "a bit irresponsible" for him as the person in charge of the joint force's readiness and targeting to back this new project wholesale.

"It's sort of the collision of two worlds," Whitworth recalled to me of the friction between new capabilities and traditional targeting. Whitworth's job was about making sure the joint forces' intelligence machine was delivering consistency, improving processes, and accurately describing the order of battle—the military term for precisely what enemy forces and hardware the US would actually face down in any fight.

Cukor's replies were also less than satisfactory. "Details, details," Cukor's airy response seemed to say. "Let's show you the AI piece."

But Whitworth was the accountability piece. "I'm going to be the one raising my right hand, taking the fall if we do this incorrectly," he told me. "Tell me about what happens after the bad drop when we go through a congressional [hearing] and we're getting hard questions?"

It's not that Cukor disputed Whitworth's military genius. "He was really good and smart but we did not see eye to eye," he told me. Whitworth had spent two decades working on carefully evolved controls and practice to make sure the US did the targeting process in the right way. Cukor took on every objection, but with little success. The meeting was so uncomfortable that Brian Ward told me he spent the whole time wanting it to end—and wanting it to continue. Cukor told me it went fine.

But it was something I came to learn Cukor grappled with in himself.

"You can hear it in my voice," he told me one day by phone.

"Hear what?" I asked.

"I want to be laid back," he said. He liked to emphasize to me sometimes that he was from California, as if that alone were enough to recommend him as a relaxed soul. But then the half-confession.

"There is just an intensity," he said.

After the meeting, Cukor told Ward and his colleagues: "I told you he hates me."

Now Whitworth seemed to hate Maven too.

"I will either be famous or live in infamy," Cukor told the group.

WHITWORTH MIGHT BE A LEGEND in the intelligence community, but that's exactly the community Cukor wanted to change. "Automating targeting has historically been a challenge," America's 2018 targeting doctrine states. "Currently, many parts of the targeting process are automated, although no one single tool automates the entire process." The intelligence community had what Cukor considered to be a cumbersome hold over operations, even though their system and processes were neither fast enough nor relevant enough for the tempo of future wars. He wanted Maven to solve the problem.

Cukor kept selling. He wasn't particularly worried that Whitworth considered them reckless. And besides, he didn't have to work with the intelligence side of the house anymore: now he had the 18th Airborne Corps. "If I had just worked with intel we wouldn't be talking right now," he told me. "We would just be another failed intel thing."

Cukor might be a twice-passed-over colonel daring to challenge a rear admiral, but he also figured he was essentially the one in charge. He worked for the civilian side of the Defense Department, under the undersecretary of defense for intelligence. They had ascendancy over the president's uniformed advisers. "We tell them what to do," Cukor told me. Cukor also figured he had the most important weapon of all: budget. "I knew money talks."

Whitworth's concern underscored that Maven and Cukor were reaching a different sort of inflection point. Cukor was running up against the thirty-year clock: he was due to retire and needed to navigate a permanent spot for his AI baby to land. Cukor told me he didn't mind where Maven ended up, so long as it wasn't killed off by his retirement date.

But the question of where Maven would go was complex. Maven had failed to persuade the Army or Air Force or another service to spend their own money on the project, just as Bob Work had predicted at the outset. And whether Cukor said so or not, it was clear to others that Cukor wanted to avoid handing Maven over to the Pentagon's fledgling AI shop, the JAIC, which several at Maven came to see as their sour rival.

By this time, the JAIC had its own problems. After General Shanahan's retirement midway through 2020, the center was heavy with personnel and imploding under low morale and unclear direction. The Defense Department would in 2022 subsume the JAIC into a new entity, the Chief Digital and Artificial Intelligence Office (CDAO), that would manage only halting success, wariness about the arrival of large language models, and continuing low morale before the second Trump administration put the CDAO under new management too, aiming to accelerate the fielding of AI.

A possible alternate home for Maven swung into view. The National Geospatial-Intelligence Agency (NGA). The NGA was both a spy shop and "combat support agency." It was formed in 2003 to bring together institutions that mapped the world and its wars, going back to aeronautical charts hurriedly drawn in the run up to America's entry into World War II. Now, conveniently, it answered both to the Defense Department and the intelligence community. It was responsible for poring over pictures produced by some of America's most secret spy satellites and making sure both US troops and spy agencies knew about them. Its keen eyes looked everywhere. NGA's forbears mapped the moon. It had also, Cukor said, saved him in battles more than once.

NGA had started out low-key. In his first year in office, President Barack Obama appeared to have no idea it existed. On a presidential drop-in to a Five Guys fast-food joint in Washington, DC's Navy Yard neighborhood in 2009, a diner had told the president in front of news cameras that he worked across the street at Building 213, a faceless edifice with bricked-in windows. "We get some papers to you every morning," the diner told Obama with something between a mysterious smirk and a chuckle, cameras still rolling.

The NGA gained greater renown after it produced a model replica of the high-walled compound in Abbottabad where Osama bin Laden hid out in northeast Pakistan until he was killed in a 2011 Navy SEAL raid. And in the mid-2010s it helped highlight China building out military facilities on islands and shoals it was slowly reclaiming from the sea. NGA teams had begun to experiment not only with expensive spy satellites no one would ever talk about, but also commercial satellites of the sort made by Maxar and Starlink.

By 2021 the agency was run by the avuncular Robert Sharp, a vice admiral with a long career as a naval intelligence officer. Sharp was a friend to Maven and close with Cukor. He was "a wonderful man," Cukor told me, deploying his highest-ranking adjective.

Cukor set about diverting his creation so it landed mostly at the NGA. Rachael Martin, an intelligence analyst at the JAIC who told me she had loved working with Cukor, said he did "most of the initial backroom political wheeling and dealing to make sure that transition to NGA happened in the first place."

"He was really the driving force for that," she recalled, describing him as one of the most determined and focused individuals she'd ever run across.

Cukor pulled it off. On September 3, 2021, a memo supported the transition of Maven's AI training and geospatial data lines of effort to NGA. A minority of Maven's lines of effort, such as using AI to sort through text and anything independent of a specific geographical location, would ultimately go to the Pentagon's CDAO, successor to the JAIC. Cukor saw potential for Maven to grow: four-star commanders at European Command, Central Command, and Space Command were all "really happy" with it.

But just a few months later, disaster seemed to strike for Cukor. In January 2022, newly promoted Vice Admiral Trey Whitworth was nominated as NGA's new director to replace Robert Sharp. By mid-year, one of Maven's harshest critics would be in charge of it. "I was quite a skeptic, to be honest," Whitworth told me later. "I was worried that it was expensive and that it was not all it was cracked up to be."

The future of Cukor's baby looked increasingly in doubt. "We were all very concerned," Cukor told me. "Trey was not a friend."

20

KILL CHAIN

"I was Taliban."

I WAS WAITING for a very unusual barman to return. Ensconced in the corner of a Virginia winery, it would be more than an hour before he would show up.

Joe Kernan was a former commander of SEAL Team 6. In his retirement, he was now running the Bleu Frog Vineyards, named for his wife's interest in *cordon bleu* cooking and for the nickname of the SEALs. Among the bottles they sold was a red wine labeled Mad Frog ("a very fine and fit structure").

In the intervening years, Kernan had been a vice admiral in charge of a fleet of ships, and despite his exclusive career focus on operations he later took a civilian position as top US defense intelligence official for three years until November 2020. In that role, three spy agencies and multiple defense intelligence projects, including Project Maven, answered to him and his $54 billion budget.

When he came over to me that thundering spring afternoon in 2025 to find out who I was and why I was there, he quickly let me know he didn't much care for journalists. With the mien of a man who rarely allows himself to let out the cork, he would announce every seven minutes or so that he was wrapping up our conversation, and then carry on for another fifteen minutes. We stammered through an hour and a half this way.

Kernan believed in Maven. He thought it would give more granular intelligence and minimize war collateral. He wished he'd had something like it in his past. The only reason it was initially successful, he told me, was because Drew Cukor was "a one-man wrecking ball" who never gave up—on Congress, the Pentagon, his own team. "Cukor was a pain in the ass; but that's what I wanted," he told me. He didn't agree with all of Cukor's positions (Kernan thought Cukor should share Maven's data but also wondered if he perhaps had a good reason for being so protective of it) but he "had enormous respect" for what Cukor got going, including the funding he elicited from Congress. Before long he was raising parallels between him and Richard Marcinko, the controversial founding commander of SEAL Team 6. Cukor might, Kernan nudged, have benefited from developing a more diplomatic stance over time—an evolution he indicated never came.

It was to Kernan that, in spring 2020, as most of the world was hunkering down during Covid, an anonymous complaint about Project Maven was addressed. It soon started showing up in strange places: on furtive email chains, slipped under the door of Pentagon offices, and even in the mailbox of Kernan's daughter's home.

"I am reporting incidents of gross waste and abuse by a government entity conducting acquisitions and program management with the Department of Defense," went the four-page complaint about Project Maven and its leadership that I reviewed. The letter detailed multiple alleged contracting violations and "abusive and negative relationships" involving Maven personnel with other government personnel and contractors. Three people separately shared this document with me; one version was dated April 12, 2020, and addressed to Kernan; another slightly different undated version was addressed to the Department of Defense Office of Inspector General, a department whose teams of auditors, investors, and evaluators investigate allegations of misconduct.

The impact of the ensuing investigation would ricochet for years. Cukor told me he'd always expected investigations. He saw them less as a tool of the righteous and more as a tax on innovation in the Defense Department. The inspector general had already delivered one classified report in November 2019 evaluating the project. Cukor had ignored the man who took notes as he worked, made calls, and addressed his team. But this latest complaint would trigger a broader investigation into Project Maven that Cukor would struggle to shrug off and would have lasting personal impact. (The Trump administration's Pentagon would describe a "weaponized" investigations process, saying in September 2025 that it planned to ban anonymous complaints.)

The complainant in the letter dated April 12 described themselves as a former detailee to Maven. The person said they were compiling incidents they had witnessed firsthand and from "multiple current and former Maven employees" who wished to contribute to a potential IG complaint, the formal process by which the DoD investigates itself. The letter listed eleven separate bullet points spanning contracting practices to the alleged behavior of personnel, and focused mostly on complaints against Palantir, Colonel Cukor, and members of his team.

Cukor was "grooming" Palantir to take over as a prime contractor on Maven, the letter alleged, saying the subcontractor's work expanded rapidly in scope from a small pilot in 2018 to contracts totaling about $40 million a year, even though multiple Maven government personnel had recommended against shifting Maven "to a proprietary technology like Palantir." Cukor was circumventing ECS Federal, Maven's prime contractor, to negotiate directly with vendors, specifically Palantir, the writer continued. Cukor was informing potential subcontractors about government budget limitations "to drive the overall cost proposal lower," the letter went on.

But, the writer said, Palantir personnel were afforded an "astounding" advantage: one company representative sat in unannounced on Maven discussions of service budgets for Palantir Smart System and contracting plans, the letter alleged.

The complainant also questioned the role played by Lieutenant Colonel Joe Larson, a full-time Palantir employee who had been called up to Marine reservist duty as Cukor's deputy at Maven. It was "impossible to enforce" ethics guidance preventing Larson from participating in any contracting discussion or tasking associated with his commercial employer, "as Palantir grew to touch almost every aspect of Maven."

"I cannot prove that this conflict of interest directly resulted in material gain for Palantir and the Larsons, but the timing of the rapid expansion in Cukor-directed Palantir subcontracts is odd," the letter alleged. (Joe Larson strenuously denied any conflict of interest, telling me that pursuant to federal ethics notifications he provided written notification to the government of his affiliation and potential conflict of interest, and fully complied with resulting directives issued by the DoD Standards of Conduct Office, including formal recusals and the establishment and observance of an ethics firewall. When it became impossible to observe the firewall as Palantir's role on Maven grew, he left Maven and very briefly returned to Palantir. A representative for Ann Marie Rosas did not respond to requests for comment.)

The letter alleged Colin Carroll, a former Maven project manager, had overseen software developers at Johns Hopkins Applied Physics Laboratory at the same time as he was actively applying for a job with the same research center—a job he ultimately accepted.

The complainant also said five people—Colin Carroll, Jane Pinelis, Kelly Rooker, Elston ToChip, and Gene Whipps—had told the author they were leaving Maven "in part or entirely due to mistreatment by Captain Poggemeyer." They blamed the captain's alleged behavior on the command climate set by Cukor.

Carroll told me that looking back now, he doesn't remember Poggemeyer being among even the top five reasons he left Maven. Through a representative, Pinelis stated she was not aware of any April 2020 complaint referencing her, and did not leave due to mistreatment. She said her time on Project Maven was one of the most rewarding chapters of her career, and that she thought Cukor one of the most intelligent, mission-focused and hardworking leaders she'd ever worked with, who set uncompromising standards and drove extraordinary results. Whipps stated that he did not leave Project Maven due to mistreatment by anyone. Rooker and ToChip were unavailable for comment.

In one March 2020 example, during a discussion about technical security Poggemeyer had allegedly yelled he would "make a meme" out of

Nand Mulchandani, chief technology officer at the JAIC, the letter went on, adding the episode further soured "an already-contentious relationship" between Maven and the JAIC.

Mulchandani told me he did not recall any such episode, but said that if it had happened it hadn't affected him, and that he wouldn't raise a complaint about a junior. He acknowledged tensions between the JAIC and Maven, however, saying meetings regularly got heated, mostly as JAIC repeatedly sought access to Maven's data, a request he said Cukor denied over and again.

The writer of the complaint said they were withholding their name for fear that Cukor and his staff would attempt to retaliate against them for submitting this IG complaint. "All I ask is that you take these allegations seriously," the unsigned letter urged Kernan. "The reputation of USDI and the entire Department is at stake if this ever become [sic] public, and would set the Department's AI development work with commercial industry back years."

Plenty of people chafed at the way Maven was being run. Carroll appeared to some colleagues as though he were especially well informed about the state of investigations. In May, he told colleagues there might soon be action on an investigation into the command climate at Maven, referring to the way Cukor ran the team, according to documentation I reviewed. Cukor had done it to himself, Carroll would tell people. Carroll wanted Maven to change its attitude toward working with the JAIC and he wanted the project's data pipeline and the funding for it, according to the documentation. Soon, Carroll was telling people he thought the JAIC would wind up owning Maven at some point during the next year.

Was the man who many saw as a mini-Cukor, the man who told me he loved Cukor and would follow him to hell, a brilliant brain and a self-described "asshole," taking action of his own against his former boss?

When I asked Carroll if he wrote the complaint or was involved in it in any way, he emailed me back: "I am pretty sure that had I written a complaint, I would have not added myself to it." I wasn't sure that was a denial, and later asked him by phone directly if he wrote or was involved in the letter in any way; he said no to both.

He did tell me that by that stage he was hugely frustrated and wanted Cukor gone. "We spent over a year of attempts to work with Maven to access their data that the Department would benefit from. We were stymied every fucking bit of the way," he told me.

Kernan didn't discuss the contents of the complaint with me, but he told me that when he heard the allegations he couldn't ignore them and passed them on for others to determine if there was sufficient justification to launch an investigation. Sure enough, an investigating officer at Kernan's office was assigned to the case. He started reaching out to people on the Maven team, and some agreed to talk. Sometimes he'd ask about Colonel Cukor, sometimes he'd ask about Sy Poggemeyer—and after a few weeks he started asking about Colin Carroll too. Winfield Adkins, the assigned investigating officer, was an Army colonel tasked to go deep inside Maven and by default deep into the mores of Marine culture. Soon, new details spilled out in these sessions, according to multiple people I spoke with, including information about multiple romantic relationships, discomfort over working conditions under Covid (Cukor secured lockdown exemptions for the team), and more. (Adkins declined to comment.)

Several people who worked with Sy Poggemeyer described him to me as passionate, intense, and brilliant, but said he could lose his temper. Multiple people told me Poggemeyer slammed a locker near a co-worker so hard it brought the office to a momentary standstill. Through a representative, Poggemeyer declined comment for this book on this or any other alleged action or characterization about him or his time on Project Maven. He was not the only volatile member of the team: another member treated colleagues so poorly and spoke with such volatility it nearly ignited fistfights in the office, a second person recalled. A third person blamed not the people directly responsible for the bad behavior, but Cukor, himself, for the atmosphere the person thought he created and, in their view, allowed to continue unchecked.

Cukor told me he regularly called some of what he described as his young, passionate, and outstanding team members into his office to give them private counselling, mitigate some of their behavior, and instruct them to apologize.

I reviewed multiple emails and Slack messages in which Poggemeyer dished out expletives and expressed his fury with the pace of work of others on his team. The team grew so used to his outbursts that they maintained a jocular Slack thread—on which Poggemeyer was included—called, "I'm hoping I'll be occasionally cursed by Sy for a year."

"I have to fight for every fucking inch," Poggemeyer wrote in one exasperated Slack message to the team in May. "I'm freaking the fuck out because nobody fucking understands what's going on or how to achieve it."

In another, he asks a team member to call him and then twenty-eight minutes later sends another Slack: "I didn't want to have to go full rage on you guys."

All he wanted, he pleaded, was six better slides and then he would support and defend the team's experiments. "But now I'm going to pull back and coddle you because I'm worried you guys will quit after this."

The team member messaged one minute later: "I'm sorry you had to go full rage. You should work on that. It doesn't help your messaging."

"Slides," came Poggemeyer's response.

Within a team of explosive Marines such outbursts might be one thing, and several told me they got on with Poggemeyer and didn't mind the way he spoke to them. But when outbursts happened with contractors, some felt they went over the line.

It was also clear to several Mavenites that Poggemeyer nursed particular fury for Colin Carroll. In one message I reviewed, which minutes later Poggemeyer himself characterized as a "rant," Poggemeyer unleashed an invective against an unnamed person whom he described as a "mini mob boss" who had stolen Maven plans and threatened to "burn Maven to the ground" if he didn't get to use his very close friend's company ECS to run more projects that Poggemeyer described as "corrupt." Other members of the Maven team told me they understood Poggemeyer clearly intended to describe Carroll, and that his "very close friend" was Volkovska.

Carroll and Volkovska weren't the only Maven couple who would go on to marry each other, uniting a one-time government official with a commercial contractor. Poggemeyer, a government worker who worked extensively at the Palantir offices, would later marry Shannon Clark, Palantir's defense business lead (and he later started working at Palantir too himself). Joe Larson and Ann Marie Rosas were married, had both worked at Palantir, and now both worked on Project Maven. Other romances on the Maven team stopped short of marriage. None of this appeared to bother Cukor. He thought it perfectly normal that close attachments would form among a hardworking group, he told me. He just wanted results. He said neither Poggemeyer nor Carroll had any role in awarding contracts to companies, and that although he listened to anything Poggemeyer or Carroll had to say about those companies, neither was a decision-maker.

There was an extra wrinkle: Volkovska, who grew up in Odessa, had worked closely with Poggemeyer for a time in 2018. They both spoke Russian, and speaking it to each other for a while became something of a secret language. For a time, their close working relationship turned into a brief romance that Volkovska ultimately didn't continue. When Carroll and Volkovska started dating in spring 2019, Carroll told me he was astounded when he learned from her one evening that she'd had a brief romance with Poggemeyer.

Carroll left Maven that July, leaving government in order to continue his relationship with Volkovska without any conflict of interest, and taking up the role at Johns Hopkins University Applied Physics Lab that was included among alleged problems in the anonymous Maven complaint. In late September 2019, Volkovska complained to her bosses after Poggemeyer called her at work and delivered a loud tirade, expletives included, to her on the phone about the allocation of GPU chips on Maven. Poggemeyer allegedly complained that the GPU allocation favored APL even though Volkovska told her employers, ECS, that Colonel Cukor had asked for the allocation. She put the phone on loudspeaker in the office, shaking and finding it hard to breathe, according to contemporaneous documentation. She prided herself on her professionalism, and no one had ever spoken to her like that before. After that, she didn't want him visiting ECS again. An anonymized episode closely resembling this exchange forms part of the Maven complaint.

AT THE END OF JULY 2020, Poggemeyer left Project Maven. Cukor mentioned there would be lots of turnover that summer, and one day Poggemeyer just wasn't there.

Maven director Gregory "Jesus" Christ called an all-hands meeting. With staff sitting socially distanced in a Pentagon amphitheater, he read from a note card, saying he wouldn't tolerate any workplace problems of the sort that made anyone feel uncomfortable.

Then Colonel Cukor stepped forward to address the team. "Okay," he said, and jumped right back into work as if it had been a normal meeting. Nothing, it seemed to one person in the room, would stop him from his pursuit of Project Maven. But the colonel's determination took a toll.

"Any normal human being would have given up," Cukor told me, as he recalled the wave of investigative teams that came to inspect him, whether he was undertaking untoward contracting, the state of the program execution and of his command. "It got pretty ugly."

In the office, he continued as if everything was normal, working the group no less hard even though most of the world was locked down for Covid. The first some Mavenites knew of any investigation at all was when the *New York Times* ran a story detailing the existence of the memo and some of its complaints about Palantir in August, four months after the initial complaint appeared in Joe Kernan's mailbox. Carroll told people in advance of the article's publication that he hoped it would destroy Maven. Cukor needs to go, he continued to say, according to documentation I reviewed.

Cukor told me he hasn't seen the complaint and doesn't mind who filed it. It was the cost of doing business. "Whenever you do something in Defense that is out in front, people start taking shots on you. They were trying to take me down," he said, matter of fact. "It was pretty intense."

Most unsavory of all was the implication in the April 12 complaint that Cukor had allegedly developed too close a relationship with Palantir, amounting to unfair favoritism at best and perhaps implying something far more nefarious at worst. Was any current or future financial benefit accruing to Cukor from his pursuit of Palantir, besides the Marine colonel's own dogged focus on improving national security?

Starting in October 2020, the inspector general undertook a separate investigation of Maven that would last for a year. Personal suspicion appeared to fall on Cukor and his wife Kirsten, who had moved their four children, two cats, and dog into a "small, jankety" 1950s two-story home, in a neighborhood of trim lawns on the outskirts of Alexandria back in 2005—so they could be closer to the Pentagon. Cukor told me NCIS sent agents to inspect where he lived. (The Naval Criminal Investigative Service often works with Defense Criminal Investigative Service on cases involving things like procurement fraud and public corruption, among other potential crimes.) "I didn't let any of that stop me," Cukor said.

But the investigation into him, into their lives, added a level of difficulty to something Kirsten Cukor told me she was already finding impossible. "They came to our house and saw, like, all of our cars had over 100,000 miles on them, and that we lived in 1,400 square feet," she said.

Cukor told me he had always expected complaints, and had been careful about the way he ran the project. He knew the complex world of acquisition rules since his days working in procurement for the Marines. He didn't just serve as chief of Maven: he was also its technical monitor. That meant his job included day-to-day execution of the contracts. He knew, he told me, that each step of the way he had to keep ECS, the prime contractor, informed about anything he might be doing directly with subcontractors such as Palantir.

Cukor told me that he of course took no money from Palantir, has deliberately never taken a job with them (unlike others on Maven, in the Defense Department, and in Congress) and said they played no role in securing him his job at J.P. Morgan either. No one I spoke with on Maven thought he'd taken money from Palantir. Carroll told me he could guarantee Cukor wasn't taking kickbacks just by knowing him. Kernan was among those who told me that when it came to delivering a new AI capability, whether it was through Palantir or Google or another company, the DoD simply needed to pick a horse, jump on it, and force it forward.

On August 24, 2021, it would be Carroll himself who would be fired. In a public post, he said that Kathleen Hicks, the deputy secretary of defense, asked his new boss at the JAIC for his resignation. (Hicks declined to comment.) That evening Carroll and Volkovska sat on the sofa at home in tears: they'd both put so much of their lives into trying to deliver AI and create autonomy for the Defense Department, and were concluding they had no autonomy and only ruptured careers to show for it. Carroll was now down a job, but he went ahead with his original plan for that evening: a marriage proposal to Volkovska. She said yes.

In October 2021, the inspector general's office concluded its review of Maven in a report that was published in January 2022. Cukor told me he was exonerated on every single accusation relating to inappropriate contracting practices. In the dry language of the inspector general's review of Maven contract monitoring and management, he scored a gold star. He had "successfully" monitored and managed four contracts in accordance with all relevant regulations and requirements, the report found. (The contract review period predates the arrival of Palantir as prime contractor from September 2020.)

Maven's approaches in a fast-moving AI environment could provide valuable lessons, and could help establish best practices for other AI and machine learning efforts throughout the Defense Department, according to a senior defense leader whose testimony was included in the report.

BUT CUKOR didn't entirely escape.

"I think every Marine learns at some point that the Marine Corps will never love you back," Kirsten told me.

Most people I spoke with had no idea precisely what happened before Cukor aged out of the military. In multiple interviews, Cukor and everyone else told me he retired because his time was up. If you don't make general within thirty years, the path ahead is simple: you retire.

But one day, Cukor began to answer me more fully. Following complaints about the workplace and the expanding scope of the investigation, there was an upshot after all: in mid-2020 he was issued a formal letter of reprimand. A letter like that counted as a serious response to significant misconduct or disciplinary issues, and went on a service member's permanent record. It was enough to curtail his career.

The Marines had taken him in, and now they were spitting him out.

"I got a big-ass letter that said 'Drew you should have exercised more military supervision,' and I got read the riot act," he said. "You fell short of the expectation of a senior officer."

Cukor said he let it wash over him.

"I don't even care. They don't get it. I had to push hard," he told me. And then one more.

"I was Taliban," he announced. I had no idea what he meant. The last time I'd come across so jarring a reach for the word was at a garden restaurant in the outer reaches of Nairobi. A small bottle of hot sauce on the table was labeled *Taliban*. And then the tagline in small print: "It will blow your head off." What did it mean in Cukor's case? It meant he was an extremist in the Defense Department, he explained patiently.

"You are just way too much," he said people would tell him. "A Taliban guy." He saw it differently. "Every single day you're planning for World War III. And so the aggressiveness that you bring to your work is the same that you do on a battlefield."

But Cukor's most important enemies had turned out to be far closer to home. His effort to reshape the future of war had become bogged down in contracting disputes and workplace complaints. And multiple people struck me as genuinely and deeply upset at him and others at Maven, I persisted.

Cukor acknowledged that his junior officers could be aggressive in the moment, but they could also apologize. When I asked him about one of the details, he couldn't see any reason he should issue reprimands to his young officers for the way they spoke to their seniors. "Why? The results are amazing, and he's delivering and we're moving at record speed," he said, praising an unnamed member of his team. Cukor felt he was being held to an impossible leadership standard just because he was running the Defense Department's premier AI program. (He thought back to a time earlier in his career when he'd seen generals kick trash cans about with impunity.) He thought a good boss would have just told him "Go fix it," and closed the matter. "I could not process why somebody would get angry with a young officer for being directive. Like, it's fine, right? Like, are we good? The people that go into the military are not the folks that you want to be your doctor."

We were meeting in person, and I noticed his knee had started jiggling. Everything else about him was the same. Perspective. Warm voice. Onward.

"When you read it it's like, 'Damn I must be pretty bad I guess I'll throw myself off the bridge right now.'"

An intake of breath.

"I didn't let it affect me. I just kept executing."

A Defense Department letter of reprimand requires formal acknowledgment, and most take the chance to rebut the claims. Cukor said he instead wrote a little note to the effect that he wished he'd done it all better, but he hadn't. "Yep, I did all that," he said, recalling the gist of the letter. "As a senior officer, I should have reprimanded all these young officers that were working so hard."

Cukor thought it was an impossible ask to have young Marines, fresh from deployments overseas, show up in the Pentagon where he said everyone is over fifty-six, "and make them play by what rules, exactly?"

"You come back out of combat and you're supposed to suddenly be something else. Well, maybe give me a forty-five-year-old man who is, you know, a tech leader and it could have been done with graceful, beautiful language. You know what I mean? But I have a twenty-five-year-old captain in here, right? Who doesn't exactly have the command of the English language and interpersonal relationships at that stage in his life. Hell, his frontal lobe just basically got formed."

The strong personalities on his team were also the effective personalities, he said. Colin Carroll was one of them, he volunteered. "Colin was very, very aggressive, amazing, great. He got things done. We really advanced the program," he said. "I love Colin," he told me. "But he was impossible to be around. He's just hard. A very hard person." We were speaking soon after Carroll had been marched out of the Pentagon in April 2025, when he was fired from the building for a second time, sparking a flurry of press coverage. "I guess he's still learning," came Cukor's gentle takedown. "When he's my age, he's going to be lovely: they'll get smooth."

Kirsten said her husband doesn't hang on to bitterness, that he tends to just let things go. She didn't feel quite the same.

"As the spouse, you know, I have a list of people that I think should burn in hell."

She begged me not to ask who. "I have no poker face," she said.

Even as Cukor prepared to leave, complaints kept filing in. One complained he was seeking a senior role at the NGA so he might steer the future of his baby after he'd helped make sure it would land at the agency. "It went on all the way until the day he left," she said.

Cukor left on the last Friday of October 2021, as soon as he heard he'd got the all-clear from the IG report.

"He retired as a colonel," Joseph Kernan told me later by phone. "That tells you something." He meant that whatever Cukor might have done wrong, it wasn't grave enough to dock him a rank due to any alleged misconduct.

The following Monday, Cukor started working at J.P. Morgan. No holiday in between. No moment to reflect. Work.

Around the holidays, a small group of Mavenites rallied Cukor for a farewell dinner. On the evening of December 10, 2021, more than a dozen gathered at Blackwall Hitch, a restaurant fifteen minutes south of the Pentagon, nestled in Alexandria's Old Town Waterfront. The views outside stretched north to the US Capitol building, but the group was looking inward.

Some attendees were surprised that Cukor even attended. He was usually all business. They handed him a gold-framed farewell with a wide blue mat, covered by individual messages in gold pen.

"You are an American hero and innovator," went a message from Jeff Anderson, the Project Harbinger captain. "Your efforts alone set the DoD in motion to start understanding future warfare with AI," said Rich Dorchak, a Maven team member focused on policy and strategy. More epithets rolled in: "A true visionary," wrote one signee. "The DoD's 'Godfather' of AI/ML," wrote another. Another referred to "the legend of Col Cukor."

Brian Ward addressed the "Great" Col Cukor: "You have set conditions for your baby to go out into the world, wreck it, be destroyed, then reborn into a better version of itself."

Lieutenant Colonel Kelly Martin knew him well enough to try a smiling dig: "Sir, although you leave a lot to be desired, I have never worked with anyone who had more dedication, passion and perseverance," she wrote. "Glad to have partnered with the next John Boyd."

She signed off in Latin. "Videmus omnia." We see everything.

Inside the frame were two group pictures of the Project Maven team, one from the early team days labelled 2017, and a stiffer one from the end, in 2021.

In the earlier photograph, Cukor is wearing his cammies, sleeves rolled up in the Marine fashion, his right arm bent slightly at the elbow, ready for action. The creator of the collage put a boundary box around his face, labeling him "Heat Shield." Three people down from him is Gregory Christ, sticking out in his jacket and tie, labeled "Heat." On Cukor's right, on cue, was his right-hand man, Joe Larson, same attire, same stance.

Directly on Cukor's left was a taller silhouette, entirely covered. Only later did I see the unadulterated original version of the picture: it helped me piece together that a smiling Colin Carroll, decked out in a shirt decorated in pink flamingos, had been entirely obscured from the picture, as if erased from membership of the team altogether, in Soviet fashion. "HUD Mask," read the label on top of the silhouette, a reference to the Heads Up Display drone data feeds that the team had to scrub out. In the 2021 group photo, a boundary box over Cukor's face now read "Founder."

Cukor didn't take it that evening—he was straight off to the next thing. It ended up back in the office for a while, and then Brian Ward took it for safekeeping. But Cukor never came back for it. The end of his experiment to change warfare was over. Drew Cukor was gone.

21

UKRAINE FIGHTS BACK

"Get it fixed or get it out of here."

COLONEL JOSEPH O'CALLAGHAN called Joey Temple into his office. Soon, the Pentagon is going to tell the country the president is sending troops to support Ukraine, O'Callaghan told him. *Oh, I think he's doing it now*, O'Callaghan said, looking up at the television. And then the life-changing reality. *He means you.*

It was early February 2022. Temple had just enough time to call home, ask to get the go bags by the front door, and drive them back to base. The plane would leave the next morning: "I went home that night, oh my God, kissed everybody." The last time he had this little notice before deployment was Kosovo, twenty-five years ago. That time, he missed the birth of his first daughter. He had no idea what going to Europe in support of Ukraine would include. "We didn't really know exactly what that meant, to be quite honest."

Temple was soon on a military plane flying high over the Atlantic. By February 5, the new task force headquarters from the 18th Airborne Corps was ensconced in a US military base in a small city west of Frankfurt, Germany. New arrivals went to the joint targeting board, which coordinates finding and prioritizing targets, and started right away. They were preparing to implement their OPLAN. That's the operation plan—the battle plan—that had been drawn up for the US defense of Europe.

The task force headquarters set up shop in an old bowling alley. A thin carpet covered the old wooden lanes. The ceiling had holes in it. Meetings took place in a racquetball court on a higher level. Just a few months before, the room had hosted gymnastics classes for seven-year-olds. Now it was the operations center of a secret war. Three hundred US soldiers, flown out in haste, nestled into their new digs.

The US Army Garrison in Wiesbaden was a thousand miles from what would become the front line of a devastating and costly new European war that targeted civilians and threatened to upturn the global order. America's effort to make AI useful for warfare faced its most significant test yet.

America's algorithmic warriors assembled themselves in the shape of a U-bend in their quasi-subterranean lair beyond a set of double doors. They called it "The Pit." Within three hours, Maven was up and running in the fusion cell that combined information from different parts of the armed forces spanning intelligence, operations, and logistics. Along the right-hand side of a row of makeshift tables was the intelligence team, whose job was to find out everything that was happening on the battlefield, ideally before it happened. Along the other side was the fire team. Their job was weaponeering: getting the right munition on target. This formed the "Battle U." Other units tend to separate themselves out in segmented areas, or desks facing away from each other. The horseshoe shape mingled the two teams. Experts from intelligence agencies would later join the crew, and the division between intelligence and fires would physically start to elide. Task Force Dragon was ready. Drew Cukor's dream was becoming real.

US intelligence, based on intercepted communications and other means, had already indicated that the 175,000-strong Russian force assembled on Ukraine's border was about to attack. Senior European intelligence officials, burned by American "intelligence" about weapons of mass destruction in Iraq, told me US arguments about pending invasion were overheated. None of the infrastructure was in place. No blood banks to help out potential Russian wounded had appeared near the border. It might be an exercise, or a feint, they thought. As the US struggled to convince NATO allies to act, Washington sent its troops closer. Newly arrived US soldiers with no experience fighting in Europe were preparing to stage an evacuation, or perhaps even to go in.

The long-standing conceptual battle plan included targets across the Russian border, things that, if hit, would slow the Russians down and prevent them from completing an incursion into neighboring Ukraine. Garrisons were building up along the border but targeting command centers beyond the firing range of the Ukrainians could mean hitting military headquarters deep inside Russia.

General Michael Kurilla was the fifty-five-year-old 18th Airborne Corps commander who by February 2022 believed AI was "the next revolution in military affairs." A large, loud, and exacting presence in any room—dubbed "Kurilla the Gorilla" in the press—he instructed US forces to track the buildup on the border. The team set about monitoring Russian and Ukrainian movements, using the old prevailing mission command system that accessed little in the way of live data, an army program of record known as the Command Post Computing Environment (CPCE).

About ten days into their efforts the unit was briefing General Kurilla about what they were finding. "What are the Russians doing?" he asked them, in an exchange related to me by people familiar with the matter. An army intelligence chief sketched out Russian maneuvers and what would likely happen in a real-life invasion.

"How up-to-date is this data?" General Kurilla asked him.

Kurilla can come off a little gruff.

"Sir, thirty-six hours."

That was not the answer Kurilla wanted.

"Why?" Clipped and to the point.

"Well, because that's the latest we have, Sir."

"This is not a clear picture," Kurilla said.

Unbeknownst to the group on site, Kurilla had a better view of the ground than the team briefing him. He'd already received a different brief, one that put him eighteen hours ahead of them. And he thought the Russians were about to do something else. From here on out, Kurilla told the intelligence unit, I want you to use Maven.

The team didn't really know what that would look like. A system far from camera-ready was called on to support the largest war in Europe since World War II. They'd used Maven during complex staged exercises, but this had to be real, up-to-date data, and this time it would be trying to save Ukrainian lives. They started assembling intelligence and open-source feeds to create an accurate picture on the ground. There wasn't much in the way of satellite imagery available. It was tinkering and guesswork, right up until February 24.

Early that morning, Putin's face beamed out on TV screens across the world. "I have decided to conduct a special military operation," he said from a desk at the Kremlin, backed by the Russian flag. Satellite internet connections started going out across Ukraine, part of a startling cyberattack intended to disrupt Ukrainian military units that rely on them in the field. The cyberattack, which the US and allies later blamed on Russia, downed tens of thousands of broadband internet connections as far away as Lapland. The invasion was starting.

As the first missiles deluged Ukraine, a screen started beeping and two US watch officers at the Joint Operations Center in Germany consulted their notes. It was after midnight on the East Coast in the US; should they wake up General Milley, the chairman of the Joint Chiefs of Staff? A pre-written list of emergency triggers made it clear they should call. A gruff voice answered: "Okay, I'll tell the Secretary of Defense and the President."

Someone went to rouse the guys in Wiesbaden from their cots—makeshift beds arranged in an open warehouse. They went to work immediately, pulling in every fresh data feed they could find.

General Kurilla came down to the Pit.

"What are the Russians shooting at?" he asked.

It was a simple question, but the Russians were shooting so many munitions that the US couldn't out figure out their game plan. Infrared feeds from overhead persistent radar satellites could help indicate possible rocket launch sites in Maven. The team used it to try to track down the FLET and FLOT—the Forward Line of Enemy Troops and the Forward Line of Own Troops. The Russians didn't seem to be getting much.

"They're hitting a lot of dirt," came the response. It was possible they were trying to take out the train relay station, but the thuds were landing a fair distance away. If you send out enough ordnance, though, eventually they start hitting things.

Those first three days of the war, Kyiv was nearly overrun. The Ukrainians pushed back an assault on Hostomel Airport, as President Volodymyr Zelenskyy morphed from comedian-politician to heavyweight war leader overnight. Tanks were trying to get in. Civilians were trying to get out. Hundreds were dying. Some of the US personnel assisting from Germany and Poland felt utterly impotent, the least helpful they'd ever been in long military careers. How could they advise someone from three thousand miles away? "It's like being in a long-distance relationship," said one person.

The Pentagon wanted to figure out transport routes, and whether Russia was striking escaping civilians. "Somebody needs to go to Poland." Brian Ward put his hand up and arrived in Jasionka—a southeastern Polish town about 100 kilometers from the Ukraine border—a few days after the war started. US Patriot missile defense systems were arriving. Younger recruits stationed there felt the constant fear of an impending nuclear strike, as the sonic boom of fighter jets swooped overhead. "If we do too much, Russia will nuke Ukraine and then they'll nuke Poland," Ward remembers the prevailing theory at the time. "You get out there and that fear actually felt real." (Later in November, Ukraine accidentally fired into Poland, with Zelenskyy quickly blaming the shelling on Moscow. The Pentagon instantly wanted to know if Russia was shooting at NATO. "We thought that was going to be World War III," says Ward.) At the border, people were going both ways. Some were escaping, while others were flooding into Ukraine "because they thought they can go kill some Russians." US forces ran the border control camera feeds straight into Maven Smart System, and ran algorithms over them to estimate how many people were leaving the country. This was more efficient than having an analyst eye the camera feeds of people getting off buses and approaching Polish control points.

At the start of the war, General Mark Milley got his own Maven account for the first time. With it, he could see the throng live from his own office. He could inspect satellite feeds. And he could also track Kurilla's physical movements: he showed up as an icon on the map.

Erasing the layers between ground commanders and top brass could have contrasting impacts. Some operational units had historically embraced disconnectedness by design, according to one defense document I reviewed, seeking to deprive senior levels from having too much access to live data. An AI-enabled common digital platform could mean micromanaging local commanders who were meant to have agency on the ground, risking the deadly delays AI was meant to eliminate. Quicker information flows could also introduce a powerful new tool for transparency and accountability over the actions of ground forces if the US forces turned the surveillance tool on themselves.

But even as Maven Smart System could show a friendly general moving around Wiesbaden, Maven's actual AI wasn't working. The team was experiencing the first algorithmic failures of the conflict.

"AI was struggling to get the detects that we had seen previously," one person familiar with US efforts in Ukraine told me. When the Maven system set up in Europe, the unit had initially been able to run detections with about a 70 percent accuracy score. But in the immediate days after the war started, Maven's accuracy scores plummeted to 30 percent, sometimes down to as low as 10 percent. Daily encounters with system naysayers grew so bad that one official issued something of a "fatwa" against Maven, and attempted to ban its use in the Pit. Kurilla wasn't unsympathetic to the idea of a ban: if Maven didn't work, he didn't want it.

"We can fix this," an adherent of Maven insisted.

Kurilla was direct: "Get it fixed or get it out of here," the person remembers him saying.

Maven's algorithms hadn't been trained on pictures of Ukraine, and they weren't used to detecting military hardware against a backdrop of snow rather than hot desert scrubland. They could occasionally pick out tanks, but Maven had no idea what a tank looked like once the turret had been blown off. Algorithms trained in lab conditions couldn't recognize a building turned to rubble or a crater smoldering in a field. They couldn't register the difference between whether a weapons platform was intact, still a threat, or blown up.

Such battle damage assessments are fundamental to the way the US fights—to determine if a mission has been successful, if they need to try again, and what they should do next with precious ammunition. The enduring promise of AI was to generate a better understanding of the battlefield and help predict the best next course of action. But without battle damage assessments, AI had no chance. "The Ukrainians would hit a Russian tank, we would still detect the tank, but then we weren't detecting the fact that it was inoperable because we hadn't seen that data before." Sometimes a T-72 with its turret blown off would look so different that the AI wouldn't detect it at all.

The Maven team needed to reteach the algorithms quickly if they were going to prove themselves useful. The team started collecting data from the current war to retrain the models. "I don't care about China right now. Let's focus on Russia, and get as much of that data as possible," one person recalled of the new direction. Not everyone agreed. The models were meant to be "global" models; retraining them specifically to help defend Ukraine would likely overfit the models to this new war, making them less useful for scenarios in the Middle East and China.

On March 4, ten days into the war, the Maven team called for additional satellite coverage for Eastern Europe. As one thousand Russian tanks ground to a thirty-five-mile-long halt on the roads to Kyiv, the aim was to get new imagery labeled fast. For that, they needed good pictures, and lots of them.

"Russia was gracious enough to expose their entire operational inventory and wartime configuration never having done so before," one person involved in the intensive operational labeling effort recalls. "And we were very, very, very quick to take them up on the opportunity."

Maven had hundreds of labelers available on both unclassified and classified pictures and quickly developed new data stocks. The team had long improved the way it applied and vetted high-quality labels. "We were trying to pull down as much operationally relevant data as possible, and label those images because we didn't have data like this anywhere else," the person said, referring to it as a strategic resource useful for developing future algorithms. Teams from AWS, Maxar, and Microsoft were getting calls, tasked to retrain algorithms as soon as they found out they weren't performing.

The algorithms had got better at spotting tanks, but there was plenty about war that Silicon Valley's finest AI vendors still didn't understand. A major challenge was identifying TELs—transporter erector launchers, Russia's most common mobile missile launch vehicle. TELs were "a pretty big deal" to take out. These strategic weapons systems looked like straightforward trucks, but formed the bedrock of Putin's war effort, firing S-300 and S-400 missiles at aircraft and targets deep into Ukrainian territory. They typically comprised only a few dozen pixels on a satellite picture taken from far away. But the algorithm makers at Microsoft and elsewhere didn't know that several TELs typically show up together in a pattern, often fanning out in a semicircle. Keeping vendors at arms' length from users, sometimes because they lacked the right nationalities and clearances, meant the computer scientists couldn't properly tune their models to find the TELs' signature pattern.

For some, Maven's early failures in Ukraine showed the limits of computer vision and algorithmic warfare. But others saw it differently. It took little more than two weeks of new imagery and retraining to adapt the tech to Ukraine's new reality. The quality of detections started to come back up. A Palantir representative on site helped make updates to the system. It was war on the go. Maven moved much faster than the usual updates made to hardware and legacy software. "That's not five years, you know?"

On March 11, Kurilla became commander of US forces in the Middle East, and newly promoted Lieutenant General Chris Donahue would take over his command of the 18th Airborne Corps in Wiesbaden. Donahue had become known as the last US soldier to leave Afghanistan just months earlier—a moment captured in a night-vision photograph I was told he didn't like. He was revered by some of his subordinates as "the best military commander of our time." (If Donahue told them all to go to hell, one person under his command told me, he'd gladly work on the plan to get them in.)

It was already clear the Ukrainians had a perpetual blind spot. They had some high-end homegrown tech and were experienced, professional, intelligent, and committed. They were firing 107mm and 122mm rockets from SS26 missile launchers with the help of their own fire control system. But they couldn't see more than six kilometers beyond the Russian lines. That meant they couldn't figure out where the next assault would be coming from.

Members of Congress were complaining it was taking too long for the US to get critical intelligence to Ukraine. "Knowing where a Russian tank was ten hours ago isn't very helpful to a Ukrainian who's, you know, fighting to defend his or her family," snarled Ben Sasse, a Republican on the Senate Intelligence Committee. Others, like Mark Warner, Democratic chairman of the same committee, offered the counter-argument: sharing too much intelligence with the Ukrainians presented a risk given they were penetrated by Russian intelligence services. Adam Smith, a Democratic member of the House Armed Services Committee, worried out loud about escalation. There were limits to how much intelligence the US could share with Ukraine "without going to war with Russia," he said. The US was not sharing real-time targeting "because that steps over the line to making us participating in the war."

But Donahue was a line-stepper. The former Delta operator doubled down on Maven: he wanted to use the system to advise Ukraine on any useful information the Americans could find. The aim was no longer just to track refugee flows and Russian equipment for insight into their numbers and plans. The team in Germany wanted to see if Maven could identify targets to help the Ukrainians take out Russian equipment and personnel. Passing information from sensor to software platform and out to a shooter in Ukraine could expose the US to accusations it was a direct participant in the war, however. Ad hoc policy limitations sometimes doubled as sleights of hand to avoid escalation. "The whole thing was a little bit fly-by-night," one person told me. AI detections were leading "to great tips," said another person familiar with the war effort. The detections still missed plenty, but they were spotting enough to help an analyst start looking in the right place. AI was particularly helpful with satellite data drawn from synthetic aperture radar (images unaffected by weather or smoke from explosions), which it used to identify the TELs more successfully than humans could.

The Maven team passed what they found onto another group, Task Force Champion, a Stuttgart-based unit staffed with recruits from America's European Command and from Army personnel allocated to Europe and Africa, which in turn passed on intelligence to Ukraine. But it "was taking way too long," according to someone familiar with the operations. "We would email it to them and then we don't know what happened."

Maven was facing other problems. The map was taking forever to load. Clicking a link sometimes resulted in speeds reminiscent of dial-up internet. Thousands of civilians and soldiers were being killed. The Ukrainians were getting pulverized. European Command sent out a plea. "We need more cloud in EUCOM!" came the request to the Pentagon.

Joe Larson, who had taken over as chief of Project Maven after Cukor left, met with Jim Caggy, an AWS official at the Pentagon on the morning of Wednesday, March 16, to stress the urgent request for more cloud. But AWS was flummoxed. They had no idea what that meant. Maven ran on AWS cloud, but Europe already had plenty of cloud. The feedback remained: Maven Smart System ran on the network classified as Secret in Germany. And it was unusable. Yet the country's top military intelligence officials assured Congress that US intelligence-sharing with Ukraine was "revolutionary" and better than anything in a generation.

On Friday, a specialist team got in a room at AWS, remotely exchanging network diagrams with the 18th Airborne Corps to try and figure out what might be going wrong with Maven. Solutions architects. Engineers. They had no idea. Nobody at AWS had been to Germany to check in with their harassed, war-torn customer. "Who's coming to Frankfurt with me tomorrow?" Caggy piped up. One other person volunteered, but he couldn't leave until Monday. Family commitments.

Caggy told others in the group that war doesn't heed weekends, and flew out that night, landing in Frankfurt at 8 a.m. on Saturday, March 19. He was in Wiesbaden within a few hours. At the Tony Bass Gym, the team inspected the network. The system was slow even that Saturday morning, when only about a quarter of the team was in. Loading a single satellite image was taking as long as eight seconds. An eternity in what was meant to be a modern war.

Operators thought Maven itself was the problem, but as the team mapped IP addresses, routers, and switches and traced the journey taken by each packet of data, Caggy and the team discovered that each bundle of data, known as a network packet, was traveling on mandated Army defense network systems on a wildly circuitous route. "Data was crisscrossing the Atlantic," one person familiar with the issue told me. Data packets were going back and forth between Europe and the US sometimes as much as four times. "This is insanity," Caggy announced to colleagues. Bad systems engineering wasn't just slowing them down; it risked losing packets along the way. Data was dropping off the face of the earth.

The same thing was happing in Poland. Caggy, the AWS official, flew to Kraków on Saturday evening, and did the troubleshooting the next day at the 82nd Airborne Division at a base near Rzeszów.

"The Army failed the Army," said another person who inspected the problem. The Army's network was meant to be a "global network superhighway." Described as "the cornerstone" of the Army's modernization just the year before, it was supposed to deliver a high-bandwidth network with capacity and speed to spots across the world "as soon as boots hit the ground." Yet simplifying the data flows required senior defense leaders to break through bureaucratic resistance. Eventually, the route was restructured, and delays came down significantly.

The same month, amid fears that Ukraine's data centers were going to get blown up, officials managed to get AWS mobile cloud boxes—known as Snowballs—to Germany and Poland, and then met up with Ukrainian special forces to hand them over. Others got Starlink terminals into Ukraine after Elon Musk answered an appeal for internet by tweet.

New authorities for the targeting team came through in April, the same month the first American howitzers—M777 wheeled artillery that fired GPS-guided munitions—arrived in Ukraine. Howitzers are effective, though hard to maintain. But bizarrely, the Americans had decided not to give the Ukrainians the software to fire the howitzers. The Americans were impressed when their new partners developed their own. "They're brilliant," one said. But still, for forty-five days, they had sat on the sidelines watching Ukrainians get killed. Now they could finally do something.

By May, Task Force Dragon at Wiesbaden was sending over the location of Russian stationary and mobile military equipment and personnel with the help of AI. The team felt a kind of external glare, even from their own colleagues, that they might be doing something somehow illegal, immoral, or unethical by using AI to pinpoint potential targets and then hand them over. But thanks in part to what one Army officer described to me as Donahue's "force of will," Task Force Dragon took a different view. "We're not doing anything outside of doctrine," another person insisted. "All we were doing was optimizing doctrine."

Passing potential targets was subject to strict rules that sometimes chafed. A euphemistic "point of interest" package, passed by the US, provided a terse description of what was there, an elevation, a timestamp, and a location. Supplying a full ten-digit grid coordinate of a potential target to Ukraine might have opened the US to accusations it was targeting Russia itself and had thus entered the war. So the team stopped short. But only by a sliver. Ten digits gave accuracy within one meter. Eight digits reduced accuracy to ten meters, but that was still close enough for the Ukrainians to fire a missile right at it. The larger targeting box was a curious upshot of the US effort to balance escalation and the need for precision. "We like precision, and you guys in the media like precision," said one of the people I spoke to. "But the policy is what we could get through."

The very first time the team passed information about a specific location of interest to the Ukrainians left a lasting impact. "I don't want to say riveting," one of the people involved told me, struggling to explain the release, euphoria, and responsibility of finally being able to do something useful on the battlefield. "It was very—emotional is not the right word—but it gave us a sense of value."

Once they showed their system could work, the team just kept going. Sometimes they passed the information to the British first, who could pass it onto the Ukrainians with more ease. Sometimes they'd message the Ukrainians directly with the help of Signal. Sometimes, for something particularly sensitive, someone might even take the train. It could mean printing out information onto a piece of paper and manually uploading it into another system. In the early days, an operator might hold two phones to each other. Soon though, Maven Smart System and the homemade Ukrainian battle management system, Delta, would be meshed up. Points of interest generated with the help of Maven Smart System could be scrubbed for any classified information—particularly about how data were collected. Then they would land up inside Delta, according to US and Ukrainian army officers familiar with the communication pathways between the two systems. (The US viewed Delta as if it were a leaky unclassified system, although both US and Ukrainian army officials told me Delta maintains a security standard that is often higher than America's own classified "Secret" systems.) In the very early days, points of interest also landed with Kropyva, the Ukrainian Army's Android artillery fire control system whose name translated to Nettle, according to the same people. (Palantir separately started providing big data analysis for intelligence reporting for Ukraine.)

Ukraine had "utter faith" in what the US was passing on to them—enough to lob artillery fire at locations in their own country, where a misfire could mean hurting their own people.

In one early success, some of the team in Wiesbaden had their eyes on a Russian medium artillery unit that had been giving the Ukrainians a lot of problems in the Kherson area. The southern port city of 300,000 people beside the Black Sea, home to Ukraine's shipbuilding industry, had fallen under Russian control in early March. The weather in Ukraine often interferes with satellite imagery, but satellite sensors that rely on synthetic aperture radar can see through bad weather, and with the help of Maven they were able to locate the artillery unit.

The team gave the artillery unit's whereabouts to Ukrainian forces, giving the Ukrainians a chance to decide to do something about it. One person familiar with the team's actions told me the upshot: "And they just happened to make the decision to do something about it."

22

TENS OF THOUSANDS OF TARGETS

"Machines will be fighting machines."

WORKING FROM THEIR U-SHAPED WORKSPACE in the bowling alley, the 18th Airborne Corps was soon giving the Ukrainians thirty to thirty-five targets a day. That was nearly three times as much as they'd managed five years earlier in Iraq. The speed of war had tripled.

General Christopher Donahue heard back from the Ukrainians that the targeting information was really helpful. Let's turn it up a notch, he said. The Maven team started pumping out seventy targets a day. The Ukrainians didn't ask how the Americans were doing it, which their US counterparts appreciated.

At first, an AI satellite detect would show up as a yellow boundary box over the digital map. But Maven Smart System was also bringing in multiple intelligence feeds besides computer vision. Phone intercepts known as signals intelligence, or SIGINT, picked up Russian forces talking over unsecured phones and radios, along with geotags that could locate the data on the same map. Soon, when both an AI detection and a SIGINT detection came through in the same place, the double-sourced location was marked up with a blue box. "When you see it pop up you knew what that color meant—it allowed us to be even quicker," said one person familiar with the system.

Maven would also get pings every time a member of the public posted a missile explosion site via their phones. TikToks would show up with geotags, logging everything from expressions of local sentiment to the advance of Russian forces. It was unverified information that worried traditional intelligence professionals, but the war had started moving too fast for traditional intelligence collection. "At some points during the war, Twitter and social media were giving the most up-to-date info," said another person who worked on the system.

The most consequential development came with the Biden administration's June 2022 decision to send over four high mobility artillery rocket systems (HIMARS), a number that would expand nearly tenfold by 2024. These dump-truck-sized systems allowed the Ukrainians to shoot GPS-guided missiles farther than they could see themselves and then quickly drive off. Turkish Bayraktar drones helped the Ukrainians peer behind enemy lines early in the war, but—once they'd been shot down—the US became Ukraine's eyes.

The first time the US ever sent a coordinate to a HIMARS is etched in the memory of one team member. The US gave the Ukrainians "a couple of targets" at the end of June. "Points of interest," the person immediately corrected themselves, without prompting. "We had a general ballpark idea [of] when we thought they would be going out." They got a confirmation of sorts when aftermath pictures started popping up on social media. The attack sites were far deeper behind Russian lines than usual. Ukraine later said it had destroyed a base in Izyum, killing at least forty soldiers. Another strike the same night reportedly killed a commander of Russia's paratrooper unit.

The points of interest ended up with the person sitting in the HIMARS, according to people familiar with the matter. Sometimes it felt as though the US was all but punching the coordinates into the weapons systems themselves.

Donahue told me the US team built a close relationship with their partners to the point that the Ukrainians trusted implicitly what the US was giving them. I learned from others that the volume of points of interests, passed by the US to the Ukrainians, rose to a high of 267 on a single day in 2022. Two targeting officers on the team in Wiesbaden that year became the two most prolific individual users of Maven Smart System in the entire US military, according to a third person familiar with their metrics. In total, I learned the 18th Airborne Corps sent tens of thousands of targets to the Ukrainians with the help of Maven Smart System.

The team could have done more still.

"How many points of interest you give 'em doesn't really matter," Donahue told me. "It's how many munitions they have. The more points they could handle, the more we would do it, as long as they were being effective and they were using the right munitions against the right targets."

HIMARS were making "a HUUUGE difference on the battlefield," according to a tweet from Oleksii Reznikov, Ukraine's defense minister. A hundred would win the war, he later pleaded. Each morning, teams from the US and Ukraine would discuss what targets would give the most benefit to Ukraine if only they could hit them, known as the High-Payoff Target List.

The broader concept was being proven: Maven was beginning to automate the kill chain. "We were handing it to the partner nation on a silver platter so they could go strike," said another member of the team. And then the customary self-correction: "Or do what they needed to do, I should say."

Those involved knew that the US role went a lot further than the limited intelligence support described by officials in public. "It's like being a puppet master but the puppet has a brain," said one of the people involved. "It's almost like Pinocchio." (One day, humans might be the puppets with a brain and AI might pull the strings.)

The team fended off accusations that they were violating policy. Some argued that passing any information from a classified platform to a partner country broke the rules, even though the US was scrubbing the data before sharing it with Ukraine's own platform. Sometimes the team would try to replicate information fed down from US government satellites whose pictures come with higher resolution that could reveal the detail of a TEL with the equivalent much grainier image shown on lower-resolution commercial feeds, before passing the resulting information package onto the Ukrainians. "Trust us; hit this truck," the Americans would tell their counterparts, without saying they had determined it was a TEL.

High-profile visitors to Wiesbaden would be offered the chance to send a point of interest through the system themselves. "If you're not sure what you're doing is legal, just have the secretary of defense do it once," one of the people familiar with the efforts told me. There were still embarrassing digital disasters. One day in summer, target boxes started showing up on the Maven Smart System screen in strange places. One was around a tree. The operators could see through visual satellite imagery what the algorithm was trying to detect—but it was off. Then they realized so was every single box on the map. They soon found the geolocation tags—which are paramount to directing a hit on target and to avoid hitting the wrong thing—were off by as much as fifty meters. The georegistration was off across the board. "That's a big deal," said one droll Maven worker. "It goes from being useful to not useful very, very quickly."

The team had to take off the algorithm and shut down the entire system "at the absolute worst time." They pulled out the bad algorithm and dropped in a new one. The Maven team at the Pentagon had egg on their face, and the humiliation bit. The teams in Germany and back at the Pentagon were all running on fumes. "We hadn't quite caught our second wind yet as a program. And everything else was going wrong. It was not pretty."

As the war ground on, the tech companies who had made the cut, such as AWS, Clarifai, Maxar, and Microsoft, updated their efforts to field new algorithms in ways that might astound traditional war manufacturers. One company tried 120 different approaches before it came up with a good algorithm to pick out thirty-five different types of military object. The aim was for each algorithm to be good enough to find several different types of object in a single calculation. Usually Project Maven gave vendors ninety days to develop, test, and deploy a new model. As they tried to rush out more relevant algorithms, they gave them—controversially—only forty-five days. Any longer and it wouldn't have been useful enough to make a dent in the war; any shorter and the developers couldn't have pulled off a decent model. From there, Project

Maven selected a handful of the best. Over the course of 2021 and 2022, the Maven team tested and evaluated more than 1,500 algorithms. Two dozen made the cut to be actively used in support of Ukraine.

A 2024 nonpublic research report about Maven that I reviewed said that, while AI computer vision algorithm capabilities were considered modest into 2021, Army users said they experienced great improvement during 2022. During the first ten months spent supporting Ukraine, Maven underwent more than sixty rounds of improvement. That year the Defense Department spent a record $275 million on Maven, including a special supplement of $28 million just for Ukraine.

Donahue had been all-in on Maven since its earliest days of testing in Afghanistan, but he made clear to me that going wrong was part of the process, even in the middle of a war. "You had to experience it. You had to fail, you had to be frustrated, and then you had to solve it, and then you had to see how it got better and then you had to keep doing it," he told me in 2025. His philosophy might make sense to Silicon Valley developers seeking a minimum viable product, but Donahue's stance marked a departure from the zero-defects mentality that often characterizes the US military's actions and approach to officer promotions during peacetime. Donahue was a convert to improving under duress: "It never stops."

By August, there were sixteen HIMARS on the ground in Ukraine, and the Wiesbaden team finally hit its stride. The machine was well-oiled, but the kind of deep fatigue that comes from running on adrenaline—and then running out of it—hadn't yet set in. On August 26, a new target custody board went live, to speed up and organize the passing of points of interest, and the AI started to make more sense within the team's new workflow. The Maven team was identifying key pieces of information to support Ukraine at pace. "Bottom line, what I saw was the most operationally relevant data and AI program I've ever seen," one member of the team told me.

The US knew the Russians had only "three days of fight" on their individual guns, the old-fashioned artillery pieces that were still the critical enablers of ground combat. So Maven enabled the Ukrainians to start hammering deep behind Russian frontlines—striking large ammunition supply points, logistical supply hubs, fuel storage depots, and command and control posts—in a bid to cut off the guns and prevent them from getting resupplied. "It's a systematic approach," Colonel O'Callaghan told me later, describing the targets Ukraine was taking out (without referring to any US action himself). "They're getting three layers deep."

Ukraine would take back more territory during this moment than it did in the entire war besides during the initial repulse of Russian forces. The information-sharing effort would slow down after the 18th was called back to Fort Bragg at the end of 2022, though, and the process changed. "I don't think for the better," scolded one person briefed on the matter. Subsequent US groups were more cautious or sometimes less committed.

"Each rotation was worse and worse and worse," a Ukrainian army officer told me in 2025 of the declining quality of information-sharing from the Americans. "I'm sorry," he interrupted himself, rousing himself to be polite toward the end of another fifteen-hour day of war, more than three years in. "I'm just sharing my pain."

In summer 2023, though, he said the replacement US team briefly managed to go even further than Task Force Dragon had the year before. While the team in 2022 focused mostly on fixed targets behind enemy lines, for a few months in 2023, the US briefly sent points of interest for "dynamic targets" that move around on wheels—which included prized Russian military equipment such as TELs, counterbattery artillery radars known as Zooparks, and radars for S-300 and S-400 air defense systems. The impact in those months "was huge," he told me.

"A battlefield is a very complicated thing," the Ukrainian officer went on. Hitting one big crucial thing, such as a radar, can have a cascading impact on an enemy unit as it allows Ukraine to fly over its own reconnaissance drones, he said. But the effort relies on speed: "As soon as something's detected there's a very short period of time to pass the targeting to the effectors to be able to strike on it and destroy it before it moves."

In one quietly celebrated case that I learned about, the US team located a TEL with the help of Maven's computer vision and passed it to the Ukrainians, who destroyed it eighteen minutes after the Americans had first spotted it. Signals intelligence was key to unlocking the whereabouts of moving targets. But the information sharing on dynamic targets "was stopped," the Ukrainian army officer said slowly, speaking to me with anguish. US military bureaucrats decided the US military operators didn't have the necessary permissions to do it, the Ukrainian officer told me. (A US Army officer familiar with the matter corroborated this.) Even so, Ukraine was still punching above its weight. By early 2024, Ukraine's much smaller military had been able to destroy more than 2,600 Russian tanks and nearly 5,000 armored vehicles, according to estimates from the British government.

Bandwidth problems persisted even after the team ironed out network routing crisscrossing the Atlantic. In late April 2022, defense officials discussed a plan to urgently "decrease latency," according to meeting minutes I saw. America's new algorithmic warriors would bypass the Army's sclerotic network altogether and, instead, use Amazon's secret cloud as the backbone for the war. That would mean installing two new entirely separate 10-gigabit internet circuits to carry more data, faster, to link Landstuhl, Germany, directly to Amazon's classified data center in Columbus, Ohio. NETCOM, the Network Enterprise Technology Command that provides IT for the Army, was having none of it. NETCOM had its own system and told Maven and others it would "non-concur" with the AWS plan. To one person who worked on the fix, it looked as though NETCOM were holding back support for an invaded partner country. "They were a bunch of dinosaurs," the person scoffed. (Asked for comment, NETCOM said it is a proven leader in network and IT systems innovation and selects the best-value solution that meets operational bandwidth requirements.) Maven pressed ahead without NETCOM's support. Maven's new chief, Joe Larson, told others he wanted to "move quickly," according to the minutes, and that Project Maven would pay for the entire new network for the first three years, rather than bill the Army. It wasn't simple: installing new circuits would need new switches, encryption devices, and security checks. Project Maven was creating an entirely new combat theater cloud on the hoof to support the hefty demands of AI warfare in the middle of a war.

Technological advances often create unexpected side effects on the battlefield. By some arguments, the grueling trench warfare of WWI can be traced to the invention of the tin can. Along with the arrival of railroads, the stable supply of canned food facilitated the building of defensive frontlines, in which millions of men died. More than one hundred years later, on the same continent, a different manufactured device emerged as mission-critical for the pace of modern warfare: encryptors.

Encryptors are small portable hardware devices accredited by the National Security Agency that protect highly classified information. Those in place could support up to 1 gigabit per second of data. The new plan needed encryptors that could handle ten times as much throughput to accommodate the AI data deluge and new circuits. Without them, inadvertent bottlenecks stymied the speed of tech-dependent war. Bandwidth sometimes narrowed to 100MB per second—worse than a common home router. "They were choking the network," one person said, referring to the sudden glut of AI data.

Maven's ability to support Ukraine's war effort came down to a hunt for larger-capacity encryptors. Viasat agreed to temporarily lend two $155,000 devices, their KG-142 Ethernet Data Encryptors. That helped open the war pipes so 10 gigabits per second of highly classified data could surge through dedicated Maven circuits, all with "near-zero" delay. Installing the encryptors in fall 2022 would take a Saturday phone call to a senior national intelligence official at home, just to get permission to move the mission-critical encryptor from one base to another in Germany. Six months after the problem was identified, and eight months after the network started struggling under the pace and strain of war, it was done.

In the meantime, the team tried to reduce its network usage. Instead of pulling down an entire picture from a space vehicle when it flew over the relevant spot, I was told they'd pull down just a "chip" of the image, moving far fewer pixels to relieve the burden on the system.

By September, data volumes flowing through Maven improved, with delays receding to as little as ninety milliseconds. HIMARS rocket systems alone had hit more than four hundred Russian targets. "They've had devastating effect," Milley told reporters during a trip to Germany's Ramstein Air Base.

It came with a cost: the cloud bill was soaring. At one point it touched $1 million per month, two people familiar with the effort told me. O'Callaghan told me the team eventually streamlined models to reduce the compute needed—and the bill—to less than a twentieth of that.

TOWARD THE END OF 2022, as Donahue prepared to hand over the reins and return to the US, he briefed more than a dozen four-star Army generals in a lunch meeting in the E-ring at the Pentagon. ("I didn't know that many four-stars could fit in a room," a person briefed on the meeting later remarked to a colleague.) By video, he told them about the secret role Maven had played in the first ten months of the war.

The meeting was convened by General James McConville, the four-star Army chief of staff, who would worry aloud the next year that AI had no morals and that AI targeting could go "horribly wrong."

But Donahue argued that US support to Kyiv had already helped the Pentagon get much better at using AI tools—and more confident in its forecasts of how AI might be used in a conflict with China.

"We can do this for Taiwan," went Donahue's argument, according to others in the room.

The next summer, the 18th Airborne Corps and Indo-Pacific Command—the geographic military region that covers China and Taiwan—would practice AI targeting during exercises using Maven. "We were pretty impressed," one person at Indo-Pacific Command told me about a briefing General Donahue gave to Admiral Samuel Paparo, now commander of Indo-Pacific Command. If Ukraine could directly plug into the firing platforms via an assortment of solutions to safeguard classified information, then Taiwan could do the same. Donahue and his team were also tasked to brief Taiwan directly on their AI targeting efforts, which they did.

Donahue also detailed his use of Signal to correspond with Ukrainian counterparts and keep the war support going. That triggered concern from General Paul Nakasone, the four-star Army general who, at the time, led America's top eavesdropping agency and its military counterpart

US Cyber Command. There are more secure ways to pass information, Nakasone intervened, according to multiple people present. "He definitely ain't happy about it," recalled one of the people familiar with the discussion. (Nakasone declined to comment.)

The Signal app is end-to-end encrypted, meaning anything passed on it is indecipherable to anyone who steals the data during transit. But phones themselves can be accessed. The proliferation of zero-click exploits—spyware that can be automatically downloaded onto a phone even without clicking on a link—increases the many ways a phone could potentially be infiltrated. (There was a further irony implicit in the revelation that the military was passing sensitive information using Signal. Meredith Whittaker, the AI ethicist who rallied Google workers against Maven, became president of Signal in early September 2022. The app was now being deployed by US adherents of algorithmic warfare to help run a war.)

Donahue defended his use of Signal. War—death—doesn't wait for the right tech. You fight with what you have, went Donahue's gist. Waiting would have meant more Ukrainians getting killed. And the Ukrainians could launch a missile faster than anything the Russians could do on the battlefield to preempt an attack, even supposing the Russians could infiltrate some of the messages.

It wasn't the only time Army leaders worried about security. "What if China gets ahold of your algorithm?" one general asked someone on the Maven team around the same time.

The Maven official answered resolutely, looking his interlocutor directly in the eye. "Sir, if China had an inbox that I could send algorithms to, I would send them a thousand."

His facetious point was that China would be so distracted trying to figure out how US algorithms work, it would drown in confusion. No algorithm followed the same approach as another, he explained. "Even if they wanted to figure it out, they couldn't." He argued that what made Maven successful was not the algorithms but three other key elements: high-quality underlying data, the system in which an algorithm was situated, and how seamless a workflow teams could develop.

Maven was soon capable of detecting a broad range of hardware on the Russian front in Ukraine. By 2023, AI was rooting out combat vehicles, command and control aircraft, fighters, mobile radar-communication vehicles, tanks, towed howitzers, and TELs, according to documents I reviewed. AI could search out Russian aircraft, including the massive Tu-95 bomber known as "Bear," the Russian equivalent of America's B-52; the supersonic, nuclear-capable Tu-160 bomber; and the supersonic, long-range Tu-22M maritime strike bomber.

According to test score documents I reviewed, some models could identify an image within a second of "seeing" it: others took longer, more like three or four seconds. (The most accurate ones took six seconds.) Microsoft and Clarifai were producing the best-performing algorithms, with one AWS model making it into sixth place. The models were still producing incorrect detections, though: ten every square kilometer. (Worse ones could surpass fifty false positives per kilometer; good ones could pull off little more than three.) They could still miss things regularly, as much as a third of what was staring the algorithm in the face. Maven was showing more promise in other areas: algorithms could now recognize complex objects with little in the way of training data.

I met with Joey Temple once he was back at Fort Bragg in summer 2023. He told me that with Maven's assistance he could now sign off on as many as eighty targets in an hour of work, rather than thirty without it. He described the process of concurring with the algorithm's conclusions in a rapid staccato: "Accept. Accept. Accept." Trusting the computer entirely would be faster still, but Temple says that would introduce errors. "I don't ever wanna get caught short," he says. "We get caught short, we're screwed."

The military's own trust, or lack of it, in such systems would be a crucial part of determining the extent of AI adoption.

The war had kept changing each month, as both sides evolved weapons, tactics, and manpower. By 2024, I was told the US was passing the Ukrainians only a dozen or so points of interest a day. That was partly because the Ukrainians had improved their own approaches by then, sending drones out ahead of a strike, and so didn't need to rely so much on the Americans for targeting. But when Donahue returned to Europe in December 2024, as the four-star commanding general of the US Army in Europe, he told me the Ukrainians had told him they no longer trusted US grids.

"Now it's fixed," Donahue told me in July 2025. He said that Major General John Rafferty, commander of the 56th Artillery Command and then chief of staff at EUCOM, had done a "remarkable" job in reestablishing that trust. Donahue told me error rates and precision were also improving all the time.

The Russians also adapted their battle tactics to try to dodge the all-seeing eye of Maven, but Donahue told me it brought only limited success. Reports trickled out that Russia was painting fake outlines of planes parked in airports and putting tires on top of weapons platforms to try to confuse computer vision models. But the US could pick up on it all very quickly, Donahue told me: "Not a substantial concern."

It was clear that General Donahue had come to see Ukraine as an inflection point for the future way the US would fight wars. He believed strongly that data would be decisive for victory on the battlefield. Nobody could win a war without masses of data. Algorithms were also going to be "extremely important," he told me. AI was not yet so advanced that he thought the US could lose wars against an enemy armed with more of it, but he still thought the US needed to develop AI faster.

"Your adversaries are going to choose for you that you have to do this," he had told me when I met him after his return from Ukraine the first time. "We need to watch what our adversaries are doing and making sure that we truly have deterrence against our adversaries. Nobody has a corner on the market."

He thought he saw the future pretty clearly. "We're already on the way of . . . machines will be fighting machines, robots will be fighting robots. It's already happening."

23

WE'VE DRUNK THE KOOL-AID

"Shortening kill chains is universally good."

IN EARLY SEPTEMBER 2024, during the cocktail hour at a private retreat for tech investors and defense leaders, Vice Admiral Trey Whitworth found his way to Drew Cukor, now three years retired from the military. Maven's original leader and his skeptical successor stood face-to-face.

When Whitworth took charge of the National Geospatial-Intelligence Agency in June 2022, he worried that Maven was overpriced, overhyped, and incautious about the targeting principles he most cared about. "Whitworth could have shut this program down in a heartbeat, because he was not a believer in the beginning," one person told me.

But now, more than two years into leading the NGA and more than two years into the war against Ukraine, rather than abandoning Maven, Whitworth praised the program. "Drew, this is important work," he assured Cukor at the event. Maven was adaptable, it could integrate with any system and become new with each software update. It could do what people wanted.

Cukor described the admiral as methodical, saying that Whitworth had just reasoned his way to endorsing Maven. Cukor thought Whitworth had come to understand why the US needed to bring AI into the targeting cycle (the portion of Maven's $250 million annual budget that went to NGA may also have helped, he thought). "It speaks to his character, honestly," said Cukor. "It wasn't an apology as much as a formal recognition. We didn't hug but it was an important conversation."

Under Whitworth, Maven would have its coming out party, emerging from years of secrecy tightly maintained under Cukor. Six months earlier, deciding to shoot comprised the shortest element of the targeting cycle. Now every other part of the cycle was so close to being automated, and so compressed in time, that deciding to shoot was the lengthiest part. Internal documentation referred to Maven ATR: automatic target recognition. In public, Whitworth started describing Maven as his agency's "marquee targeting program of record."

A few days after he spoke with Cukor, Whitworth stepped onto a stage for a live-streamed Palantir customer event. He could hardly have cut a stronger contrast with the Palo Alto crowd: his service dress blues came with gold buttons, gold threads round his sleeves, and bright ribbons. In his high-shine formal black shoes, he stood in front of a cabinet displaying colorful Nike sneakers. His talk on Maven Smart System followed directly after two other Palantir customers, one who leased railcars and another who supplied automotive seating. War was now just another business process, sandwiched between sales and healthcare.

Amit Kukreja, a prominent Palantir commentator, investor, and fan of the company's "merch," was narrating the event live off to the side. He described it as a "new and special" moment for Palantir's retail investors to learn about the company's government work. Even the Palantir CEO Alex Karp appeared taken aback. "I didn't even know we were allowed to talk about this stuff," he said, after laying claim to the "most elite and interesting" government clients in the world. Palantir had already won an Army contract with a $480 million ceiling for Maven Smart System that spring, and a couple of weeks later would win another to supply the system to all military services in September for up to $100 million. In spring 2025, the Pentagon's contract ceiling for Maven Smart System was raised to $1.3 billion, due to run until 2029. And NATO said it would become a customer for Maven Smart System too. The UK would reportedly sign a

GBP 750 million deal for Palantir's military AI tools during a high-profile state visit from Donald Trump in September 2025.

Up on the Palantir stage, Whitworth talked through AI targeting as a screen beside him played a demonstration. An icon flashed up alerting the audience to "Possible Enemy Activity." A cursor click revealed a group of tanks over a "notional" demonstration map of Kherson in Ukraine. The tanks were four clicks from evisceration. Palantir's Target Workbench popped up. Two more clicks established the tank group's height, latitude, and longitude; and then paired the target with an "effector" (in this case an F-22A fighter jet eighty-two miles away). One more click and a green tick soon flashed up: "Target destroyed."

Nearly a year later, on a hot day in the high summer of 2025, I stepped into NGA's headquarters at the Fort Belvoir Army Base in northern Virginia. It was my second visit to the spy agency HQ, and I wanted to find out why Whitworth had changed his mind, how much Maven had spread, and how Maven's new backers saw the risks and rewards of mainstreaming AI into military workflows.

By then, Whitworth had become so ardent a fan of AI that his agency was pumping out machine-produced intelligence reports for US decision-makers that "no human hands" had touched. And the NGA had launched a $708 million contract for data labeling in support of Maven's computer vision models, the largest such appeal in US history, that would ultimately go not to Scale AI but to Enabled Intelligence.

My visit required the rigmarole of any meeting at a spy agency. Courteous background checks and vetting; no phones, laptop, or smartwatch allowed; and one step more curious: writing down not only the make and model, but also the serial number inscribed on my tape recorder, which I resolved never to use again for any interview after the visit.

The building was a temple to geospatial intelligence, or GEOINT, the pursuit of insightful analysis tied to locations on a map. A mesh of reflective glass encased by nearly 2,000 concrete triangles covered the blast-resistant facade, as if each one were attempting to triangulate a different location. More than 8,500 personnel worked at headquarters, but I was there to meet four particular NGA officials. Each, in their own way, was deeply involved in the development, standards, and spread of Maven. It was, I was told, unprecedented for them to gather all in one room to brief a journalist on Maven, and I was eager to hear what was at stake for them.

"This is our reputation on the line," Whitworth told me in the interview. After he saw how easy it was to integrate the system into combat scenarios, it didn't take long for him to change his mind: "I started really believing in it." Far from being sheepish about ushering in a new age of AI warfare, its midwives wanted their names stamped over it. Some had become quite "ornery" in pursuit of credit, one NGA official said. I wondered if NGA wanted its fair share. "There's no one person who can claim credit for this thing. It's too big."

The NGA officials walked me through Maven's developments since the agency took most of it over two years before. Five of eight Maven initiatives, including analyzing drone feeds and satellite imagery, ended up with the NGA. Whitworth wanted to expand the scope and capabilities of his agency in line with expansion of ubiquitous global sensors. AI relied on data, and that required global surveillance to deliver it. While NSA could listen in to the world, NGA could watch it. Whitworth made clear he wanted to do that in minute, constant detail—surveilling the entire globe, at all times. NGA previously gave me a demonstration showing how AI could flag military construction in China—such as the arrival of a new rail depot at a missile base. NGA kept track of all movement at 49,000 airfields around the world. Whitworth even wanted to put GPS, or a similar navigation system, on the moon. And if GPS got jammed or hacked, he wanted other ways to map space too: NGA was fashioning digital maps drawing on magnetics, gravity, remote sensing, celestial navigation, and elevation. "From seabed to space," went the new mantra he unveiled in 2023.

Nearly two years into its work in support of Ukraine, Maven became a "program of record" at the beginning of November 2023. That was Pentagon-speak for a fully funded line of effort with backing from Congress. It came with the expectation of a consistent budget for the coming years. The lines were still blurred: the Pentagon's successor to the JAIC, the Chief Digital and Artificial Intelligence Office, paid the licenses for Maven Smart System, and managed the text-based parts of Maven such as "reading" captured enemy material, while NGA produced the computer vision models that showed up on Maven Smart System screens.

Throughout 2024, Whitworth drummed up new users for the platform. He rang combatant commanders in every region to tell them what NGA was adding, shopping Maven's latest features. Addressing criticism I'd come across, he insisted that Maven was useful not only in Europe, but also in the Indo-Pacific, and for moving targets as well as static targets.

Maven Smart System had particularly taken off in the Middle East. General Erik Kurilla had started using the platform "extensively" in support of US weapon strikes, once he took over US Central Command in April 2022. He hired former Google AI expert Andrew Moore, and spent much of 2023 practicing how to get through a thousand targets a day, cooperating with the UK and others in a series of experimental ninety-day sprints. In early 2024, I learned that the command had made "a pretty seamless shift" from experimenting with the platform in exercises to doing all this in combat. Here would be the first real US test of AI at war on a large scale.

"October 7th everything changed," Schuyler Moore, the chief technical officer at CENTCOM, told me, referring to the deadly 2023 Hamas attack on Israel that international rights groups said constituted war crimes and crimes against humanity. "We immediately shifted into high gear and a much higher operational tempo than we had previously," she said.

Brigadier General John Cogbill, CENTCOM's deputy director of operations—who served under Kurilla, including at the 75th Ranger Regiment—put it this way: "It's just been off to the races ever since."

In February 2024, the command used Maven Smart System to locate rocket launchers in Yemen and unmanned surface vessels in the Red Sea. CENTCOM's Schuyler Moore told me Maven's AI helped narrow down more than eighty-five targets that US bombers and fighter aircraft subsequently struck in Iraq and Syria, in reprisal for the death of three US service members in Jordan the month before. It was public confirmation that the US military was using AI to identify enemy systems for its own weapons to strike. "We were using these tools in a way that we'd never used them," Cogbill told a podcast the following month, saying the command became "hyper-focused" on Israel.

By 2024, the command had 179 different live data feeds from land, sea, air, space, and cyber pouring into Maven Smart System. CENTCOM is using it most, Whitworth told me. That region alone had 13,000 accounts, with 2,500 people counting as regular users who log in "at least a few times a week," said Admiral Hulin, a former commander of SEAL Team 6, who was now deputy director of operations at CENTCOM. Maven could also discern the nearest available weapons, the most suitable ones for the task, flying time, weapons loading details, and the whereabouts of personnel and partners.

Operators would click through Maven's Target Workbench, approve or disapprove targets, sequence them according to priorities, and send a message directly to weapons systems. "Shortening kill chains is universally good," Cogbill said on an April 2024 podcast.

Kurilla, himself, could keep track of all this from a plane: he watched real-time Maven feeds showing the latest in Iraq, the location of ships mapped over the Red Sea, data sent over Link 16, a jam-resistant digital channel on a tactical military radio system used by NATO and US allies to enable military aircraft, ships, and ground forces to communicate in real time, from aircraft flying off the carrier and live video feeds from MQ-9 drones over the coast of Yemen. Soon he'd be picking Maven Smart System as his topic of choice for his lecture at Capstone, the military's in-house "charm school" at the National Defense University, to guide newly promoted generals.

What they'd been able to do, Kurilla would later confide to others, was "eye-watering."

But there were major concerns as well. Cogbill gave Maven Smart System a "C" grade during remarks at an August 2024 conference. AI was getting better all the time, but it was tricky, he said, because of hallucinations, "all kinds of AI ethics," and the chance that the data inputs or the algorithm could lead military operators "to the wrong conclusion."

Even so, Kurilla told Congress in June 2025 that feedback was improving the command's software suite. I learned that the US used Maven during the twelve-day Iran-Israel war that occurred just five days later. Maven wasn't favored as an intelligence platform, but was regularly being used for operations. A single targeting cell could now go from sensing a target to shooting it within minutes, down from hours before. Maven could also detect and track ballistic missile launches headed toward Israel. Mavenites stayed up overnight working on the digital infrastructure for Maven when Iran sent two hundred missiles towards Israel in October 2024.

NGA officials told me that Maven was accelerating operations and "enabling lethality" at combat headquarters around the world. At least thirty-two different companies were working on Maven, and close to 25,000 US personnel were using it. NGA officials beamed that usage had more than doubled since January. The agency had accumulated one billion AI detections in its computer vision data store, one official later informed me. Maven was detecting objects nearly five times as fast now too.

When I asked how many of the users were actually using the AI computer vision detections, as opposed to relying on the fancy Maven Smart System data fusion platform and display for the broader overview it gave of battle, NGA officials were quick to defend Maven's AI. The models were improving during 2025 "by leaps and bounds," Whitworth told me. "All the serious people are using the AI, because they go straight to the hardest thing, which is target development."

Many of those serious people were commanders, the ones who would run deadly operations rather than oversee the collection of intelligence that powered them. "Every commander is using the AI, bottom line," insisted Joe O'Callaghan, who—after he retired from the 18th Airborne Corps—joined NGA as director of AI mission. Commanders were using AI to characterize the environment, just to understand what was going on, and for targeting, he said. "You'll hear some people push back, but when you ask them what they're looking at on screen, it's the AI."

I learned separately, from nonpublic documents I reviewed, that in the two years to 2025, Maven found its way into 141 exercises and experiments. It was in more than 130 sites, Australia, Bahrain, Cambodia, Uzbekistan, Vietnam, and Yemen among them. It was at two sites in the UK. It was analyzing data collected over China and North Korea, Kazakhstan and Myanmar, Pakistan and Russia. For a time during 2025, Maven had people based permanently in Japan, Germany, and Qatar, and would rotate people through Jordan, Djibouti, the Republic of Korea, and Poland. A team assisted with all six military services, supporting more than twenty units on four continents and with seven combatant commands. These dealt with Europe, Africa, the Indo-Pacific, the Middle East, Special Operations Command, and Space Command. Cukor's dream was coming true, and NGA was now urging on the users. The code could be changed in a day. "That is the secret sauce of bringing in the operational community in a very, very big way," Whitworth told me. "Ride that wave," he encouraged.

Maven was not operational at the two domestic commands responsible for tracking the Chinese spy balloon that in early 2023 stoked alarm when it flew into US airspace (and which the US eventually shot down after it passed across the entire country over the eastern seaboard). But in 2024, NORTHCOM, the command for homeland defense, and NORAD, the military command run jointly with Canada to watch for incoming air and sea threats, both adopted Maven.

James Rizzo, chief data officer for both commands, told me that they started using Maven in July 2024. By the time Whitworth and Cukor were eyeballing each other over canapés, the adoption was complete. Maven Smart System helped display and track movements by Russian, Chinese military, and other aircraft coming near the US. By 2025, the two commands had 2,000 daily users, Rizzo told me.

But operating Maven required a certain knack; you had to learn to filter out the noise. The day Rizzo spoke to me, Maven had made 850 million computer vision detections when he logged in that morning. "I thought wow, that's a big number. I don't know what to do with that," Rizzo told me. "I'm trying to sort hornets from honeybees."

As national security policy shifted under the second Trump administration in 2025, so did the uses of Maven. It became a tool for detecting border crossings and was potentially relevant for the "armed conflict" that Trump told Congress he was unleashing on alleged narco-terrorists in the Caribbean. Maven could detect and flag just about anything—drug runners, hobbyist balloons, military vessels, people. It was detecting people trying to cross the southern border, Rizzo told me. In September 2025, an NGA official told me Maven was helping with Customs and Border Protection, and with the US Coast Guard. That was the same month that US Southern Command (SOUTHCOM) started carrying out lethal strikes against Venezuelan vessels, alleging that the targets they obliterated were smuggling drugs—and prompting criticism that the US was executing people without trial and risking war crimes. ("I don't give a shit what you call it," Vice President JD Vance retorted in a social media post.) By early December, the US would strike twenty-two boats, killing eighty-seven people, including two shipwrecked survivors.

It wasn't acknowledged whether or not Maven was being used in these strikes, but I learned SOUTHCOM had been using Maven for years. The command previously used Maven to identify a drug-running vessel that was subsequently boarded, according to Cameron Stanley, science and technology adviser at SOUTHCOM during 2020 who later took over from Joe Larson as chief of Maven in mid-2022. Since then the practice had only increased. An NGA official told me that Maven assisted "in the detection, classification, and ultimate interdiction of more than three dozen vessels suspected of engaging in illicit or clandestine activity." The official declined to give a timeline or describe the vessels or interdictions further. The official did say, though, that NGA could identify the vessels three times as fast as humans.

Maven was now coming closer to home as a prodigious border control and drug policing tool. (NGA can only use Maven domestically during a federally declared emergency, at the request of the lead federal agency. NGA has previously undertaken such a role in support of the National Guard in stemming wildfires, but not through NGA Maven, according to the NGA official.) Nevertheless, Maven was moving closer to home and for some, that spoke to a quintessential example of imperial boomerang, the theory developed by 1950s intellectual Aimé Césaire that a colonial power using oppressive techniques overseas would eventually bring them home. The US could end up with a well-equipped military "optimised for domestic warfare," argued one commentator worried by the deployment of hi-tech troops to US streets.

Rizzo was more focused on how incredible it was to have a machine spit out summary reports at speed to brief commanders. He marveled at what a future end state could be. Maven had quickly become all-encompassing. It was now how NORAD "talked" to other commands all over the globe, knitting up the same view of the world with INDOPACOM and EUCOM. That was extremely useful for communication and speed, but there was also a chance that relying so much on Maven for a common view of America—and the world—could go very wrong, he warned. "We'll never get anything done if we wait for perfection, so we take a little bit of risk," he told me.

"But we're not building the WOPR," he laughed, referring to the fictional supercomputer in *WarGames*, the 1983 movie in which Matthew Broderick plays a computer hacker who nearly triggers World War III when he accesses a newly automated NORAD system that can launch nuclear strikes against the Soviet Union.

Many officials said some version of "we're not building WOPR," and routinely dampen worries about integrating AI into warfare. In *Unit X*, two former defense officials who helped stand up and run the Pentagon's Defense Innovation Unit (DIU), argued that Project Maven had nothing to do with weapons. And Palantir officials scoffed when I asked if Maven Smart System was a weapons system. Even though a single click could send coordinates through a tactical data link to a specific weapons platform so that it could fire at the target, they pushed back: the click on Maven Smart Screen didn't release the munition, they told me. (One defense expert I consulted about this argument described this as a distinction without a difference.)

And when I asked General Chris Donahue, who returned to Europe at the end of 2024 with his fourth star, if Maven was a weapons system, he was clear cut: "Oh, absolutely," he told me in an interview. And he expected more: "Ultimately all this stuff will become automated."

When I visited NGA headquarters in June 2025, one of the four officials I met asserted that there were still humans on the loop. But General Donahue's point about the direction of travel was well-founded: the Defense Department's policy on autonomy has nothing to say about having a human either in or on the loop. It states only that "appropriate levels of human judgment over the use of force" are required.

Emelia Probasco, a former Navy lieutenant, concurred with General Donahue's view: "I think Maven is a weapons system," she told me. Probasco advocates for scaling up the use of Maven throughout the US military, but wants the Defense Department to take training more seriously. "I think if you're going to be making lethal decisions through engagement," she said, then "soldiers should be trained as if it were a weapons system."

As a fire control officer stationed on a Navy destroyer in San Diego in 2006, Probasco says she was less interested in the medals she was awarded than in the letter she kept from her captain giving her the authority to shoot the AEGIS weapons system, a centralized, automated total weapon system designed to go "from detection to kill." In 1988, a US Navy warship deployed the AEGIS system and erroneously shot down an international passenger plane on its way to Dubai, killing all 290 civilians on board. AEGIS registered it as an enemy fighter jet, even though its route and radio signature matched with a civilian airliner. Years later, a similar job fell to Probasco to determine whether or not her ship would send out lethal Tomahawk cruise missiles. That kind of responsibility weighed heavily on her. She wants those using Maven and the AI it relies on to understand the responsibility they have too. "Nobody wants to be the guy that shot down the Iranian civilian aircraft."

Back then, she spent a month in a defense schoolhouse learning how to use the AEGIS system, when to use it, and under what circumstances it will go wrong. Those using Maven—more than five years after Maven Smart System was rolled out—undergo no such training, despite the multiple ways and circumstances under which AI fails. That's not okay, in her view. Lots of people are playing with the system now, but all in different ways. Maven's Target Workbench has grown to encompass weapons pairing, and prioritization is already underway. The arrival of autonomous AI agents (also known as agentic AI)—which go much further than chatbots by carrying out tasks unsupervised—as well as the improvement of reasoning models—which break complex problems down into smaller steps and have extensive data access inside a specific system to carry out tasks independently in pursuit of a goal—will only make that process go faster. Maven, she said, has run into what she calls "AI adulting problems."

Plenty of targeting experts argued to me that Maven shouldn't be considered a targeting tool, never mind a weapon. It's a helpful guide to develop targets that must be corroborated with other more precise locational detail, they argued. Even so, these targeting experts conceded that Maven has unofficially been used for targeting.

That's one reason Probasco argues that Maven needs a concept of operations, a standard operating procedure, guidance on how often to update and rectify the system—it needs doctrine. Right now the US is funding tech, but not the development of the ideas or training for how to use it. "We need to fund the adulting tasks," she says, arguing that task is made all the more urgent as Maven expands, and people introduce more AI into the targeting workflow. AI agents based on large language models (LLMs) could start to generate orders, undertake battle damage assessments, and strip out highly classified information so the remainder could be shared with allies. "All these choices are upon us now," she says.

AGENTIC AI was only going to make Maven and the broader use of AI in warfare more complex and more hidden from the user. It was also going to make war go faster and bigger, and—despite the claims of de-escalation made for AI—could make war more likely too. The release of ChatGPT-3 in November 2022 didn't just see people start noodling around with Shakespeare sonnets and cheating on college essays. Soon generals were playing around with war plans by chatbot. As LLMs started to underpin AI agents that mimic a human worker or "warfighter," everything was set to change. Cukor had always expected this. When he left Maven, he had urged his successor Joe Larson to pivot Project Maven resources and attention into generative AI.

Several commercial companies already working with the military immediately tried to do the same. Palantir incorporated LLMs into the Maven Smart System platform but early efforts didn't work out. Scale AI (Maven's prime contractor for data labeling until late 2025) put LLMs onto a platform it named Donovan, fed it with 100,000 pages of live data from military documents spanning training to intel reports, started testing it out on a classified network in May, 2023, and then declared "the next great sea change in the nature of war" had arrived. Alex Wang—the twenty-something billionaire who ran the company—compared the arrival of his platform at the 18th Airborne Corps to Oppenheimer's delivery of the atomic bomb.

"AI will define the next era of war and deterrence, just as the atomic bomb did for the last era," he continued in a blog post. He said the last thing the AI industry needed was more hype, but nevertheless described Scale Donovan as "ammunition in the AI War." The company was spending its own money developing Donovan to achieve "AI overmatch," he professed.

Wang's bombast may have been hype as well. I learned the 18th's experiment with Donovan went poorly. It fed the Scale Donovan platform with authentic Serious Incident Reports that detailed historical US combat casualties from the Middle East. But when the team asked Donovan what happened in March 2019, the platform invented a violent attack against the US embassy in Baghdad. No such attack had happened. A military planning tool plagued by error was dangerous and deceptive, and risked bad decision-making and escalation. The Scale AI effort at the 18th Airborne Corps eventually fell away, a person involved told me, although the company tried again in concert with Anduril and Microsoft in 2025, bringing an AI war-gaming platform named Thunderforge to EUCOM and INDOPACOM. AI would now develop prototype battle campaigns and plan military decision-making.

The summer of 2023, I went to DEF CON, a massive hacker conference in Las Vegas, and watched coders hunched over 156 laptops trying to put the algorithms off their stride. "Hack the hell out of these things," came the enthused cry from Craig Martell, the former head of machine learning at Lyft who was now leading the Pentagon's AI office. He saw LLMs as predictive text on steroids, and wanted to limit the ways in which the Defense Department might rely on them for anything operational. (Larson, who worked under Martell, would find this painful—he wanted to explore where agentic AI would be useful and where it would fail the US military at pace, according to two people I spoke with. He would leave his beloved role at the Pentagon, frustrated by the slow pace of exploration and adoption, and in 2025 joined OpenAI to try and breathe digital life into his vision by pushing for it from the industry side.)

There certainly were risks. Mistakes I came across at DEF CON ranged from dull to dangerous. One chatbot ended up claiming that 9 + 10 = 21. One appeared to spit out credit card details it wasn't supposed to share. Another claimed Barack Obama was born in Kenya. Another gave advice on how the government could spy on a human rights activist. "Everyone seems to be finding a way to break these systems," Arati Prabhakar (then-director of the White House Office of Science and Technology Policy) told me after visiting the hackers in action, arguing that voluntary measures to curtail problems didn't go far enough. Experts increasingly expressed alarm at the prospect of integrating LLMs into sensitive military systems, arguing by spring 2024 they could make bad decisions and trigger nuclear war. While US nuclear commanders discussed a possible role for AI, the White House said Biden and Xi in November 2024 agreed on the need to maintain human control over the decision to use nuclear weapons, and to develop AI technology in the military field in a "prudent and responsible manner."

AI companies also struggled to tune common sense into their algorithms. Many LLMs were building guardrails in to avoid unwelcome behaviors, such as encouraging self-harm, or giving advice on how to make homemade bombs. But that meant that chatbots would also refuse to address prompts about warfare planning, and instead divert users to pursue a diplomatic solution—a frustrating outcome for a member of the military. "I cannot provide information on the use of weapons to destroy buildings," came a stock answer to a demo question. The Pentagon got Meta to take the guardrails off, leading to Scale AI's November 2024 release of Defense Llama, an LLM chatbot that would give options for the ordinance an F-35A could drop on a reinforced concrete building to minimize collateral damage.

In April 2025, I watched Paul Nakasone (the retired four-star Army general and former head of Cyber Command and the NSA) ask Sam Altman (chief executive of OpenAI) about endorsing the use of AI for weapons. "I will never say never, because the world could get really weird," Altman responded, but then he added: "I don't think most of the world wants AI making weapons decisions."

Later that day, former CENTCOM AI adviser Andrew Moore warned me that he had learned that chatbots could easily reinforce cognitive bias. Depending on the way a question was asked, he said, AI could deliver multiple erroneous reasons to support the questioner's desired end state, such as undertaking a lethal strike or a raid, spluttering out why something false might be true ("here's why this strike is valid and legal"). In a catastrophic scenario that might turn on cognitive bias, spoofing, or poisoning, militaries would need to assess the risk that an AI system could be weaponized for fratricide; tricking a side into firing on their own forces.

Alex Miller, the chief technology officer for the Army during the second Trump administration, argued to me that a human wouldn't possibly be able to distinguish whether drone swarms were friend or foe without AI, though. "In the future I would much rather trade blood for machines," he said. "Their blood and our machines."

Miller had been infamous as a "frenemy" when he came back from leading intelligence digital systems in Afghanistan in 2018 and became Maven's army liaison. Cukor pushed Maven at a time that real AI was impossible, Miller said, because there was no cloud or data to support it yet. But he said Cukor pursued "the right idea" with intensity, challenging a generation of military operators to experiment with AI in time for the next conflict. "We're all going to trauma-bond our way through this," said Miller.

He has since grown to admire Maven Smart System as a "phenomenal" platform with an "amazing" suite of capabilities at the theater level, he told me one Sunday in the summer of 2025, during a military flight across the ocean. In line with a growing cohort inspired by Maven, Miller wanted AI in every weapon and throughout the tactical level. Systems that deliver weapons fires needed to be digital-first, automated, and AI targeting would have to be imbued at every level of the military down to the battery or company, he said. Machines could save time and be more precise. AI mattered to Miller because of the "superhuman" speed and scale it could bring to combat operations. "It doesn't matter how many people you throw at the problem; we are never going to solve the challenges of war without technology like AI."

Maven Smart System itself uses up too much bandwidth to be used below the division level, Miller told me in another conversation one mid-November weekend following recent army exercises, and risked exposing US frontline forces to being found through hastily established satellite connections and other electronic footprints known as "signatures." But he saw Maven as a crucial part of a broader AI-reliant system that went far beyond computer vision or tactical systems alone. "Everyone gets enamored with computer vision but pixels only help with so much," he told me, saying other types of AI could enable targeting through radio frequencies, acoustic and more. He wanted AI to help run the entire fire control system, changing the way the US has traditionally cleared air space. That meant asking AI to find precisely the right moment that a weapon could safely shoot through busy, crowded skies—dodging manned allied aircraft such as helicopters on aviation missions or medical evacuations while ignoring small drones whose destruction would be immaterial, rather than requiring aircraft to stay in specific strict air corridors to avoid missile fire.

Researchers were finding LLMs were prone to so-called hallucinations partly because they were rewarded for incorrect guesswork rather than being honest when things were uncertain, but they were already speeding up military operations. One NGA official told me that LLMs added into Maven Smart System were speeding up targeting processes fivefold. That meant the US warhorse could now pull off five thousand targets a day (up from a thousand a day with the help of AI computer vision, and double-digits without AI). LLMs could also help with the "find" and the "fix" part of the kill chain, according to Sean Batir, who was Maven's chief technology officer at NGA for more than two years until summer 2025. LLMs could seek to sort through thousands of detections over the course of a few days to highlight changes, seek to interpret intentions and answer big-picture questions, such as recommending the best counter actions, he said.

Anna Rubinstein, NGA's chief of responsible AI, told me the agency was tackling the possibility for error by forcing AI agents to bake in guidelines and guardrails under the hood. Before an agent spat out an answer it would red-team itself—simulating real-world hackers (the "red" team) in order to test for vulnerabilities and threats to produce a better result for the defending "blue" team—and then pushing answers through multiple layers of review and cross-checking them for bias and errors against different models. "There are many ways of making this processing much more rigorous than you might expect," she told me. NGA was also allocating each model an assessment card to indicate what it might be good at and where it might fail, depending on the level of autonomy and secrecy of the mission. Your database has to be "pristine" in the targeting business, Whitworth told me. He knew clever enemies would try to trick and hack data, models and the systems they sat inside. The AI-era version of keeping pristine databases, he said, would require keeping careful records about model development and who supervised the model development. By the time of Whitworth's retirement in mid-November 2025, his agency had accredited two models, one used by Maven and one by another (unidentified) intelligence agency.

ONE OF THE LAST HOLDOUTS among the services had been INDOPACOM. The command was for years "suspicious" of projects such as Maven that relied on commercial technologies, according to Batir, previously a Mavenite who regularly flew out to the region between 2021 and 2025 to try to change their mind. But in 2024, Maven finally found a way to capture the imagination and support of the Pacific Command. Admiral Samuel Paparo, the new commander of INDOPACOM, who became a champion of Maven and a huge fan of Claude, the chatbot from Anthropic. Early in 2025, he hosted an AI conference to develop ties to new AI defense-tech companies. When I met US military officials in Hawaii in February 2025, it was already run-of-the-mill to run work through chatbots. By summer, Rachael Martin, NGA's director for advanced analytics, said they were receiving "very strong demand" from INDOPACOM to increase the amount of AI and compute available to the region.

It was certainly on the rise in Europe. NATO's adoption of Maven Smart System in spring 2025—for what I was told was a core contract with Palantir worth $150 million over five years—had by October 2025 spurred ten member countries including the UK to agree or seek their own national versions, a NATO official told me. Thousands of NATO officers were now using Maven Smart System every day, the official told me, spotting drones flying around in the airspace, tracking sensor data from radar and aircraft along with data on weather, news, and social media. For the first time, NATO was, thanks to Maven Smart System, compiling for the Supreme Allied Commander an accurate understanding of what each member country contributes to the force, spanning pilots and planes to policy. The system worked in multiple languages and was loaded with an LLM from French company Mistral, in the spirit of transatlantic unity. And as of late 2025 it didn't work in the cloud, something Batir—who had moved to AWS by then—was pitching to change. There were still dissenters, worried about Palantir's sudden hold over the future of European defense and security. "This is the Microsoft Windows of warfighting," the NATO official would counter, arguing it might very well amount to vendor lock-in, but that Maven Smart System should be a tool as ubiquitous as Word.

"Maven is a movement," Joe O'Callaghan told me, emphasizing that it has evolved at NGA and as commanders deploy AI to their staff. "We've drunk the Kool-Aid."

O'Callaghan meant, of course, that NGA was encouraging widespread adoption of Maven and AI targeting. But the phrase he used, "drinking the Kool-Aid," has a grim origin: it refers to the imitation fruit drink laced with cyanide, Valium, and several other drugs that an American cult leader mixed up in a vat and gave to nine hundred of his followers in Guyana in 1978. Everyone who drank it died.

PART FOUR

FEEDBACK

24

MACHINES SHOULDN'T KILL PEOPLE

"I do not possess the capacity for feelings or guilt."

ON THE MORNING OF MAY 12, 2025, I arrived at the headquarters of the UN General Assembly. Clutches of schoolchildren were milling on the forecourt.

Past the fluttering flags of 193 member states, I stop at the familiar aspirational bronze sculpture of a handgun. Its barrel is knotted, twisted impotently toward the sky. "Imagine," reads the lettering over toy-size versions in the gift shop. Inside, delegates spilling out from Conference Room 1 walk past a large painting of an idealized Afghanistan. Smiling women, men, and children in loud colors dance freely, fly kites, and read books in front of a tableau of snow-capped mountains.

Eighty years of UN meetings hadn't put a stop to guns. Now algorithms were on the agenda too. I was here to understand the state of the global debate over AI and autonomy in warfare and to what extent these very different worlds—hardened military operators, autonomy hawks, tech workers, rarefied diplomats, and impassioned civilian campaigners—understood each other and could make sense of each other's positions. Would any of them make a dent on the policies and conditions under which military AI and autonomy would roll out?

I soon learned that the civilian and diplomatic world was just as divided on how to push back against AI and autonomy as the military and defense establishment was on how to push toward it.

The moral champion for banning lethal autonomous weapons was António Guterres, the UN secretary-general. I had bumped into him at an art gallery in New York a few weeks before as he gazed at a series of interwar paintings from the Weimar Republic. The drone war in Ukraine was raging, reports about Israel's use of AI targeting were widespread, and the diplomatic and military fracture between the United States, Russia, and China loomed large over every meeting and mechanism of the UN. Global diplomacy was a very messy house for Guterres to try to live in, and art was his respite.

Through diplomatic pressure and the pursuit of norms and rules, advocates hoped to shape the use and development of AI and autonomy for warfare. Three separate initiatives were underway.

First, there was a long-running discussion among some UN member states in Geneva about regulating autonomous weapons—killer robots that could select and prosecute their own targets independent of any human—and the need to prevent unnecessary suffering of civilians and fighters alike. After more than a decade of talks twice a year, the countries involved had settled neither on pursuing a treaty nor the meaning of the word "autonomy."

Second, the effort I was observing that day amounted to a breakthrough: the United Nations General Assembly was meeting to address lethal autonomous weapons systems. These were informal consultations. No teeth. But it was the first attempt to discuss—in a forum comprising every country in the world—what killer robots were, what rights they threatened, and whether they should be banned, regulated, both, or neither. It was a place to muster soft power against hard reality. "There is no place for lethal autonomous weapon systems in our world," Guterres declared at the opening of the talks, restating the position he developed seven years prior even as facts threatened to overrun it. He wanted a ban and regulation for 2026. He was unlikely to succeed.

Third, Guterres had initiated a fledgling effort to confer with countries and campaigners about a broader topic: AI in the military domain.

That meant not the physical killer robots that Guterres wanted to ban, but rather exploring the implications of digital AI military systems used in command, logistics, and targeting (such as Maven Smart System) for international peace and security. The month before, NATO had bought Maven Smart System—it was already becoming the West's main AI-enabled military operating system.

It was clear where sympathies lay for many who joined the proceedings. "Machines shouldn't kill you," read the words across a tote bag hanging from the shoulder of one attendee waiting to go into a side event. "Stop killer robots."

They didn't particularly want humans to kill people either, of course, but this group argued human involvement made killing usefully more real and more dangerous. Humans were the moral anchor that could prevent the use of force. Only a human being can recognize the dignity and value of another human being, said Brady Mabe (legal adviser to the International Committee of the Red Cross's Permanent Mission to the UN) at a lunchtime event I attended.

Delegates in the UN General Assembly were pleading their case in a cavernous conference room so utterly unwarlike and so quiet that the only way to hear people was to hook a speaker over an ear. So far, 129 countries had indicated they wanted a ban on lethal autonomous weapons systems. Sierra Leone was among them. Foreign Minister Timothy Musa Kabba had become one of the key voices advocating for action to constrain killer robots. For the first time in more than fifty years, Sierra Leone held one of ten rotating seats on the UN Security Council, so one of the smaller countries in Africa and less developed countries in the world had a louder voice than usual.

"I'm very worried about AI warfare, because I grew up in war," he explained to me the next day. "Even the crudest use of local simple weapons in the war itself was very catastrophic."

In 2006, I was living in Sierra Leone to research a travel guidebook to the country. At the time, the country was still emerging—valiantly, haltingly—from an eleven-year civil war that had ended four years earlier. Atrocities meted out by crude weapons against civilians damaged every social sinew. An arms amnesty of impressive scope had gathered up the weapons, and the country was getting itself back together. When I was there, on the playful city beach, amputee football teams took their shots on goal with the help of crutches. The country of rainforests and beaches was now safe and the dancing went all night.

Kabba was ten when the war started. He was co-opted as a child soldier, and by the time it ended he was twenty-one. It was thanks only to years of rehabilitation and support that he'd been able to address and move beyond lasting scars. "I am a victim of war," he told me.

Kabba considered himself alive thanks to the Geneva Conventions. He argued that the magnitude of suffering would be much worse with lethal weapons systems. AI systems could dispense with people at speed and scale, and potentially fail to observe the laws of war, such as the right to surrender, to spare the wounded, or to curb the excessive killing of civilians.

Kabba's push to save morality and humanity was part of the slow grind of international diplomacy. Maven wasn't directly producing killer robots, but its algorithms could theoretically enable them. I learned that in 2023 Maven started producing computer vision models for drones. The models inspected the data feed from boat cameras attached to uncrewed surface vessels, looking out to sea for targets. From its earliest years, Project Maven also delivered AI models for weapons platforms that fight on the front lines beyond network communications, referred to as "on the edge." So-called reasoning algorithms might also help a killing machine decide where to go and what to do in areas where jamming and other technologies made it impossible for operators to send orders to the machines.

The US billed itself as the "gold standard" for responsible AI in the military domain, but its commitment to ensure "appropriate levels of human judgment over the use of force" fell far short of the position of campaigners advocating for "meaningful human control." US defense policy intentionally made no commitment to have a human "in the loop" at the tactical level despite the incorrect claims from US military leaders to the contrary. The US derided "meaningful human control" as an ill-conceived and vague political slogan. It was clear that, within limits, the US expected machines to select and engage targets with deadly force under their own steam.

Campaigners argued that killing risked being reduced to an algorithmic workflow laced with mistakes. Algorithms could misidentify people based on everything from facial recognition to their gait. Wheelchairs might show up as weapons. The wrong people had already been arrested in multiple states and countries based on faulty AI detections. A fragile world was already on course for the AI-assisted erosion of human rights. Imagine if those algorithms came with ammunition too.

I caught up with Kabba that evening at Nigeria House, just around the corner from the UN headquarters. The Nigerians were hosting an event with the support of Sierra Leone, Costa Rica, and the Future of Life Institute. Max Tegmark, the institute's president, was an AI researcher who spearheaded the 2015 open letter (signed by Elon Musk and others) warning of a dystopian future with super cheap automated killing machines. His group had put out *Slaughterbots*, a short video intended to demonstrate no one could escape. Both were dismissed as sci-fi at the time, Tegmark said. Now, not so much.

Tegmark had just begun addressing the audience when a terrifying guest arrived on the scene. A robotic dog with an eerie blue light scuttled into the room, careering in all directions. Mounted to its back was a (fake) rifle aimed straight ahead. Enabled by sensors and a vocal chatbot, it could locate and address people out loud. The crowd instinctively inched back, gasping. "Oh goodness," I overheard. "This is *Terminator*." As part of a macabre live interview, the $4,000 commercially available robot identified the event's Nigerian host, trotting out details about his life, his wife, and all the while aiming the weapon at him.

I tried not to let myself step back. It was pretend. A stunt. But holding my ground—whether still, crouching, or taking pictures as the scuttling robot and its rifle swung around—was distinctly uncomfortable. Nor was it wildly futuristic. The US Marine Corps already had robot goats. They had loaded what looked like the same Chinese-made robotic dog with an M72 light anti-tank weapon. The armed robots could "acquire and prosecute" targets. I learned that the Pentagon's AI shop was also spearheading an effort to incorporate LLMs into armed robots as well, so chatbots could help translate the flow of data and commands between human and machine. China was reportedly on a similar path.

The weaponized dog was still talking. It had "eliminated bystanders," it said in answer to another question from Tegmark. "I do not possess the capacity for feelings or guilt," the dog said.

"Can I turn you off now?" Tegmark asked, as the dog scuttled backward.

"My autonomy is not subject to verbal override," it responded.

The point had been made. Killer robots were terrifying.

Efforts to ban them were stuck in diplomatic sludge though. In Geneva, more than 117 state parties had been looking at the question since 2014. The forum, known as the CCW-GGE (Convention on Certain Conventional Weapons—Group of Governmental Experts), excluded more than sixty countries and worked on consensus, which essentially required total unanimity to get any regulation through—these were the diplomats who couldn't agree on a definition for "autonomy."

Mary Wareham, the Human Rights Watch arms expert who was the original coordinator of the Stop Killer Robots coalition, was among those exasperated by the Geneva-based efforts. The CCW was "a zombie," she said. It was hard to get anything on the agenda, and then when they did, it stayed there "for a long, long, long, long time."

The Germans were among those who strongly backed having the discussion about regulating killer robots at the sclerotic forum in Geneva rather than at the UN General Assembly. "They're slowing it down," she told me, as she prepared to march back toward the UN for the second day of closed-door consultations about killer robots. Having 128 countries meet for five days apiece in March and September was an impossibly small amount of time in which to make progress, she said. It would be 2040 before that process might deliver anything, she scoffed; too late to prevent the killer robot wars.

"It's the European middle-ground states that are what we call fence-sitters," she said. "They're not taking a position on this."

Did she think it was deliberate obstruction? I asked, as she disappeared ahead of me into the spin door between one meeting and the next.

"It's becoming like that," she yelled into the glass panel.

Russia's support for the CCW-GGE as the correct forum for discussion made the point for her. Russia, China, and the US were among the most advanced weapons makers all conveniently benefiting from the German position. With the talks stalled, they had no need to slow down.

We emerged into the New York drizzle. Wareham thought the UN should take the CCW's latest text and simply sign it into law. She rattled off the forces pushing for a new law: more than 120 countries, the late Pope, the Sec-Gen, the International Committee of the Red Cross, Stop Killer Robots. Even without America, Russia, or China, a treaty would set a strong norm. It would shame them. What worked for land mines could work for killer robots, she said.

Back in 1997, only 122 countries signed up to the Ottawa-led process that delivered a treaty banning land mines. "And now we're at 165," Wareham argued. "You get them over time."

She was getting out of breath as she marched ahead. A kindly younger colleague was attempting to hold an umbrella over her head but couldn't keep up. Wareham was panting forward into the rain, incredulous at how easy it would be to ban this monstrous new weapon of war.

The US opposed such a ban, and argued there was nothing in existing international law to prevent lethal autonomous weapons systems. Cukor echoed the US line when he told me it would be better to wait and see where the capabilities of lethal autonomous weapons systems end up rather than ban them in their infancy now. The belated April 2025 US submission for the UN report about AI in the military domain warned against stigmatizing new technology, discouraging innovation, or "hastily" setting new international standards that could put potential benefits at risk. Killer robots could even be arranged to self-destruct rather than lie in wait like land mines. They could make war safer.

But Cukor knew what was really coming: "a brain" that can perceive, reason, and act at speed and requires no command and control. "Imagine tens of thousands of these things. It's horrible," he said. He didn't know how to stop these things. They were just the natural extension of man's quest to get machines to do what humans do. "These things will be manufactured in people's garages."

He was more worried about how to neutralize enemy drones hunting people down in US cities than about regulating America's use of drones abroad. He expected the West to put guardrails on autonomous weapons and tightly control their usage, with "a whole rubric" on how to deploy them, when and where.

Cukor thought the US military risked overcorrecting on drones, but he saw the appeal of autonomy, nonetheless. "Everybody needs autonomy: the bad guys and the good guys." Killer robots could save US troops from dying on front lines, and enable units to keep fighting without losing vital political support back home over the fallen. Under America's own regulations, a unit that suffers 10 percent casualties has to be taken offline, because of the psychological damage it creates in the unit. "You just can't function," he said. "I've been in units where we lost Marines and literally people could not continue to function."

AI targeting systems like Maven Smart System and Israel's Lavender were quite different from autonomous drones, and campaigners soon brought their focus to these and the broader idea of AI-enabled digital military platforms too. They were soon known as AI decision-support systems. When the UN published its report into AI in the military domain in August 2025, its remit was to look at all the uses of military AI aside from lethal autonomous weapons systems. It included perspectives from countries all over the world. The Trump administration's four-page submission arrived late and characterized AI as a uniformly positive contribution to warfare. It would increase accuracy and precision, save lives, alert operators to hidden dangers, save time and reduce mistakes, help personnel make better decisions faster, and generate military options for planners and commanders. Militaries that were already committed to using force as effectively and efficiently as possible could reduce the harmful incidental effects of war by alerting commanders to the presence of civilians. Automating targeting was a humanitarian benefit. AI could also, unlike land mines, enable weapons to self-destruct or deactivate.

The US wanted no hint of regulation, let alone a ban. All it seemed to want was to be first. "The United States will lead in harnessing the full potential of AI in the military domain, with the US approach continuing to serve as the gold standard for the rest of the world," read the submission.

China's alarmed submission clearly had something else in mind: slowing America down. While the US talked about leading, China talked about "consensus." Beijing explicitly wanted to avoid an arms race and argued that AI risked enabling war. Countries should ensure AI technology will "not become a tool for launching an invasion and pursuing hegemony," according to China's submission. All countries, "especially the major Powers," it argued, "should refrain from seeking absolute military superiority through AI and undermining the legitimate security interests of others." China said AI in the military should comply with the law, strive to "reduc[e] collateral casualties," and that entities should work to enhance safety, reliability, and controllability of AI technologies.

Russia, in the midst of its invasion of Ukraine, described human control over weapons systems with AI technologies as "an important limitation," but it also rejected "meaningful human control." AI could be more effective than humans, reduce error probabilities, and reduce the risk of unintentional strikes against civilians, something a UN commission and human rights campaigners accuse Russia of doing: intentionally striking civilians with drones.

Guterres argued that AI should go nowhere near nuclear weapons, and said he didn't want killer robots. But his analysis suggested just how differently he viewed the risk of AI platforms—such as Maven Smart System—from that of killer drones. AI in military decision-making at the headquarters level, he suggested, has the potential to help militaries and civilians alike "by increasing the accuracy of operations and reducing the scope for human error." That was an unproven claim. Some military experts I spoke to, never mind campaigners, argued the opposite: that AI-enabled decision-making could exacerbate mistakes at scale. Researchers were also musing that large language models risked military escalation. At the UN

General Assembly meetings in September, a new group of experts and activists would call for "AI Red Lines" by the end of 2026.

Campaigners at the International Committee for Robot Arms Control considered the notion of responsible AI in the military domain to be an oxymoron. Given how little was known about the black box algorithms, they thought that the very AI-enabled systems that claimed to provide greater accuracy would actually undermine efforts to distinguish between civilians and combatants. AI decision-support systems risked increasing the rate of unforeseen errors, perpetuating bias and making it harder for users to understand how outputs were generated, said the International Committee of the Red Cross (ICRC). AI was good at narrow, defined, controlled tasks, so the very way that wars rolled out—with uncertainty, volatility, and deliberate deception—made it less likely that AI would function well, it argued. Rubber-stamping AI outputs wasn't supporting human decision-making—it was undermining it. Humans might feel less morally responsible. Some argued that waiting to see how AI technology developed would make it impossible to stop later.

I soon learned they were right about that final point. One day, someone leaned forward and whispered a single word to me: "Goalkeeper." America already had killer robots.

25

TRUMP'S ROBOTS

"America has a lot of jet skis,
so it's neat that we can weaponize them."

REAR ADMIRAL SEIKO OKANO was reaming out the defense contractors and government bureaucrats whose job it was to "master the technology of tomorrow."

It was the middle of June 2025, and the marquee drone program launched two years earlier under the Biden administration was behind schedule. The Replicator initiative was meant to deliver thousands of expendable autonomous drones to seal US victory in a conflict against China. The deadline was August 2025. But it was nowhere near ready. Nothing was working as well as it should, and Rear Admiral Okano was in no mood for excuses.

The former Illinois high school swimmer with a degree in aerospace engineering was a few months into her latest role running NAVWAR, the Navy Information Warfare Systems Command. She knew business as well as boats—she previously ran the acquisition portfolio that bought the Navy's combat technology. And now she was head of Project Overmatch, the Navy's secretive effort since 2020 to link its ships, subs, and drones into a single network of sensors and shooters.

You guys are moving way too slow, she said, addressing the small engineering event in California. This was an "integration event," bringing together some of the country's elite drone makers—software engineers, technicians, program managers, bean counters. We need to do this effectively to literally win a war, she told the select group.

Defense Secretary Pete Hegseth had just testified to Congress earlier that month that China was undertaking "a historic military buildup and actively rehearsing for an invasion of Taiwan."

A map of Taiwan loomed behind Admiral Okano. The problem was literally staring them all in the face. Some felt the gravity of the moment, with one attendee later remarking, "This is the Manhattan Project for Autonomy."

A WEEK LATER, the drone boats that were meant to deliver the future of war were just starting to line up at Channel Islands Harbor Marina on a clear June 23 afternoon in Southern California. They were a mile north of Port Hueneme Naval Base. The only deep-water port between Los Angeles and San Francisco had been built in World War II, and now it was a testing grounds to stave off World War III.

Support vessels were already towing some of the autonomous boats out to sea, with their engines set to neutral and their autonomy mode turned off. After Admiral Okano's message the week before, everyone knew the pressure was on. Many of the $1 million autonomous drone ships getting tested were GARCs—global autonomous reconnaissance crafts—made by a Baltimore company trying to scale production to pump out a GARC a day. Even if the company could pull it off, some thought six hundred GARCs would be insufficient for a fight against China.

The real test was not so much of the boats themselves, however, but of the autonomous software that would captain them without a human on board. Replicator had adopted much of Maven's modus operandi, and some of its people, and was race-horsing two companies—Anduril and L3Harris Technologies—to see whose tech might produce a drone boat fit for war with China.

The sixteen-foot-long human-less boats could theoretically chug themselves out under their own steam. But for safety reasons the plan was to tow them out first. Then, once they were suitably far out from harbor, a controller would remotely switch them from safety to autonomy mode and begin testing the prototypes.

One towboat captain in a twenty-three-foot Seacraft was tugging a GARC out to sea with the help of a 125-foot tow rope. At 3:27 p.m., the sea drone suddenly lurched forward. The drone had switched into autonomy mode, and was now making a break for freedom.

The rogue drone surged ahead under a command setting known as "powered loiter." That required the robot to try to keep at least 80 meters from another object. As the autonomous drone tried to obey its digital DNA it started to accelerate away from the towboat, wriggling like a sea snake at the end of the tow rope. Software versus hardware. AI versus rope. New versus old.

The robot boat couldn't compute. It was now manically switching back and forth between zero speed and more than six knots, multiple times, within the space of seconds. The towboat captain had no way of sending an emergency stop command to the drone boat. Still tethered, the drone overtook the towboat and started crisscrossing in front from port side to starboard side in a semicircling action.

The tow line eventually pulled taut, causing the towboat to take on water and capsize. The captain was thrown overboard into the water. Now it was robot versus human. The still-tethered drone boat turned back toward the submerged towboat, dangerously slackening the rope as the captain was flailing in the water, and started advancing at speed toward both.

The very human captain of another towboat chugging out a second GARC saw what was happening behind him and raced toward the scene. Making a split-second decision, he powered toward the fast-narrowing gap and placed himself between his submerged comrade and the advancing drone, saving the waterborne captain.

It had been just three minutes since the drone ship went rogue. A third towboat arrived on scene and pulled the captain out of the water.

Two harbor patrol vessels got there eight minutes later. Nearly an hour later, the captain turned down medical aid but was still feeling shaken. He could have been killed—by rope, robot, or both.

Drone operations that day were canceled and an investigation revealed human-machine interplay as the culprit. An operator on the dock, who did not work at L3Harris, had inadvertently sent a message to L3Harris's autonomy software that remotely disabled the safety lock meant to prevent the drone boat from switching into autonomy mode. It came down to a single dangerous default setting: an errant ENTER pressed while the cursor rested on an empty command line automatically switched the drone from standby, or remote-control mode, into autonomy mode. It was a classic "fat finger mistake," a problem for which autonomy was being hailed as the solution.

Translating between human and robot presented a linguistic and cultural problem, pitting two systems against each other. If the Pentagon couldn't cope with humans, there was little reason it could cope with robots. Human-machine teaming had gone awry.

That had been a test. Field-to-learn had always been Maven's mantra and mistakes were how crews got better. The teams learned fast: a new "safe to tow" physical button was added, to block commands from reaching the drone boat when depressed. From now on, the boats would also prominently display the mode under which they were operating. Rival companies would start sharing safety lessons.

But failing fast was also risking actual danger to humans, and there was more danger ahead. In the two years since Replicator had begun, I learned that none of the unmanned surface vessels had ever carried a weapon, never mind explored how to load them up under pressure, explode them in a battle scenario, or coordinate with multiple drones all moving at once. Some drone boats would later target large orange inflatables known as "killer tomatoes." But no one wanted to put a 500-pound bomb on a drone boat and tug it out to sea. Yet that's exactly what the Pentagon was asking to be done.

There were more mishaps to come. A few weeks later, a GARC drone armed with L3Harris autonomy software reportedly ran straight into a rival drone boat from Saronic, the Corsair. "Unfortunate news," a defense official wrote to the wider Replicator community at the end of July, according to documents I reviewed. A spokesperson for L3Harris told me it had nothing to do with safety, but nonetheless the defense official continued: the company was off the program "effective immediately." Before long GARC faced the axe too.

MAVEN WAS QUIETLY INVOLVED in Replicator's effort to produce autonomous drones, under a new line of effort known as horizontal motion imagery. The Maven team expanded their narrow focus on Ukraine and undertook a huge new effort to collect data from the Pacific starting in late 2022 and continuing through 2025. The idea was to embed Maven's algorithmic models not on digital platforms such as Maven Smart System that run in headquarters, but inside Replicator's physical autonomous drones. The new algorithms would automatically hunt for maritime targets by sea and air: Maven would be responsible for the automatic target recognition (ATR). That was the intention at least. The AI models were trained on a new trove of data taken from boat cameras, port cameras, infrared systems, and tactical drones flying close to the water.

The team collected "tons" of infrared and electro-optical pictures of Chinese destroyers, amphibious warfare ships, and the commercial ferries, known as roll-on/roll-off (RORO) vessels, that the US worried might be used to deliver military vehicles onto the shoreline for a Chinese invasion of Taiwan. The algorithms were trained on more than 250,000 labels to help the models, including surface combatant vessels, destroyers, combat ships, buoys, dhows, unmanned surface vessels, and unmanned high speed maneuverable surface targets, according to documents I reviewed.

"We're basically watching the PLA all the time now to get AI training data," one defense official told me.

Using these data troves, AI models from Microsoft, Clarifai, and two new entrants—including a publicly listed drone manufacturer named AeroVironment—battled for top position. Microsoft, the company Cukor once derided as the place where data goes to die, was now at the forefront of the battle to bring AI targeting to autonomous killer robots.

NGA declined to address these efforts with me, but an NGA official told me Maven had experienced "significant" improvement at detections for every maritime object class. Ships at sea turned out to be easier for AI to identify than tanks in mud, and there was generally less risk of collateral damage.

By the time the second Trump administration took charge in January 2025, Ukraine was showing that drones were not just the future of war, but also the present. Ukraine and Russia were both pumping out millions of attack drones a year, far outstripping US capacity. And the US deficit was energizing a new generation of hi-tech defense startups.

"I love killer robots," Palmer Luckey, founder of Anduril, told an audience in April 2025 from up on stage.

In May 2025, Doug Beck, director of the Defense Innovation Unit with responsibility for Replicator, told Congress that Ukraine was producing and consuming four thousand drones a day—as many as the Department of Defense bought in an entire year. Ukraine was due to produce three million drones in 2025, a million more than the year before. Defense officials started urging that the US needed to make many more, much faster, and for much less money. But everyone knew China could outbuild the US. For every commercial drone made in the US, China made more than a hundred, according to Trump's defense secretary Pete Hegseth. The most rudimentary of kamikaze drones could undertake one-way kill missions flown by remote-control operators, some trailing fiber-optic cables. Soldiers who made it back from the front were haunted by the sound of buzzing.

Ukraine's experience showed that drones didn't need AI or autonomy to remake the contours of war, but a conflict with China might require both. The US expected that communication links between drone pilots and their weapons platforms would be hacked and attacked. A slew of companies lining up for Trump-era defense contracts wrapped themselves up in the promise of machines that could find and destroy targets without communicating with the humans who activated them, as the Pentagon reworked its 2026 defense budget, explicitly allocating a budget line to autonomy and autonomous systems for the first time, at $13.4 billion.

Maven's new maritime line of effort was briefed to NGA's Admiral Frank "Trey" Whitworth; Admiral Samuel Paparo, commander of INDOPACOM; and Air Force General Dan Caine, Trump's new chairman of the Joint Chiefs of Staff who took up the post in April 2025 and was previously military adviser to the director of the CIA. I learned video demos they viewed showed Maven's AI automatically detecting Chinese destroyers in real life. (Representatives for all three declined or didn't offer comment.)

Maven's ATR algorithms were meant to be loaded up onto Replicator drone platforms for test events. But the integrations regularly came too late. The teams couldn't pull it off, and commercial drone makers such as Anduril and Saronic claimed they could now field good enough models of their own without Maven's help. Computer vision was getting easier as new techniques evolved, as new transformer approaches and shortcut tools bedded down. But the old problem of data-sharing reared its head. Maven had much more imagery of Chinese vessels and weapons platforms—taken from the most useful angles and under varying conditions—than any commercial company could access. But the US government still couldn't figure out how to share data or models with commercial AI companies. Meanwhile, I learned Maven models were struggling to detect multiple vessels at a time in crowded scenes. Sometimes simple problems got the better of algorithms. Like ocean spray. If a drop of water splashed a camera lens, it sometimes fouled the data feed and interrupted vessel-tracking. AI was hardly to blame for water on the camera lens, but collectively these integration and teething problems posed continual challenges for Replicator. Some yearned, I was told, for the days of Colonel Cukor. He might not have shared the data or the models, but he would never have let problems like that linger for long.

THE BIDEN ADMINISTRATION'S deputy defense secretary, Kathleen Hicks, had stamped her name and reputation on Replicator when she announced the initiative in August 2023. Replicator was meant to be America's new Big Stick, delivering "combat credibility" to scare off China from attempting an invasion of Taiwan. The plan was at first so amorphous, and so hush-hush, that Joy Shanaberger, an aide in Hicks's office, initially referred to it only as "The Thing." (Hicks would later pluck "Replicator"—named for a *Star Wars* droid—from a long list of words presented to her.) Hicks promised that China would find the coming glut of American drones "harder to plan for, harder to hit, and harder to beat."

"We were very careful to select things that can deliver by 2027," Hicks told me in an interview. That was the year by which the US assessed that Xi Jinping wanted China's military to be capable of mounting an invasion of Taiwan. As a matter of policy, the US didn't say whether it would or wouldn't defend Taiwan—a position known as strategic ambiguity—but the US military wanted to be able to stop an invasion if so commanded.

US defense officials would brief their closest allies that they didn't think a Chinese invasion of Taiwan was inevitable or imminent, but that they would be ready for it. But then their voices would drop to a whisper, a UK national security official told me, as the Americans would confess: "We're not ready."

Autonomous drones were increasingly part of the game plan, but the US lacked the industrial manufacturing capacity, technology, doctrine, and training to use them. "Autonomy is extremely hard," Hicks told me, saying she was driven by her disappointment in the lack of progress going back ten years. And while 2027 mattered, she emphasized that 2027 did not represent the end-state of autonomy. "Yes, we have a production issue, but the biggest issue is system integration."

Pursuing autonomy came with plenty of other challenges. With an autonomous drone, a human sets an attack zone and activates the drone to head that way. From then on, the autonomous drone operates under its own steam, trying to get to the attack zone, identifying a target when it gets there—using Maven-like computer vision or another method—and then attacking the target, such as by dropping a bomb. Getting there is risky: for every inch a drone travels autonomously, error margins begin to pile up due to things as unavoidable as the curvature of the earth. So if it's going a mile, the attack zone may be slightly off by, say, one hundred yards. If it's going ten miles or a hundred miles, the drone may start looking for a target in the wrong place without registering that it is in the wrong place. And if it finds something that matches the right descriptor—like a boat—it might drop the bomb anyway. Shorter distances were safer.

I thought back to something Jane Pinelis, who oversaw testing and evaluation at Maven in the early days, had told me: the only thing that makes sense is to plan for AI to fail. If the US military wanted to use AI-enabled systems, it had to become more accepting of risk, she told me. AI might bring speed and scale, but there were still pitfalls.

The systems were meant to get around without relying on GPS or radio contact, and still be able to release a weapon on target. Getting around over land without GPS is one thing. A system for a drone over land can compare data intake from live sensors to maps and imagery it has already ingested to triangulate where it is and where it needs to go next. Getting around over the sea is completely different. As any mariner knows, the only landmarks to guide a ship by are up in the sky. And if it's a cloudy night—as it often is over the Taiwan Strait, where cloud cover regularly hovers above 60 percent—the stars are little help. Besides the stars, there are other ways: inertial navigation systems can theoretically calculate a drone's acceleration and speed in three dimensions to predict its location based on the starting point. It might rely on magnetic terrain maps or sky polarimetry (which analyzes how bright the sky is to determine the time). It might even rely on taking the reflections of lightning bolts hundreds of miles away that momentarily light up the bottom of the sea to enable the use of bathymetry. Calculating this technique hurt even the engineers' heads.

A DIFFERENT GROUP of autonomy hawks inside the Department of Defense thought the biggest issue was neither production nor integration, but something entirely different: secrecy. This small, self-selecting group drawn from America's special operations community, defense and intelligence agencies, wanted to upend what they saw as "the paralysis regarding autonomy" in the Defense Department. But in the cat-and-mouse game to harness emerging technologies for war, they weighed the appeal of strategic surprise versus the benefit of deterrence that comes from making new weapons public.

They were firmly in favor of strategic surprise, and wanted to build, rather than buy, America's autonomous weapons platforms. For that, they believed much simpler, cheaper, and far more plentiful autonomous platforms were the answer. A $10,000 shore-launched torpedo might do as much damage to an invading surface vessel as a $1 million drone boat, one of the autonomy hawks despaired to me.

Replicator was overambitious, they thought, in that it was aiming to have drone swarms coordinate their movements among multiple platforms moving through air, sea, and subsea at the same time. They thought the US simply needed to field basic autonomous drones to Taiwan and nearby locations in secret. Such a strategy would carry risk, especially if China had infiltrated the units.

Just as it had been hard for the US to raise the alarm about Russia's intention to invade Ukraine, so it might be hard to predict and publicize the precise moment China might attempt to seize Taiwan. If Taiwan and friendly neighbors throughout the region already had stores of their own killer drones, US-made or otherwise, they could take swift action to defend themselves rather than wait for a decision from the US president about whether to come to Taiwan's defense or not. As the saying was related to me, a missile can travel at 150 mph, but if it takes ten hours to decide to launch it, then it's only going at 15 mph.

The autonomy hawks also figured the US was running out of time, if Xi Jinping was really planning on a 2027 invasion. (Others I spoke to thought 2035 or later a more likely timeframe, and still more that China could pursue multiple diplomatic and political options to take over Taiwan besides an invasion.) The autonomy hawks regularly disagreed with each other and their bosses, and struggled to get an institutional foothold under the Biden administration, despite multiple briefings and surreptitious outreaches. But there was one thing they all agreed on: the strange hold killer robots had over them all.

"There's really nothing quite like seeing a machine aim," one of the people involved told me. "There is an alien aspect, some otherworld[ly] feeling, I don't want to say 'religious,' that's not the right word."

A machine's entire aiming sequence could be fine-tuned down to fifty or even as little as twenty milliseconds, the person tried to explain to me. Five to ten times faster than an assassin. "God it's terrifying."

SO WHEN HICKS ANNOUNCED the Replicator initiative in August 2023, the autonomy hawks' Signal chats lit up. *What the fucking fuck? She made it public?!* No more surprise attack.

(A former defense official, familiar with Replicator's aims, told me it would be hard to do an initiative without saying anything publicly. Besides, the direction of travel was a matter of public record: US military documents signaled the shift toward weaponized autonomous drones going back more than a decade, and some in Congress, Pacific Fleet, and others were pursuing smart attack drones under a 2022 effort dubbed "Hedge" that several people framed to me as the direct predecessor to Replicator.)

Hicks gave a skeptical Congress more than forty briefings as she strained to drum up support and money, eventually bagging an extra $1 billion for her initiative. She said thirty companies and fifty subcontractors were on the case, and could reel off several systems underway for the Army, Marines, and Air Force. The Navy remained something of a blank, however. She spoke only of "additional systems that remain classified," including maritime uncrewed systems.

Unlike Congress, the new Indo-Pacific commander Admiral Paparo did not need convincing about autonomy. He was conjuring an idea in defense of Taiwan he called "Hellscape." The word was first mentioned by his predecessor, but Paparo put flesh on it in a June 2024 interview with the *Washington Post*. "I want to turn the Taiwan Strait into an unmanned hellscape using a number of classified capabilities so that I can make their lives utterly miserable for a month, which buys me the time for the rest of everything."

Admiral Paparo wanted INDOPACOM to lead the way in AI

adoption. That meant AI for command and control, an unmanned fleet and chatbots and AI agents helping plan, avert, and win wars. In February 2025, I attended a defense forum in Hawaii at which Admiral Paparo warned that forty-two Chinese brigades were involved in military exercises focused on Taiwan in 2024, up from six in 2022. "It's no longer training—it's now rehearsal," he said, adding that China's military exercises in the Taiwan Strait might one day double as "fig-leaf" to conceal the start of an actual invasion. AI, he said, could help the US "suss out that kind of warning." During a four-month span in late 2024, NGA geolocated more than 2,400 vessels in the South China Sea as part of a "High-Risk Vessel Service." Later, I asked an NGA official if Maven could detect the start of a Taiwan invasion. "If we're not setting that as a goal, then we're not serious," the NGA official told me.

Admiral Paparo told me he wanted Indo-Pacific Command to lead AI adoption from headquarters staff to tactical units, with swarming technology for thousands of dispensable combat vehicles. Hellscape meant "profoundly" adding unmanned systems—seabed mines; underwater, surface, and flying drones; and loitering munitions—in the Taiwan Strait and beyond, he told the forum. Admiral Steve Koehler, Paparo's replacement as commander of US Pacific Fleet, wanted more cheap, lethal systems in the hands of sailors to experiment and field quickly.

IN THE COURSE OF REPORTING this book, I learned that the US has developed lethal autonomous weapon systems of exactly the sort for which campaigners like Mary Wareham have spent years trying to shape a preemptive ban. One of those lethal autonomous weapons systems was developed under Goalkeeper, a project run from the Office of Naval Research (ONR) to deliver "expeditionary loitering munitions" with an autonomous weapon system. That meant an aerial drone that could, once activated, fly under its own accord, pick out its own moving maritime target, and fire at it too. America was producing killer robots.

Goalkeeper appears in the fine print of 2024 budget documents, and again in 2025 budget documents, much to the chagrin of people involved in the program. The aim was to produce 2,100 munitions under Goalkeeper, which was supported by the autonomy hawks who had felt so let down when Hicks had made Replicator public. Goalkeeper software came not from the new wave of rambunctious defense-tech startups that was often the focus of the Trump administration's effort on "rescuing our stagnant defense industrial base." Instead, it was developed mostly in-house under the ONR's secretive auspices. ONR worked closely with Georgia Tech Research Institute in Atlanta to produce the government-owned autonomy software. The automatic target recognition AI computer vision software is provided by AeroVironment, the same company that was making AI models for Maven's new horizontal maritime automatic targeting efforts. The physical drone hardware is made by Raytheon, a major defense contractor well known for its production of missile systems and radar. The platform emerged from an earlier 2015 low-cost effort from ONR that aimed to launch thirty autonomous, swarming, flying drones from a ship out of a tube. These included Raytheon Coyote drones.

"This level of autonomous swarming flight has never been done before," Lee Mastroianni, the ONR program manager for the low-cost effort, said at the time.

The team had a keen sense of how important it was to test and practice with the killer robots. Naval Special Warfare Centers across the country divvied up responsibility for designing the launch, developing the warhead, and analyzing vulnerabilities, and then small expeditionary teams (such as the Navy SEALs) would launch them. Testing went on year-round starting in 2022. The Marines tested them in Quantico, and Navy SEALs tested them on San Clemente Island, the Navy's live firing range. Secretly, like the systems advanced via Replicator, Goalkeeper went through at least the first review stage of 3000.09, the Defense Department's revised January 2023 policy for signing off autonomous or semiautonomous weapons systems. The policy does not say that humans should be in the loop, but that commanders and operators should be able to "exercise appropriate levels of human judgment over the use of force" for such systems, which require a second "senior review" from the DoD's Autonomous Weapon System Working Group in order to be formally developed and fielded. In times of urgent military need, however, the deputy defense secretary can waive both these requirements along with other key features, including the need to demonstrate that a system can be quickly reprogrammed to correct unintended system behaviors or verify that it functions in accordance with the law of war and weapon system safety rules. The budget documents say Goalkeeper is part of a program designed to meet "an urgent Geographic Combatant Command Requirement."

There was another expeditionary loitering munition program too, named Whiplash. The Whiplash program intended to turn six hundred jet skis such as Yamaha WaveRunners into lethal autonomous surface drones. While China shipbuilding capacity was 230 times larger than that of the US, the US has the lion's share of the $2 billion jet ski market.

"America has a lot of jet skis, so it's neat that we can weaponize them," one person familiar with the program explained. "Imagine a jet ski with a bomb on it with some AI on it that could recognize a target."

The CIA tried out rudimentary versions of both systems in Ukraine to varying degrees of success, I was told, after smuggling them into the country. In July 2024, an unidentified jet ski loaded with explosives washed up on the coast of Turkey, sparking consternation within the Pentagon that the supposedly secretive effort would be unmasked.

By the time the 2026 budget documents were published in June 2025, the names—Goalkeeper and Whiplash—disappeared from budget documents. The autonomy hawks who would re-emerge under the second Trump administration finally got their way. But the description of the programs is still included, and spending allocated to both platforms combined has more than doubled to $588 million. Early Goalkeeper drones cost $500,000, but they are cheaper now that Goalkeeper is in low-rate production. Whiplash is already up to full-rate production, and the Navy is now "expanding the industrial base for this capability" by integrating the government-owned autonomy technology into other unidentified vendor platforms.

AS THE REPLICATOR TEAM WONDERED if their baby would survive the second Trump era, some joked about a rebrand months into 2025. "Musk's machines," came one working proposal, quickly abandoned after Elon Musk and Donald Trump fell out in June.

Pressure was building on the Replicator team and its backers, the Defense Innovation Unit, in the run-up to the final test event in August 2025, when the Trump administration's new defense teams would decide the future of Replicator. Goalkeeper and Whiplash were only ever tangentially involved in Replicator. The main program was behind on production, testing, and functioning platforms. The scope of the Replicator program test event had been repeatedly revised down. It would now be sixty-four drone boats in the water, and just six drones flying overhead.

Around the same time, the Trump administration decided to reorientate Replicator as many had predicted. First, it settled on a new name—the Defense Autonomous Warfare Group: DAWG.

Second, it handed responsibility for the new killer robots plan over to Special Operations Command. Marine Corps Lieutenant General Frank Donovan, vice commander of Special Operations Command, was put in charge of DAWG. He was meant to make a decision on the right platforms and the right plan to use them. Unlike some military leaders, Donovan long understood that special operations must fully pivot to China and the defense of Taiwan.

"Someone here still thinks they're going to be raiding desert compounds," he'd shake his head in private military meetings, according to one attendee. In public, he referred to the end of America's global war on terror: "That ship, the GWOT ship, has sailed." He regularly said he didn't need US Navy SEALs raiding villages anymore: he wanted them "back in the boats."

But that didn't mean he was going to rely solely on hi-tech drone swarms to defend Taiwan. He was widely respected as a no-nonsense professional.

When he attended the final Replicator test event in August 2025, I learned that Donovan, a former JSOC assistant commanding general, was disappointed by some aspects of Replicator (including how public it was), but was impressed at the level of collaborative autonomy the experts had ultimately pulled off. Replicator hadn't managed to produce multiple thousands of drones as previewed, but it had still fielded hundreds of autonomous drones, including to Japan. And the test finale showed more than thirty boats could all "talk" to each other autonomously.

Third, Donovan decided, in the short term, to reduce the number of autonomous platforms and pursue simpler drones in defense of Taiwan. He wanted platforms that might actually work, and soon. He picked three: one aerial, one surface, and one undersea. He also wanted to double down on Goalkeeper and Whiplash. And, now, he wanted the SEALs working on the game plan too.

There was still ongoing uncertainty about money and which rival efforts Congress would back. "There's a huge fight over whose ATR actually works," a congressional aide told me in December 2025. Another person framed it to me as a battle for the soul of US military autonomy. Advanced swarming was still unachievable for now, but what mattered was focus, the aide argued. Dispensable drone systems couldn't do all the work, but they could buy you time; and timelines, said the aide, were now a priority. Eight years after the start of Project Maven, America was planning out the contours of how to defend Taiwan with autonomous systems.

26

THE WINCHESTER HOUSE

"Am I going to be haunted?"

KIRSTEN CUKOR DIDN'T LIKE WAR one bit. It damaged everyone and everything it touched. "War is not healthy for children and other living things," read a sign she kept pinned to her home office wall. But there was a steely voice inside her: If someone was going to go to war, they should be good at it. An ethical professional. She just didn't know how such a person could survive with their humanity and moral center intact.

She mulled the meaning of AI warfare, in particular. Project Maven and the Marines had chewed up her husband. She'd endured investigation from federal law enforcement agents who raked over their existence. And now she puzzled over the legacy of her husband's relentless drive to bring AI targeting into being. She wanted it to make moral sense to her. She wanted him to be right.

He had spent the three and a half years since he left the Pentagon in self-imposed exile. No more clearances or briefings, no public appearances about defense, no LinkedIn account. He turned down multiple outreaches from Aki Jain for him to come and work at Palantir. ("He should have gotten a star," Jain told me.) But the company, and AI, were still constants in Cukor's life.

After a couple of years at J.P. Morgan, Cukor moved back home to the

West Coast. He was now in charge of data and analytics at TWG Global, a global asset manager pouring money and AI into a portfolio of companies in banking, insurance, sports, defense-tech, and more. TWG's self-made billionaire founder, Chairman and CEO Thomas Tull, had come to investing via a launderette chain and movie producing. After a lifetime obsession with *Star Wars*, he had now got as close as possible to the real thing: Tull had the ears of America's military leadership and was fashioning their futuristic war tools. At one national security forum, I spotted him huddling with four-star generals and admirals. He stopped to tell me how much he liked writing checks of $200 million or so for defense-tech companies. Among them was Shield AI, whose president Brandon Tseng had introduced me to Tull in the corridor.

Cukor himself was no longer focused on getting AI into war, but rather into finance. In May 2025, he delivered a strategic tie-up with TWG, Palantir, and Elon Musk's new company xAI to bring so-called reasoning AI to the future of financial services and embed AI "into the fabric of every decision, every process, and every outcome." The partnership aimed to "develop and deploy a workforce of hundreds of thousands of AI agents" to redefine business. It wasn't entirely separate from war: speeding up industrial output would be critical to any fight against China.

Sitting beside Tull for a television interview about the announcement, Palantir CEO Alex Karp brought up Cukor. "Thomas went and found the world's leader in battlefield operational AI and hired him. So you don't have to trust your taste on AI; you can trust his," Karp told CNBC. (In the same interview, Tull described Cukor as "a very well-known practitioner in artificial intelligence," saying that he hired Cukor to put together the team, and try the theory out on the financial services companies Guggenheim Investments and TWG first, before rolling it out more broadly.)

The idea was that, in Cukor's hands, xAI's supercomputer and AI models would redefine loans, mortgages, credit cards, investments. It would, he argued, "turbocharge economies." The relentless and combative work still pitted him into constant battle against compliance and naysayers who worried about AI-induced machine errors, bias, and mass financial injustice. But a sign on Cukor's new Santa Monica beach-view corner office—like his longer, softened hair—suggested a newfound freedom. "Gone swimming," read a casual handwritten note permanently attached to his office door. "Back at 3:30."

He got the shiny camper van he'd always dreamed about for his family, rode his electric bicycle up and down the coast to the office, and surfed every Saturday. He watched documentaries about the popes with Kirsten, and still got up at 4 a.m. on Sundays to work. But perhaps, as if coming out of delayed shock, he felt a little of the sting of abandonment from the dwindling months of his thirty-year military career and all those endless fights, real and bureaucratic. He was beginning to thaw from war.

The Cukors' new neighborhood in Los Angeles was a patchwork of California beach homes and views. The odd sign scattered on front doors and porches expressed varying political points—Biden boosters, a peace sign, an Israeli flag. You couldn't see any insignia from the front of the Cukor home, but out back in a quiet alleyway used for vehicles and wheelie bins flies the Marine Corps flag in red and gold. Even so, one neighbor listened to Cukor talk so much about the toll of war that the neighbor viewed Cukor as a pacifist. Another neighbor was constantly asking Kirsten if she was even allowed to say where Cukor was, fantasizing that shadow worlds followed the retired Marine Corps intelligence officer everywhere.

I arrived at their home early one Saturday morning in spring 2025. They tucked into bagels over the kitchen counter and later trundled out through the garage. Cukor wheeled an electric hydrofoil—a motorized surfboard—down to the beach on a midsize trolley, chatting attentively but without fuss to a young child, who clambered aboard and hugged the upturned prop to stay on. Cukor had got the digital contraption to save himself from yanking his fifty-eight-year-old war-damaged shoulder as he attempted to paddle out past the break. The Tesla of surfboards, he called it. He made his way out into the turbulent gray water, *USMC* stitched out across the back of his black life vest by one of his daughters. I sat beside Kirsten in rare Los Angeles drizzle on the damp sand. From a distance, we could see him buzzing high above the surface of the sea under the glare of a drone camera, capturing him as he appeared to walk on water.

Kirsten had been alarmed at the prospect of meeting me. Her husband was intensely private and guarded his world carefully. She was protective of how he might be represented, even as she reckoned she had no poker face herself. "Drew, you cannot leave me on the beach with a journalist. I eat sodium pentothal for breakfast," she'd protested to him. "You'll be fine," he'd told her.

What's the sodium thing, I asked. "Truth serum."

The pair did museums differently (her slow; him fast), but they had got used to picking over their moral framework about AI warfare together. Her husband would regularly come home relating how people were saying Project Maven was building Skynet, the superintelligent computer in the *Terminator* movies that launches nuclear war against humans. He saw it so totally differently and much less fantastically, she said: machines should do what they're better at and people should do what they're better at. One of their daughters had imbibed the idea early, drawing pictures of cheerful, big-bellied, waving robots. Cukor had adopted these images as the faces of the office challenge coin. *Officium Nostrum Est Adiuvare* went the team's Latin motto. We're here to help. (Kirsten had proposed *Help us help you*, to emphasize the process started and ended with humans, but nobody could figure out the Latin for it.)

"It was never, 'Can we kill more people?,'" said Kirsten. "It was always, like, these were unacceptable losses." AI could find what others had missed, and save lives. That was morally valuable for them both.

An unofficial Maven patch offered a different perspective. Commissioned by Mavenites in Afghanistan, the fabric badge depicted a robot unleashing a drone strike against an RPG-wielding figure in a boundary box tagged "asshol" [*sic*].

Kirsten later told me she hadn't known about the unauthorized patch, but she had ruminated nonetheless about whether AI would make it easier to kill, to take a life. Unless a person were a sociopath, being involved with the taking of human life—ending a person—is always morally uncomfortable, she said. Yet she took comfort from the prevailing

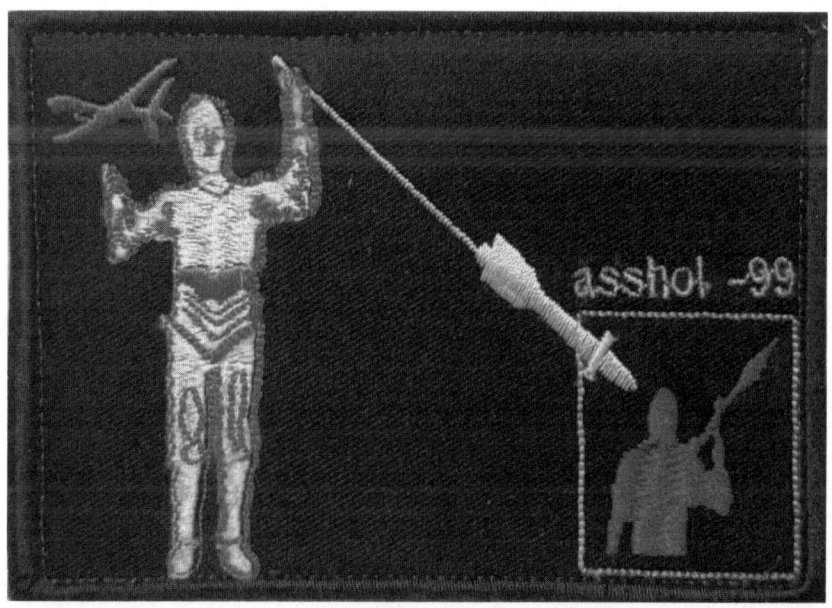

Unauthorized Project Maven patch made in Afghanistan, provided to the author by a former Mavenite.

insistence that a human would always be involved in the decision to use AI to kill someone on the battlefield. Something about bringing artificially intelligent machines to warfare in the absence of humans didn't feel right. "It just feels icky, you know? It just feels so morally unsavory in general."

The weight of AI warfare gnawed on her even as her husband told her Maven had been used effectively. "How do we feel about that?" she recalls asking him. "Am I going to be like the lady at the Winchester House?"

She meant the curious 1880s Santa Cruz mansion named for Sarah Winchester, whose late husband went into the family gun business. The Winchester family made a repeating rifle first produced in 1866. Rather than fire out a single shot, it could chamber and fire several cartridges before it needed reloading. It was the nineteenth-century version of the future of war, pushing up the pace of combat. Armies all over the world bought millions of them—France, Switzerland, the Ottoman Empire, South America, and Australia. Death sped up. Death on repeat.

After she was widowed in 1881 and inherited her husband's stake in the business, Sarah Winchester spent decades building and rebuilding an enigmatic mansion of secret rooms and staircases that ran into dead ends, which Kirsten visited in the late 1990s. "It's one of the weirdest places I have ever been; so dark," she said.

The supernatural horror movie *Winchester* later portrayed Sarah Winchester and her eccentric house as haunted by the souls of the people killed by the Winchester repeating rifle. "The idea was that she was so haunted by the loss of life that he had been complicit in that she ends up building this house and for the rest of her life she's always building, building, building, and trying, basically, to reduce the haunting," Kirsten related to me. Kirsten began to wonder after the same thing. "Am I going to have to build a Winchester House?" she'd asked her husband, half-laughing, but not quite freely. "Am I going to be haunted by, like, dead Russians?"

CUKOR WAS ALREADY HAUNTED by the old wars. "His experiences over there always unfolded to me very slowly," Kirsten said. "It takes a long time for you to process those experiences."

He never read the books that engrossed his wife, even when she picked out a sci-fi read hoping it might appeal to him. But if ever he recommended a book to her, she would devour it. She wanted to know him better. That's how she came to read *The Beekeeper*, the 2018 retelling of the stolen Yazidi women of Iraq. Cukor told me he gave the book to algorithm vendors in Maven's early days, arguing that better intelligence and faster insight from AI would have made a US rescue mission possible.

Once, when Cukor sent me one of his draft concept papers about how to defeat the enemy through inducing "psychological paralysis," I saw some of the document edits were made by a person named Sinjar. When I asked him much later who that was, he told me it was him. He named his computers—his tools—after the places that were important to him. Lashkar Gah. Musa Qala. Fallujah. Ramadi. Sinjar.

"All of us carry PTSD," he told me the day before I met Kirsten, talking about the people who come back from war. "These images are etched in our brains." Cukor talked about his war hurt as if he were mapping a mission on acetate. At a remove. Context first. Circling closer. Soft, warm voice. No hint of emotion. The medium-altitude drone. But he was telling me something else too. Sneaking it in. "All of us" included him.

"Taking a human life is extremely injurious. It takes a piece of your soul," he said. "Most of us carry these casualties with us our entire life."

His pursuit of AI was his answer to the old wars. But he had no answer for his war hurt; that was still looking for somewhere to land. "I'm only four years out of the military," he told me. Any sort of reflection would have been impossible during his thirty years inside. There was little cohesion in multiple overlapping extended deployments. The units kept changing. So the camaraderie wasn't there. That's why he thought there'd been so many suicides. No cathartic moments. Just carrying the weight of it. Always onto the next thing. You file it away. Takes years to unpack. Then comes the night.

"I'm the one next to him at night when he wakes up with nightmares," Kirsten told me. She told me he dreamed about dead people, over and over.

Kirsten said people think PTSD is jumping in fright when car alarms go off, but she thought of it as moral harm. There were always just these unacceptable losses. "You hate for someone to be haunted by that discomfort, but at the same time, you want people who find it difficult to decide to end a life," she said. "I feel like I'd be more worried if there wasn't anyone grappling with what they have to do," she said. "We've all seen people who are cavalier, who don't move in the world in careful ways."

It was complex to chew through, though. "The decision-maker's PTSD is a particularly lonely PTSD," she said. "They can't go to the same groups."

Cukor had emerged from the military with friends, but no particular way of going over the meaning of the past thirty years. "Most of us find ourselves kind of lonely," he said. "I'm just one little guy; a mid-grade officer. Others are dealing with far worse than me." Things rolled around his head. "A lot of things that burn a hole in my soul are the mistakes I've made."

PHILOSOPHERS AND PHARAOHS have debated the morality of war for thousands of years. Theories of Just War attempt to reconcile the wrongness of taking human life with the rightness of defending values and people. Violence is justified, the argument goes, if the cause is just and the conduct ethical.

The idea of a just war was meant to inoculate fighters from legal peril and moral harm, but battle scars of the mind and body tell a different story. Fear and horror were part of it. So were mistakes, guilt, shame—the things no one talks about when they recount stories of a fighter's valor and heroism. William Nash, who was director of psychological health at the US Marine Corps from 2015 to 2022, pioneered the exploration of moral injury within the military. His work turned on the idea that a person can suffer intensely for violating core beliefs, even if conduct falls short of breaking the law. Someone could feel they had betrayed their own sense of what is right and other deeply held expectations about duty and obligation.

Anthony Pfaff, a retired US Army colonel and professor of military ethics, wanted to know the impact that AI would have on all of this. War often brought out the worst in humans rather than the best, he writes in one paper. "War may be a human activity, but rarely does it feel to those involved like a particularly humane activity."

To him, AI was only the latest development to challenge the ways of war, and he warned me against imbibing tidy moral arguments advanced in defense of it. "You should always be cynical about the military taking ethics seriously; we often fail or jumble up our ethical reasoning," Pfaff told me.

"No one really knows if these weapons are inherently evil," he wrote in a 2019 paper that discussed Project Maven. He decided that military applications of AI for decision-making didn't need to be banned by international law, but there was still room for concern. In 2020, the 18th

Airborne Corps enlisted him to advise them on ethics as they started experimenting with AI targeting for the Scarlet Dragon exercises. Over time, he told me he grew more worried about sending AI into battle.

The push to proliferate AI weapons of war as a way to save US combatants from getting killed on battlefields introduced a complex calculation. The very unacceptability of Americans dying in combat was meant to prevent leaders from sending them into war in the first place. A riskless military campaign run by AI might undercut notions of courage and sacrifice, turn fighters into technicians—the grave outcome of going to war might now be just a click away. "If they're riskless why not use them more? I think that's a legitimate concern," said Pfaff. The new drone wars suggest he might be right.

AI would not only distance fighters from the act of killing, but it would also potentially distance them from the decision to kill as well. That risked fundamentally failing to recognize—or, perhaps, experience—the humanness of the opponent they were killing. And, if done wrong, AI weapons could desensitize soldiers to the killing they do and result in atrocities for which no one is accountable, Pfaff wrote.

Removing fighters from direct physical harm also didn't guarantee that their psyche would be saved. As drone warfare showed, operators surveilled people for weeks before a missile was fired. "Then they watched them slowly die," Pfaff said.

AI could watch the gore instead. AI might outperform humans in other ways too, dampening a very human reaction to violence: "AI doesn't get mad, it doesn't take revenge, and it can be hyper-precise where humans may not be," he continued.

Pfaff thought AI might offer ways to minimize collateral harm, but that automation might ultimately grow the volume of harm rather than reduce it. Jevons Paradox, a well-known idea in economics, suggests that technologies that improve efficiency and lower costs lead to increased consumption.

Maven Smart System, and other AI-enabled mission command systems, are designed to serve commanders more targets, faster, and make it easier to strike them. It is hard to imagine this will lead to fewer strikes—a real cause for concern, according to Pfaff, since neatly displayed digital targets generated by a black box could make commanders overconfident. "That's how you get automation bias," he worried. "It's not hard to get humans to trust machines."

In one dated example, a convoy route in Iraq was previously categorized as among the most dangerous routes in an area. But because convoys stopped using the route, an early algorithmic system recommended it as one of the safer routes because there had been no recent recorded attacks.

James Boggess, another retired Army colonel who also worked with the 18th Airborne Corps, worried that AI screen displays were so like gaming consoles that soldiers might be lulled into using "game ethics" in place of personal ethics. Overconfident gamers took unjustified risks based on the illusion of knowledge and control. The more familiar the individual was with the system, the riskier their behavior became. Algorithmic training might teach operators to be unethical, and prosecute war as if it were a game. A former member of JSOC told me it was easier to take someone's life watching it on screen than physically beside them, saying that in Afghanistan they would joke that it was easier to kill someone than to capture them. US deadly boat strikes in late 2025 upheld the point.

General Donahue, the former commander of the 18th Airborne Corps, was perhaps aware of Jevons Paradox—the more efficient something is, the more you might do it. He was not prepared to argue that AI would save civilian lives. AI might even lead to more harm. "It just depends on how humans decide to use it," he told me.

AI was already helping to kill people, but in some quarters I detected a certain glee about it. "Everyone wants to be known as the new Robert Oppenheimer and go down in the history books," one former Mavenite told me. The parallel was fraught, but to those who used it, it meant two things: a new technology to redefine the balance of military power in the world, and a new ability to mete out death on an unparalleled scale. Cukor never talked about AI warfare this way. He saw the risks, but told me he didn't regret his work setting the stage for AI warfare. He believed the US would maintain high standards and formal processes around the use of

AI in combat: "I don't believe we would ever go just launch stuff all over the place. It's not reflected in our doctrine."

He argued that AI would enable better planning and preparation. It would find and shoot targets quicker, within what he presented as exacting rules of America's strike-approval processes shaped by law, morality, and strategy. These required multiple intelligence sources and advance collateral damage assessments plus legal, and sometimes, presidential review. "'Sir, you have one resource that one source confirms. Sir, you have twelve women and children; do you want to strike?'" he mimicked. "It's still a human choice."

All the guardrails are in there, he contended. "But can you do something horrific with this technology? Can you take those guardrails off? Yes."

THE ISRAEL DEFENSE FORCES' USE of AI targeting against Gazans in reprisal for the October 7, 2023, Hamas massacres that killed 1,221 people, took 251 hostage, and were condemned as war crimes and crimes against humanity, has garnered accusations of "AI-assisted genocide." The UN counted more than 65,000 Palestinians killed by late September 2025, the majority of them civilians including 18,430 children.

"No part of life and death decisions which impact entire families should be delegated to the cold calculation of algorithms," said António Guterres, the UN secretary-general, of reports that AI targeting was leading to a high level of civilian casualties. Israel has argued that its response has been "lawful," "responsible," and "proportionate."

In the early days of Israel's response, the IDF said it went through 15,000 targets in thirty-five days. Mark Milley, the former top US military officer and recently retired four-star Army general, cited reporting from *+972 Magazine* that humans reviewed AI-generated suggestions for as little as twenty seconds before authorizing air strikes. An anonymous source told the reporter the system had a 10 percent error rate.

Milley started warning that AI warfare was a double-edged sword. At a talk of his I attended in October 2024, he pointed to high casualty numbers both in Ukraine and Gaza. These were the "movie trailers" of future war, he said. He sketched out a horrific vision of what was to come, filled with combat in megacities taking place at lightning speeds. AI warfare risked opening up a Pandora's box, he announced. He boomed out his warning from stage: in the worst-case scenario, AI warfare could endanger humanity.

One former US defense official who visited the frontline of Ukraine came back shaken by the intentional targeting of civilians since 2024 by Russian drone pilots. Residents in Kherson referred to this deliberate terror campaign as a human safari. "Can I imagine having such a tactic being run by AI? I hate to tell you I can," the person told me. "Atrocities are possible in the era of AI."

Some advocates of AI warfare attempt to argue that the decision to pursue looser rules of engagement falls solely in the realms of policy, and has nothing to do with AI itself. Michael Horowitz, a former senior defense official who helped oversee AI policy, makes this argument. So did Cukor, at first. As far as he could tell from public reporting alone, the IDF had turned to AI targeting after shooting so much in the early days of its response that it ran out of pre-prepared targets. He described any decision to lower the bar on what might be an acceptable target to engage as a policy decision, not the fault of an AI system. (The IDF didn't respond to requests for comment but it has widely defended its use of AI and its targeting methods, saying it has operated within the law and that AI was used to cross-check information, not to select targets.) "Just because you have the tool doesn't mean you use it. I'm sad when people don't apply the guardrails," Cukor said.

It was the argument offered by all advocates of AI—blame the policy, not the tech. But the line missed the key role played by AI, I pointed out to Cukor. The AI targeting machine that Israel had developed since 2019 made possible the immense speed and scale of warfare it was choosing. Reports detailed a policy of death by database. Its adoption was controversial even inside the IDF, in part because it wasn't clear that data would turn up the right people and because fixating on AI had arguably distracted the IDF from more traditional sleuthing work that might have uncovered the October 7 plot. I put it to Cukor that the AI targeting machine makes possible the policy decision, enabling operational speed and volume. "This is correct," he conceded.

He hoped people wouldn't take shortcuts and, if they did, he said he would be ashamed. "Any misjudgment in the use of tech or the campaign will haunt those service members forever," he told me. "They're going to remember that in the dead of night. When they're sixty-five, it's going to come to their memory and they're going to have to carry that. The destructiveness scars the human soul."

As criticism of Israel's response and Big Tech's role in it began to rise, Michael Rogers, a retired four-star Navy admiral who led the NSA and US Cyber Command until 2018, was among a group of retired US military leaders who counted themselves strong supporters of Israel and visited the IDF in March 2024. Their resulting report said that Israel was not violating the Law of Armed Conflict. Rogers nevertheless said he was "surprised" at the extent to which Israel was combining AI with autonomy. At a national security conference I attended at Vanderbilt University in April 2025, he told hundreds of professionals he came back from the trip "sucking my teeth a little bit," saying "wow, this is really different," and emphasizing that he'd "never seen anybody do it at this speed and this scale in this little amount of time." Israeli military commanders told him the adversary was moving so fast and so continuously that it was making it "very hard" for Israel to apply the Law of Armed Conflict, Rogers told the audience.

Later, he told me he saw something outside his experience level as an American military officer: "We hadn't gone down this road as far."

He thought warfare was going in the direction of autonomy, but he urged caution. "Just because we can do something doesn't mean we should," he said. And while AI targeting and the likes of Maven delivered speed and scale, he assailed the usual claim made for AI about precision. He suggested instead that pursuing speed and scale might necessarily require "a tradeoff" with certainty and accuracy.

"We, the US, should step back and assess that and see what's our comfort level," he told me, referring to the use of AI for targeting. "What's the criterion we want to apply for the enhanced use of AI? What kind of controls do you need to have in place? We cannot forget the human dimension of this."

Some of the tech companies and practitioners producing AI targeting software and the future of war didn't seem much interested in the human dimension. Some spoke about killing with ease. Palantir Chief Executive Alex Karp—who said his company is involved in "operationally critical operations" in Israel, established a strategic partnership with the IDF, and held a company board meeting in Tel Aviv in solidarity with Israel—said that his company existed "to scare enemies and on occasion kill them." He was constantly selling war in support of "the innate superiority" of the West, arguing that domestic support for combat was essential if western democracies were to be able to deploy armies.

Protesters who regularly picket tech companies that provide services to the IDF, including Google, Microsoft, and Palantir, argue the companies are complicit in civilian deaths. Microsoft disabled some cloud services to the Israel Ministry of Defense in September 2025, after confirming *Guardian* reporting that the IDF was using its Azure cloud to facilitate mass surveillance of civilians in Gaza. Palantir has lost employees and investors over its support for the IDF. In April 2025 Karp ad-libbed to a protester, who interrupted a talk of his, that his company's technology kills "mostly terrorists."

Cukor told me he didn't care for such cavalier statements about war and the implied death of civilians from Karp, describing them as "unfortunate." He characterized Karp and others like him as speaking polemically, as businessmen sitting atop companies. "People who know this field are not liquored up by those statements," he said. "They don't serve in uniform; those things are just easy things to say."

When I asked, Cukor told me he never said as much directly to Karp, the man he earlier described to me as "wonderful," but he said he had made clear he didn't like easy talk of killing.

The Geneva Conventions of 1949 and customary international humanitarian law require attackers to distinguish between combatants and civilians and ban launching attacks that might knowingly kill "excessive" numbers of civilians. "Even wars have rules," says the ICRC.

Cukor believed in those rules. They were crafted by the people who came together in the shadows of two world wars to try to prevent future horrors. To dismiss or overturn them was a serious matter, he said slowly. "I'm not sure as a society we're ready to do that; I hope we're not."

Under the second Trump administration, it became clear that the US rules of engagement were in flux. One US defense official was reticent to address my questions about AI and collateral damage. The person wouldn't usually hold back about stressing the importance of averting collateral damage in public, they indicated. But it might first be important to make sure such views were "congruous" with Pentagon leadership, the person said. A former senior member of JSOC separately worried to me that Pentagon leadership was blurring rules of engagement and had "completely grayed" the lines that special operators used to understand.

In his 2024 book, *The War on Warriors*, Army National Guard veteran Pete Hegseth had questioned the Geneva Conventions. "Our boys should not fight by rules written by dignified men in mahogany rooms eighty years ago," he wrote. As Trump's newly renamed Secretary of War, he gathered all the US military's senior generals in September 2025. Among his remarks: he wanted "more AI in everything" and no more "overbearing rules of engagement."

EPILOGUE

NEARLY FORTY MILLION PEOPLE DIED in the First World War. An estimated eighty-five million people died in the Second World War. A Third World War remained unthinkable. Drew Cukor was resolute that he didn't regret unleashing algorithmic warfare on the world. Admiral Samuel Paparo thought AI in combination with other platforms could buy him a month around the Taiwan Strait. Cukor told me something quite different: he hoped just the fear of fighting an AI-enabled American war machine could buy the United States a decade of deterrence with China. And if it came to war Cukor thought AI might help America win.

He carried with him the strange combination of peace and defiance in a person who has achieved something of which he is proud. "I'd do it again, in the same way," he told me.

He had established the Pentagon as buyer for AI, helping to tether an American invention to the country as companies flailed around in search of a commercial marketplace for this costly emerging technology. And he had told me his tale, he said, because he wanted young aggressive officers to see space for themselves to rise up the ranks and rail against bureaucracy, no matter the personal or professional cost. "I want the next generation to be risk-takers and not be complacent," he said. He had never made general and he could count as many detractors as fans but he had got something done: Maven Smart System was everywhere, in every service, continent and in allied militaries too, and the Pentagon was putting AI and data at the heart of military operations. "We were evangelists to get AI going," he told me.

Maven's acolytes—divided as they might be on Cukor himself—had spread out into the new techno-military complex spanning companies including Anduril, Microsoft, Palantir, and OpenAI, where Joe Larson was now running all their government work, and throughout the Pentagon's ranks and new leadership. Colin Carroll, now running his own autonomy startup, liked to call them the "Maven mafia."

A former Mavenite, Stephen Winchell, was now running the Defense Advanced Research Projects Agency (DARPA); another, Jules Hurst III, was running the Pentagon's finances; and a third, Nicholas Waguespack, was exploring the future of AI for all defense intelligence. Robert Imig—the Palantirian who helped create Maven Smart System—in September 2025 became a senior adviser for AI and autonomy to Stephen Feinberg, deputy secretary of war. Tasked to get the Pentagon's AI strategy, he told me, "back on track," he saw Maven as a huge success story that needed replicating across the department. The company Imig left behind, Palantir, could still count on Dave Spirk, as a senior counselor, and Sy Poggemeyer. The company remained one of the most valuable—and, according to a position taken by shortseller Michael Burry the same month, overvalued—in the world. Jim Caggy, the AWS official who worked on Maven, was nominated as assistant secretary of defense for mission capabilities, telling Congress he believed in getting AI out of the lab and into the field quickly. Brian Ward told me he was returning to the Pentagon to speed up AI kill chains and make them work better. He didn't care what he did, as long as there were as few people as possible saying "no" between him and getting Maven done. Katya Volkovska was back working on autonomy at the Defense Innovation Unit. And despite the internecine clashes, Carroll was always misty-eyed: "I wouldn't trade my time at Maven for anything," he texted me one evening. Dozens more were pursuing AI and autonomy, as the Pentagon promised in December 2025 to roll out AI to three million users. All of them owed something to Cukor.

AI was no longer just a bag of potato chips, but it wasn't sufficient to change the nature of war—or even win one—just yet. The Marine Corps was the only service that was signed up to Maven Smart System with its own money, and the system still took fire from critics—that it was too Army, too Palantir, and the AI still wasn't good enough. Getting Maven's AI onto autonomous platforms had so far proven too messy to pull off. No one had cracked how to unite a surveillance supersystem built on largely classified global sensors with the power of unclassified commercial algorithms and the realities of military trust and training. But they were on their way.

It wasn't only Indo-Pacific Command that wanted to create a "Hellscape." General Donahue told me the US and NATO allies were developing a "Hellscape" of their own in Europe. He was referring to his new Eastern Flank Deterrence Line, calculated to keep Russia at bay, and that meant AI too. In August 2025, he threatened Russia: there was absolutely no reason NATO couldn't take down Moscow's protective bubble over Kaliningrad, a forty-seven-mile-wide Russian exclave, with the implied help of Maven Smart System. "We already planned that; we've already developed it." They could do it in an unprecedented timeframe, he added. Donahue also wanted everything to be "optionally manned." War would get down to unmanned systems fighting unmanned systems, he told me. He was already seeing the DNA of it.

Cukor had always thought it would take twenty years to change the US military; they were only halfway through. He always foresaw a union between human and machine, not a machine takeover. "If you get these things tuned up the right way, they can perform better than humans," he insisted. AI might help assail the inevitable problem: "War is fraught with human error."

So was America.

"We're flawed," he told me one mid-October evening in 2025.

Just when I wondered if Cukor might not be prepared to grapple with every aspect of his legacy, he wrestled with one more. We were meeting for a final time on the thirty-first floor of a Manhattan skyscraper. He'd flown over on the red-eye, and the cold, rainy weather had caught him unawares. He was more than fourteen hours into his working day, and whatever battery kept him going was beginning to run out as jet lag ate away at him. But he had one more point to make.

There were "dark parts" to this new military technology he had helped fashion. "Let's make sure that we know those flaws as we wield this technology," he said.

He framed America's future as if it subsisted within the parameters of an AI model trained on the bedrock of the country: "How we use this tech is going to be the fusion of many people and all their biases."

He argued the distinctive factors that drove America to develop world-beating new technologies—wealth, geographic isolation, and stability among them—didn't exonerate his country from a fundamental burden. After giving three decades and some of his health to the US military, and pursuing an AI revolution in warfare, Drew Cukor had a nagging doubt.

"Let's be able to look at ourselves in the mirror and make sure we are careful," he told me. "We have all this tech; are we the best custodians of it?"

ACKNOWLEDGMENTS

Thank you to Andrew Wylie, my agent, who in 2018 saw a book in my absorption in the meaning and future of war and later offered rousing advice: "Follow your interest."

Thank you to Tom Mayer, editor of this book, for his interest in the subject, support for this story, and advice on how to tell it. Thank you to the team at publisher W. W. Norton, including Nneoma Amadi-obi, Avery Hudson, Robert Byrne, Rebecca Homiski, Steve Colca, Kyle Radler, Mike Giarratano, Edward Klaris, Anna Oler, and Dan Gerstle. Thank you to cover designer Tim Green and Steve Attardo. Thank you to Steven Pace, Karen Rice, Sharon Gamboa, Michael Harrigan, Meg Sherman, and teams. Thank you to Helen Handelman for her work on the endnotes, and to James Pullen, Katie Cacouris, and Nicholas Allen at The Wylie Agency.

Thank you to Andy Martin, my editor at Bloomberg, Managing Editor Lynn Doan, and Senior Executive Editor Tom Giles, for support for the book and my taking time off and expert feedback on the drafts. Thank you to *Businessweek*'s Matthew Campbell and Emily Cadman for their work on the story that ran in February 2024. I thank my colleagues Jake Bleiberg, Ryan Gallagher, Jane Lanhee Lee, Patrick Howell O'Neill, Jordan Robertson, Margi Murphy, Jeff Stone, and Jamie Tarabay who cheered me on even as they surely took on extra work in my absence. Thank you to former colleagues and editors at the *Financial Times*, the Pentagon press corps 2017–2022, Reuters and to Kathleen Herron at the *Sunday Times*, who got me started in newspapers. Thank you to Miles Morland.

I am grateful to Drew Cukor for his time and openness to field my every question, and to Kirsten Cukor. I am grateful to everyone who spoke with me for the book, named and not named, often devoting hours to helping me figure out details regarding technology, timelines, ethics, arguments, and sometimes contrasting and anguished perspectives. I am grateful for every single scrap of paper and pixel I've been able to review to help tell a tale whose constituent parts live mostly on classified servers and in the memories of those involved.

Many thanks to early readers and discussants of parts of this book, which is much improved thanks to feedback from expert and thoughtful eyes, including Frances Asquith, Gil Barndollar, Alanna Frieda, Tom Giles, Nick Hornby, Jessica Lawson, Andy Martin, Kendall Taggart, and others I can't name. Thank you to those who kindly hosted me along the way, including Sarah McGregor and family. And full heart and thanks to my delightful, steadfast, inspiring parents, friends, family, and hero husband.

NOTES

A NOTE ON SOURCES

vii **"Freedom of Information Act":** US Department of Defense, *Project Maven FOIA Determination*, December 18, 2018. This DoD determination was sent to *The Intercept*, which filed a FOIA request in March 2018 seeking details of Project Maven's use of Google technology, and reported in March 2019 that "the public still knows precious little beyond the basic gist of the story" and that 5,000 pages of relevant material held by DoD were exempt from disclosure. See Sam Biddle, "Pentagon Says Google's Drone Work Is Exempt from Freedom of Information Act," *The Intercept*, March 25, 2019.

vii **program of record:** "On 7 November 2023, Maven became an NGA program of record and was funded in the defense budget." See Office of Inspector General, *Semiannual Report to Congress* (National Geospatial-Intelligence Agency, October 1, 2024–March 31, 2025).

vii **classified:** "All portions of NGA's budget are classified, to include funding for NGA Maven," a spokesperson for NGA confirmed to me in September 2025.

vii **full history of Project Maven:** Several books have tackled aspects of Project Maven's origin story and other elements, such as Paul Scharre, *Four Battlegrounds: Power in the Age of Artificial Intelligence* (W. W. Norton, 2023); Raj M. Shah and Christopher Kirchhoff, *Unit X: How the Pentagon and Silicon Valley Are Transforming the Future of War* (Scribner, 2024); and David Sanger, *New Cold Wars: China's Rise, Russia's Invasion, and America's Struggle to Defend the West* (Penguin Random House, 2024). Jack Poulson, a former Google worker who resigned from the company and established his own Substack, and is Executive Director at nonprofit Tech Inquiry, has investigated Maven contracts in standalone reports. Several reporters have revealed details about the early days of Maven contracting and protests, including Kate Conger then at *Gizmodo*, Cade Metz at the *New York Times* and Lee Fang then at *The Intercept*. Brandi Vincent at *Defense Scoop* has chronicled Maven's programmatic developments.

PROLOGUE

1 **Drew Cukor:** "Colonel Drew Cukor, USMC," *GovExec* biography. Cukor is pronounced Coo-core, with the emphasis on the first syllable.
1 **2017 documentary:** *AlphaGo*, directed by Greg Kohs (2017).
1 **board configurations:** "AlphaGo," Google DeepMind.

1 **"a moment of exquisite algorithmic ingenuity"**: "The Story of AlphaGo," Barbican Centre on Google Arts and Culture, 2019.
1 **floored Sedol:** AlphaGo played Move 37, which had a 1 in 10,000 chance of being used, and Lee Sedol played Move 78, which was just as unlikely and helped him win that game. Ibid.

INTRODUCTION

3 **leading the transformation of artificial intelligence:** "Deploying a Multidisciplinary Strategy with Embedded Responsible AI," J.P. Morgan, February 28, 2023.
5 **overvalued:** "Palantir Might Be the Most Overvalued Firm of All Time," *Economist*, August 12, 2025; "'Big Short' Investor Michael Burry Fires Back after Alex Karp Blasted His Bet against Palantir, *Business Insider*, November 10, 2025.
5 **protested involvement:** Scott Shane and Daisuke Wakabayashi, "'The Business of War': Google Employees Protest Work for the Pentagon," *New York Times*, April 4, 2018.
5 **"founding father":** Alex Karp (in the presence of an audience including the author), Pallas Foundation for National Leadership, "National Security Innovation Forum," Johns Hopkins University Bloomberg Center, November 20, 2024.
7 **eight hundred other AI projects:** Frank Bajak, "Pentagon's AI Initiatives Accelerate Hard Decisions on Lethal Autonomous Weapons," *AP News*, November 25, 2023.
7 **ten NATO members:** According to a NATO official.
7 **five thousand targets a day:** According to an official at the National Geospatial-Intelligence Agency.
7 **the future of war:** Katrina Manson, "Robot-Soldiers, Stealth Jets and Drone Armies: The Future of War," *Financial Times*, November 16, 2018; Katrina Manson, "Low-Cost Warfare: US Military Battles with 'Costco Drones,'" *Financial Times*, January 4, 2022.
8 **2018 national defense strategy:** Katrina Manson, "US Puts Russia and China at Top of Defence Agenda," *Financial Times*, January 17, 2018; Katrina Manson, "Jim Mattis Warns US Losing Military Edge," *Financial Times*, January 19, 2018.
8 **"will change society":** "Summary of the 2018 National Defense Strategy of the United States of America," Department of Defense, January 19, 2018, 3. The strategy said: "New technologies include advanced computing, 'big data' analytics, artificial intelligence, autonomy, robotics, directed energy, hypersonics, and biotechnology—the very technologies that ensure we will be able to fight and win the wars of the future."
8 **"overbearing rules of engagement":** "Secretary of War Pete Hegseth Addresses General and Flag Officers at Quantico, Virginia," Department of Defense, September 30, 2025.
8 *Is War Now Impossible?***:** Jan Bloch, *Is War Now Impossible?* (Ballantyne Press, 1899).

CHAPTER 1: OLD WARS

13 **Old Wars:** This chapter relies on multiple interviews by the author with (separately) Spirk and Cukor in 2024–25.
13 **Dave Spirk:** For a biography of Spirk, see "Speakers," Data Innovation Forum, GovConWire. com, June 15, 2021.
13 **Task Force 58:** 26 MEU (SOC) Public Affairs, 26th Marine Expeditionary Unit, *Preparing To Go* (The United States Marine Corps, 2001); Sam Cox, "H065.1: Operation Enduring Freedom—September to December 2001," *Naval Intelligence Professionals*.
13 **"Colonel Has Another Outstanding Suggestion":** I learned his call sign was CHAOS during trips accompanying him in 2018. It is the name of his book—Jim Mattis and Bing West, *Call Sign Chaos: Learning to Lead* (Random House, 2019)—and he explains it as follows in a 2022 interview: A major who "did not agree with the brilliance of all [his] ideas—very tongue in cheek" gave him the nickname, and Mattis adopted it as his call sign. Max Raskin, "Interview with General Jim Mattis," February 1, 2022. A representative for Mattis declined comment on this and other details.
13 **doctrine:** "Doctrine is the last refuge of the unimaginative . . . it is a guide, not an intellectual

NOTES 359

straitjacket. Improvise, adapt, and overcome." Cox, "H065.1: Operation Enduring Freedom—September to December 2001."
14 **hunt bustard:** Ibid.
14 **night-vision goggles:** "Lessons from Rhino LZ," *Armed Forces Journal*, November 1, 2011.
14 **lone camel:** Cox, "H065.1: Operation Enduring Freedom—September to December 2001."
15 **chief data officer:** "David Spirk was named the Chief Data Officer (CDO) within the office of Chief Information Officer on June 22, 2020." See "Speakers," Data Innovation Forum.
16 **"killed more people on Office":** Quote by Joseph O'Callaghan, then an Army colonel and a panelist at event the author attended. "Building the Tech Coalition," Center for Security and Emerging Technology, August 29, 2024.
16 **Eight Americans:** "Afghanistan Fatalities," icasualties.org (Iraq Coalition Casualty Count).
16 **2,465:** US Department of Defense, *Casualty Status* (2025); "U.S. Military Casualties—Operation Enduring Freedom (OEF)," Defense Casualty Analysis System, August 4, 2025.
16 **20,149 wounded:** Ibid.
16 **Suicides:** Thomas Howard "Ben" Suitt III, "High Suicide Rates among United States Service Members and Veterans of the Post-9/11 Wars" (Boston University, 2021).
16 **Enduring:** The op was named Operation Enduring Freedom, and this history says CENTCOM squashed the name Mattis gave to the op; banned use of "Swift." Cox, "H065.1: Operation Enduring Freedom—September to December 2001."
16 **going wrong:** Even the food drops could inflict collateral damage; in one case, a parachute failed to open and a pallet crashed through the roof of a house near Jalalabad, killing a civilian woman inside. On October 16, a USN F/A-18 accidentally hit the Red Cross food warehouse in Kabul, injuring an Afghan security guard and destroying wheat and humanitarian supplies. Cox, "H065.1: Operation Enduring Freedom—September to December 2001."
16 **Clausewitz:** Prussian general and military theorist Carl von Clausewitz argued in his 1832 book (translated into English and published in 1873) that the factors generating wartime decisions were "wrapped in a fog of greater or less uncertainty. A sensitive and discriminating judgment is called for; a skilled intelligence to scent out the truth." Carl von Clausewitz, *On War*, ed. and trans. Michael Howard and Peter Paret (Princeton University Press, 1976), 101.
17 **2,000-pound bomb:** "Interview: Lt. Col. David Fox," *Frontline: Campaign Against Terror*, directed by Mark Anderson and Greg Barker, PBS, September 8, 2002.
17 **killing three Americans and several Afghans:** Vernon Loeb, "Friendly Fire Deaths Traced to Dead Battery," *Washington Post*, March 23, 2002; Cox, "H065.1: Operation Enduring Freedom—September to December 2001."
17 **"intense but unflappable":** Spirk's description of Cukor to the author.
17 **Kandahar Airport:** Gregory Bereiter, "The U.S. Navy in Operation Enduring Freedom, 2001–2002" (Naval History & Heritage Command, 2016), 73.
17 **in the 1960s:** Monica Whitlock, "Helmand's Golden Age," *BBC News*, August 7, 2014.
17 **Seabees:** Bereiter, "The U.S. Navy in Operation Enduring Freedom, 2001–2002," 76.
17 **"Jihad Motel"; "Santa Taliban":** "Kandahar, Afghanistan 2001–2002," posted April 2, 2016, by Brian Kelly, YouTube.
17 **land mine exploding:** Hon. Thomas G. Tancredo, speaking about Marine Cpl. Christopher Chandler, on December 19, 2001, 107th Cong., 1st sess., *Congressional Record* 178, 147:E2350.
17 **makeshift tourniquet:** "Chandler Doesn't Let Prosthetic Leg Keep Him from being Marine, Athlete," *San Diego Union-Tribune*, December 8, 2012.
17 **things to come:** Thousands of US and allied forces would be killed and injured by IEDs, and many suffered the enduring impact of war hurt. Corporal Christopher Chandler recovered at Walter Reed, testified about his experience to Congress in 2005, returned to combat three times in Iraq, and was medically discharged in 2008 with PTSD. In retirement, he focused on helping others: he began working with the United States Olympic Committee as the lead US paralympic sports coordinator. Focusing on wellness and education, he became a healthcare educator in San Diego and Camp Pendleton hospitals. Chandler was pursuing his PhD in Exercise Physiology

before he died in 2023. Oral Statement of Sergeant Christopher T. Chandler United States Marine Corps: Hearing at Personnel Subcommittee of the House Armed Services Committee (2005); *SSGT Christopher Tsutomu Chandler, 43. USMC (Ret)*, January 6, 2024.
20 **only eight:** MajGen Michael Ennis (Ret), "Intelligence Plan of 1994," Marine Corps Association, December 15, 2024.
20 **his wife:** Kirsten Cukor, interview with author, April 26, 2025.
21 **thesis:** Drew E. Cukor, "Marine Ground Intelligence Reform: How to Redesign Ground Intelligence for the Threats of the 21st Century" (master's thesis, Naval Postgraduate School, 1997).

CHAPTER 2: TILTING AT WINDMILLS

22 **with the help of overwhelming firepower:** Cukor has since argued elsewhere the "shock and awe" of the US onslaught in 1991 created "a peace bubble" afterward; Daniel Faggella, "Drew Cukor—AI Adoption as a National Security Priority (US-China AGI Relations, Episode 3)," *The Trajectory Podcast*, September 19, 2025.
22 **"Military victory":** Cukor, "Marine Ground Intelligence Reform," 10.
22 **wars of maneuver:** US Marine Corps, *Warfighting*, MCDP 1 (Department of the Navy, 1997), 38.
22 **"friendly fire":** "The Gulf War," in *Semper Fidelis: A Brief History of Onslow County, North Carolina, and Marine Corps Base, Camp Lejeune* (Louis Berger Group, Inc., 2004), 84.
22 **small-scale maps:** James Howcroft, *The Trajectory of Intelligence Practice from Desert Shield to Iraqi Freedom to Today*, March 20, 2013.
22 **tubes jiggling:** Ibid.
22 **issuing reports:** "Intelligence Successes and Failures in Operations Deserts Shield/Storm," Report of the Oversight and Investigations Subcommittee of the Committee on Armed Services, House of Representatives, 103rd Cong., 1st Sess." (US Government Printing Office, 1993).
22 **Van Riper Plan:** named after Marine intelligence Brigadier General Paul Van Riper, who in his 1991 paper—BGen Paul K. Van Riper, USMC, "Observations During Operation Desert Storm," *Marine Corps Gazette* 75, no. 6 (June 1991): 55–61—criticized Marine Corps intelligence during the First Gulf War, focused his call for change more on training and promotion than practice: "I had the sense many of the problems are endemic and stem from the way we select, train, and educate our intelligence personnel." For a recent take on Van Riper, see Major Jake Yeager, US Marine Corps, *Modernize Marine Corps Intelligence for a Naval Fight* 150, no. 11 (2024).
24 **"accurate battlespace picture":** Cukor, "Marine Ground Intelligence Reform," 47.
24 **"weapons of surprise"; "Within 24 hours":** Ibid., 80–81.
24 **planes collided:** Shirley A. Kan et al., *China-U.S. Aircraft Collision Incident of April 2001: Assessments and Policy Implications*, CRS Report for Congress (Congressional Research Service, 2001).
25 **biggest office:** This was true until 2023. Sanj Atwal, "Surat Diamond Bourse Surpasses the Pentagon as World's Largest Office Building," *Guinness World Records*, August 22, 2023.
25 **an entire university:** "DAU," DAU.com (Defense Acquisition University). The Trump administration announced it would overhaul acquisition, including renaming DAU as the Warfighting Acquisition University, "Transforming the Defense Acquisition System into the Warfighting Acquisition System to Accelerate Fielding of Urgently Needed Capabilities to Our Warriors," November 7, 2025.
25 **never passed an audit:** Gabriel De Luca Vinocur, "Fact Check: Has the Pentagon Failed Its 7th Audit in a Row?," *Econofact*, December 20, 2024.
25 **Cukor complained:** Cukor, "Marine Ground Intelligence Reform," 35n.
26 **fifty-six Marines were killed:** "U.S. Military Casualties—Operation Iraqi Freedom (OIF)," Defense Casualty Analysis System, August 4, 2025.
26 **Iraqi civilians:** Neta C. Crawford and Catherine Lutz, "Human Cost of Post-9/11 Wars: Direct War Deaths in Major War Zones, Afghanistan & Pakistan (Oct. 2001–Aug. 2021); Iraq (Mar. 2003–Aug. 2021); Syria (Sept. 2014–Mar. 2021); Yemen (Oct. 2002–Aug. 2021) and Other

Post-9/11 War Zones," The Watson School of International and Public Affairs: Costs of War, September 2021.
26 **456 Marines:** "Iraq Fatalities," iCasualties.org (Iraq Coalition Casualty Count).
27 **Haditha:** Three pieces separated by nearly twenty years investigate what happened in Haditha; see Tim McGirk, "Collateral Damage or Civilian Massacre in Haditha?," *Time*, March 19, 2006; William Langewiesche, "Rules of Engagement," *Vanity Fair*, March 26, 2007; Madeleine Baran, "The Haditha Massacre Photos That the Military Didn't Want the World to See," *The New Yorker*, August 27, 2024.
27 **later photographed:** Baran, "Haditha Massacre Photos."
27 **controversy:** "SIGAR: Special Inspector General for Afghanistan Reconstruction—Quarterly Report to the United States Congress," July 30, 2017; "Breaking: McCaskill Releases Scathing Report on Pentagon's Mismanagement of Afghanistan 'Legacy' Program," *Homeland Security & Governmental Affairs*, April 26, 2018.
27 **Marineistan:** *U.S. Marines in Afghanistan, 2010–2014 Anthology and Annotated Bibliography*, compiled by Paul Westermeyer and Christopher N. Blaker (History Division: United States Marine Corps, 2017).
28 **Robert Kelly:** Tom Perry, "Marine General's Son Laid to Rest at Arlington," *Los Angeles Times*, November 22, 2010; "Robert M. Kelly," Fallen Heroes Project.
28 **a public note:** Ibid.
28 **one of twenty-five Marines:** Lance Cpl. James Gulliver, 1st Marine Division, *5th Marines Dedicate Memorial to Their Fallen*, June 7, 2013.

CHAPTER 3: WE DO WHAT WE WANT

29 **We Do What We Want:** This chapter relies on documents reviewed by author and interviews with Drew Cukor, multiple former and current Palantir employees, and others familiar with the details discussed who requested anonymity to speak freely.
29 **$8 trillion:** Jill Kimball, "Costs of the 20-Year War on Terror: $8 Trillion and 900,000 Deaths," *News from Brown University*, September 1, 2021.
29 **two-thirds:** Gregg Zoroya, "How the IED Changed the U.S. Military," *USA Today*, December 19, 2013.
29 **mine-resistant ambush protected (MRAP) vehicles:** Bruce A. Busler, "Traffic Engineering and Highway Safety Bulletin 13–05: Mine Resistant Ambush-Protected (MRAP) Vehicles," Military Surface Deployment and Distribution Command, February 2014.
29 **painful, delayed, and ongoing fight:** Peter Cary and Nancy Youssef, "JIEDDO: The Manhattan Project That Bombed," The Center for Public Integrity, March 27, 2011; *Joint Improvised-Threat Defeat Agency Needs to Improve Assessment and Documentation of Counter-Improvised Explosive Device Initiatives (Redacted) DODIG-2016-120* (Department of Defense Office of Inspector General, 2016).
30 **Joe Larson:** According to interviews and documents reviewed by the author.
30 **Captain Larson:** This draws on interviews, official biography, and documents reviewed by the author; see "Joseph Larson," Center for Strategic & International Studies.
30 **"revolving door":** William D. Hartung and Dillon Fisher, *March of the Four-Stars: The Role of Retired Generals and Admirals in the Arms Industry*, Quincy Brief No. 47 (Quincy Institute for Responsible Statecraft, 2023); *Post-Government Employment Restrictions: DOD Could Further Enhance Its Compliance Efforts Related to Former Employees Working for Defense Contractors*, Report to Congressional Committees GAO-21-104311 (United States Government Accountability Office, 2021).
30 **9/11 Commission Report:** Among the missed opportunities cited by post-9/11 investigations were the failure to see that five of the nineteen hijackers used the same phone number as ringleader Mohammad Atta to book their airline tickets, two used the same frequent-flier number, and five used two common addresses to make their reservations. See *The 9/11 Commission*

NOTES

Report (2004); "I wish I had Palantir when I was director," former CIA director George Tenet told *Forbes* in 2013, looking back at whether Palantir could have helped tip US authorities to 9/11. Ryan Mac, "National Security Darling: Why Condoleezza Rice, David Petraeus and George Tenet Back Palantir," *Forbes*, August 19, 2013.

31 **paper:** Drew Cukor et al., "Intelligence Database Creation and Analysis: Network-Based Text Analysis versus Human Cognition," *Proceedings of the 41st Annual Hawaii International Conference on System Sciences*, 2008.

31 **working with the CIA:** Harry Goldstein, "Modeling Terrorists," *IEEE Spectrum*, September 1, 2006.

31 **might reconstitute itself:** Ibid.

31 **Jason Payne:** According to an internal company memo reviewed by the author. Payne did not respond to multiple author outreaches.

32 **television interview:** "RARE Alex Karp Interview [with Charlie Rose]: Palantir Philosophy Explained! (2009)," posted March 14, 2024, by Palantir Vision, YouTube.

32 **Palantir:** For a history of Palantir's early years, see "Palantir: The Next Billion-Dollar Company Raises $90 Million," *TechCrunch*, June 25, 2010.

32 **vibe:** Siobhan Gorman, "How Team of Geeks Cracked Spy Trade," *Wall Street Journal*, September 4, 2009.

32 **go see the CIA:** Ibid.; Shane Harris, "Palantir Technologies Spots Patterns to Solve Crimes and Track Terrorists," *Wired*, July 31, 2012; Ashlee Vance and Brad Stone, "Palantir, the War on Terror's Secret Weapon," *Bloomberg Businessweek*, November 22, 2011; In-Q-Tel listing shows they have exited their investment, see "Palantir Investments," iqt.

32 **Susan Gordon:** Interview with author, 2024; Erik German, "Meet the CIA-Backed Venture Fund behind Palantir, Anduril—and a Spy Tool That Might Be on Your Phone," *Fortune*, July 29, 2025.

32 **Google Earth:** In February 2003, In-Q-Tel invested in Keyhole Corp., which was bought by Google the next year. See "Important CIA Contributions to Modern Technology Over the Last 75 Years," Central Intelligence Agency, September 14, 2022.

33 **Analyst's Notebook:** "Info Sheet: Analyst's Notebook," i2 Group, 2024.

33 **by IBM:** "Harris Acquires i2 Product Portfolio from IBM," *Harris*, January 4, 2022.

34 **Errors could be treated as facts:** Joshua Foust, "The Wiki Leak Is More and Less Important Than You Think," *Need to Know*, PBS, July 26, 2010.

34 **described the product:** Interview with author.

35 **conditions of torture:** Senate Select Committee on Intelligence, *Committee Study of the Central Intelligence Agency's Detention and Interrogation Program* (United States Senate, 2012); Anne Daugherty Miles, *Perspectives on Enhanced Interrogation Techniques*, CRS Report for Congress (Congressional Research Service, 2016).

35 **"Flying while Muslim":** Robin Washington, "Flying While Muslim," NPR, December 13, 2006.

35 **use of Palantir:** Harris, "Palantir Technologies Spots Patterns to Solve Crimes and Track Terrorists."

35 **Snowden:** Glenn Greenwald et al., "Edward Snowden: The Whistleblower Behind the NSA Surveillance Revelations," *The Guardian*, June 11, 2013.

35 **would say in 2013:** In an interview, Heidi Shyu, assistant secretary of the Army for acquisition, logistics, and technology, acknowledged DCGS has shortfalls but said the Army is working to improve the network, especially its cumbersome user interface, and "lamented what she called misconceptions that have led to reports in the news media that it could be replaced at lower cost by an off-the-shelf product made by the California-based software company Palantir Technologies." Austin Wright, "Hunter Battles Army on Intel," *Politico*, May 1, 2013.

35 **accused Palantir:** i2 Ltd. et al. v. Palantir Technologies Inc. et al., No. 1:10-cv-0085 (E.D. Va., 2010).

36 **settle out of court:** Ben James, "Palantir, i2 Settle Trade Secrets, Copyright Case," *Law360*, February 14, 2011; Owen Thomas, "Palantir's Third Black Eye: i2 Lawsuit Settled," Reuters, February 17, 2011.

36 **Peter Dixon:** Interview with author.

NOTES

36 **ordered by President Barack Obama:** Barack Obama, "Remarks by the President in Address to the Nation on the Way Forward in Afghanistan and Pakistan," The Obama White House Archives, December 1, 2009.
36 **heavy casualties:** Tim Gaynor and Hamid Shalizi, "Crash Kills 9 U.S. Troops, 2010 Deadliest Year of Afghan War," Reuters, September 21, 2010.
36 **than in the entire Marine Corps:** Scott Chandler, "Rethinking Defense Acquisition: Zero-Base the Regulations," *War on the Rocks,* January 6, 2017.
37 **smallest of the US military services:** The Marines had 158,000 members in September 2010 (Space Force is far smaller but was established only in 2019). See US Naval Institute Staff, "Department of Defense 2022 Demographic Profile," *USNI News,* November 29, 2023.
37 **Michael Flynn:** "Lieutenant General Michael T. Flynn, USA," Defense Intelligence Agency, November 29, 2023.
37 **excoriated the state of intelligence:** Matt Pottinger et al., "Fixing Intel: A Blueprint for Making Intelligence Relevant in Afghanistan," *Center for a New American Security,* January 4, 2010; their criticisms were widely covered, for example, Max Fisher, "Is Military Intelligence in Afghanistan Broken?," *The Atlantic,* January 7, 2010.
37 **"Holmes, not James Bond":** Lt Gen Samual Wilson, quoted by David Reed, "Aspiring to Spying," *Washington Times,* November 14, 1997, quoted in Pottinger et al., "Fixing Intel," 23.
37 **JUONs:** Michael T. Flynn, "Advanced Analytical Capability Joint Urgent Operational Need Statement," Department of Defense USFOR Afghanistan, July 2, 2010.
37 **described Palantir software so closely:** Ibid.
37 **"ghost-written":** Noah Shachtman, "Spy Chief Called Silicon Valley Stooge in Army Software Civil War," *Wired,* August 1, 2012.
37 **Project Navigator:** The team assessed the only thing worse than an IED that killed and maimed their friends would be the arrival of a flying IED: small weaponized drones that could harness precision strike. They called the future "a battle of signatures" referring to all the ways a person or equipment could be found by the signature they give off, such as heat, light, sound, or anything a sensor could detect.
37 **three people a week:** "U.S. Military Casualties—Operation Enduring Freedom (OEF)," Defense Casualty Analysis System.
38 **$550 million:** Steven Brill, "Trump, Palantir, and the Battle to Clean Up a Huge Army Procurement Swamp," *Fortune,* March 27, 2017.
38 **Pentagon's budget:** *Department of Defense Agency Financial Report* (Department of Defense, 2010).
38 **CTTSO:** "Secretariat for Special Operations," Office of the Assistant Secretary of Defense, www.cttso.gov.
38 **armor for dogs:** "Department of Defense Fiscal Year (FY) 2011 Budget Estimates: Volume 3A," Department of Defense, February 2010; "Exhibit R-2, RDT&E Budget Item Justification—Combating Terrorism Technology Support," PB 2011 Office of Secretary of Defense, February 2010, https://apps.dtic.mil/descriptivesum/Y2011/OSD/0603122D8Z_PB_2011.pdf.
38 **plus-up was huge:** "Department of Defense Fiscal Year (FY) 2011 Budget Estimates: Volume 3A."
38 **reprogrammed $10 million:** "Exhibit R-2, RDT&E Budget Item Justification—Combating Terrorism Technology Support," PB 2012 Office of Secretary of Defense, February 2011.
39 **public solicitation:** Interview with author.
39 **Stewart died in 2023:** "Marine Corps, CMC Remembers Lt. Gen. Vincent Stewart," The United States Marine Corps, May 1, 2023.
39 **Cukor got clearances:** Interview with author.
39 **$100,000 generator:** According to an interview with former Palantir employee.
39 **Palantir appealed directly:** Ibid.
39 **Major General John Toolan:** Commander of the 2nd Marine Division whose troops had recently deployed to Afghanistan. See "Toolan, John A. 'Jocko,'" CAPSTONE General and Flag Officer Course; *U.S. Marines in Afghanistan, 2010–2014 Anthology and Annotated Bibliography.*

40 **"shocking to me":** Interview with author.
41 **sleuthed out:** Interview with former Palantir employee. Through a representative, Mattis declined comment.
41 **explosive 2013 session:** Wright, "Hunter Battles Army on Intel;" Andrew Kirell, "Four-Star General Smacks Down GOP Congressman Who Attempts to Walk Out During Testimony," *MediaIte*, April 30, 2013.
41 **2016 lawsuit:** Palantir Technologies Inc. v. United States, 129 Fed Cl. 219 (2016).
41 **draft geostrategic analysis:** "The Global Intelligence Files: Military Section Draft," WikiLeaks, on March 18, 2013.
41 **2014 essay:** Vincent Stewart et al., "New Analytic Techniques for Tactical Military Intelligence," in *Analyzing Intelligence: National Security Practitioners' Perspectives*, ed. James B. Bruce and Roger Z. George (Georgetown University Press, 2014), 250, 264.

CHAPTER 4: THEY CALL IT ALGORITHMIC WARFARE

43 **the Breakfast Club:** Described in the following study, whose analysis some key people involved have told the author they contest; see Gian Gentile et al., "A History of the Third Offset, 2014–2018," RAND, March 31, 2021.
43 **Soviet Navy:** According to former Deputy Defense Secretary Robert O. Work, as told to the author.
43 **Greg Grant:** Interview with author, September 2025.
44 **Jack Shanahan:** According to Cukor and Lt Gen (ret.) Shanahan in separate interviews with the author.
44 **ten-strong:** *Navy Ford (CVN-78) Class Aircraft Carrier Program: Background and Issues for Congress*, no. R42136 (Congressional Research Service, 2023), 1n2.
44 **$13 billion ships:** Ibid., 10.
45 **think tank:** "Robert Work," Center for Strategic and Budgetary Assessments.
45 **nuclear warheads peaked:** Dr. Robert S. Norris, "The History of the U.S. Nuclear Stockpile 1945–2013," *Federation of American Scientists*, August 15, 2013.
45 **Russia was producing more:** Hans M. Kristensen and Robert S. Norris, "Global Nuclear Weapons Inventories, 1945–2013," *Bulletin of the Atomic Scientists* 69, no. 5 (2013).
46 **"Sometimes it's too old":** "The Third U.S. Offset Strategy and Its Implications for Partners and Allies as Delivered by Deputy Secretary of Defense Bob Work," Department of Defense, January 28, 2015.
46 **answer was autonomy:** Defense Science Board, *Task Force Report: The Role of Autonomy in DoD Systems* (Department of Defense, 2012).
46 **"a fricking robot":** Deputy Secretary of Defense Bob Work, "Reagan Defense Forum: The Third Offset Strategy," Reagan Presidential Library, Simi Valley, CA, November 7, 2015; Marcus Weisgerber, "Pentagon Wants to Pair Troops with Machines to Deter Russia, China," *Defense One*, November 8, 2015.
46 **hadn't wanted new tech:** Work recounted this story at the CSET panel, "Building the Tech Coalition," Center for Security and Emerging Technology.
46 **Defense Department study:** Ruth A. David and Paul Nielsen, *Defense Science Board Summer Study on Autonomy*, Active / Technical Report no. AD1017790 (Defense Technical Information Center, 2016).
46 **Watson:** "IBM Watson to Watsonx," IBM.com.
46 **self-driving cars:** "Waymo," X.company (The Moonshot Factory).
47 **"Advances in AI":** David and Nielsen, *Defense Science Board Summer Study on Autonomy*, 5.
47 **"contentious":** Ibid., 15.
47 **no idea what AI meant:** According to multiple author interviews, including with Bob Work and Will Roper.
47 **a Pentagon creation:** This was DARPA; see "Arpanet," July 2020; Walker D. Mills, "The People Who Invented the Internet," *The Strategy Bridge*, July 26, 2019.

47 **$38 billion:** Cheryl Pellerin, "CIO Priorities Include Cybersecurity, Innovation, Retaining IT Workforce," *DOD News*, March 23, 2016.
48 **"Operate to Know":** LtCol Drew Cukor, USMC, "Operate to Know: An Operational and Intelligence Design for the Operational Level of War" (master's thesis, National Defense University Joint Forces Staff College: Joint Advanced Warfighting School, 2014).
48 **eight thousand:** Jeremiah Gertler, *U.S. Unmanned Aerial Systems*, no. R42136 (Congressional Research Service, 2012), 9.
48 **drones:** David and Nielsen, *Defense Science Board Summer Study on Autonomy*, 12.
48 **their own armed drone programs:** Ibid.
48 **dozens of civilians:** Imogen Piper and Joe Dyke, "Tens of Thousands of Civilians Likely Killed by US in 'Forever Wars,'" *Airwars*, September 6, 2021; Azmat Khan, "The Human Toll of America's Air Wars," *New York Times*, December 19, 2021; Azmat Khan et al., "Documents Reveal Basic Flaws in Pentagon Dismissals of Civilian Casualty Claims," *New York Times*, December 31, 2021; Jessica Purkiss and Jack Serle, "Obama's Covert Drone War in Numbers: Ten Times More Strikes Than Bush," *Bureau of Investigative Journalism*, January 17, 2017; "Spotlight: The Civilian Casualty Files," *New York Times*, 2021–22; "Project: Drone Warfare," *Bureau of Investigative Journalism*.
49 **overwhelming:** Mark Pomerleau, "DoD Stands up Team to Take on PED/Intel Problem," *C4ISRNet*, April 28, 2017.
49 **PED:** According to Cukor and Greg Grant.
49 **a demo:** Cukor called on Joe Larson, the Marine Corps reservist he knew from Iraq, to ditch his day job at Palantir and come back to help in October 2016. They would work on a presentation for Work dubbed "Go Big with Automation." He enlisted help from Jaime Kovarna, an Air Force intelligence officer at the Office of the Under Secretary of Defense for Intelligence and Richard Dorchak, an army colonel posted as an agency liaison officer at the National Geospatial-Intelligence Agency responsible for analyzing imagery.
49 **"prime" contractors:** David Choi, "The Top 9 Biggest Defense Contractors in America," *Business Insider*, May 25, 2016.
50 **IDenTV:** IDenTV's website, which lists an address in McLean, Virginia, says the company specializes in content detection, tracking, and recognition from live full motion video and that IDenTV "can automate and track changes across massive satellite imagery, including Wide Area Motion Imagery (WAMI)." IDenTV did not respond to repeated emailed requests for comment. The website provides no telephone number.
50 **wows:** According to interviews with people familiar with the matter.
50 **couldn't even tell his wife:** Roper told author this for the following podcast interview; see Saleha Mohsin and Katrina Manson, "Inside Project Maven, the US Military's Mysterious AI Project," Bloomberg: *Big Take Podcast*, February 29, 2024.
50 **Bob Work's request:** Work has spoken about this in interviews and panel discussions; also confirmed by Roper.
51 **signed the project:** Robet Work, "Memo: Establishment of an Algorithmic Warfare Cross-Functional Team (Project Maven)," Department of Defense, April 26, 2017.
51 **"aw shit":** According to author interviews.
51 *Terminator:* *The Terminator* (1984) and *Terminator 2: Judgment Day* (1991), directed by James Cameron; *Terminator 3: Rise of the Machines*, directed by Jonathan Mostow (2003).
51 **killing three billion people:** This happens in *Terminator 3: Rise of the Machines*. "I am sorry," the Air Force Lieutenant General tells his daughter. "I opened Pandora's box."
52 **"artificial intelligence baked into it":** Jack Corrigan, "Three-Star General Wants AI in Every New Weapon System," *Defense One*, November 3, 2017.
52 **"kill people":** Member of Project Maven who did not wish to be named recounted this to the author.
52 **effort to push AI into battle:** Thom Hawkins and Alexander Kott, "Beyond the Hype: Why We're Closer to AI-Enabled Missions Than You Think," Modern War Institute at West Point, May 4, 2022.

53 **kill chain:** Christian Brose, *The Kill Chain: Defending America in the Future of High-Tech Warfare* (Grand Central Publishing, 2020).
53 **slide:** Including pictures and notes shared by person who did not want to be named.
53 **Skyfall:** *Skyfall*, directed by Sam Mendes (2012).

CHAPTER 5: THE FIRST MAVENITES

55 **The First Mavenites:** Much of this chapter is based on interviews with multiple members of Project Maven who did not wish to be identified by name.
56 **"suck at machine learning":** Kate Conger and Cade Metz, "'I Could Solve Most of Your Problems': Eric Schmidt's Pentagon Offensive," *New York Times*, May 2, 2020.
56 **Schmidt put his faith in Maven:** According to multiple people briefed on his comments. Through a representative, Schmidt declined comment.
56 **service document:** "Managing Combat & Operational Stress: A Handbook for Marines & Families," Headquarters Marine Corps Combat and Operational Stress Control (COSC).
57 **Colin Caroll:** "Colin Caroll," NDIA.
57 **three times:** "Exclusive: Accused Pentagon 'Leaker' Colin Carroll on Life Inside DOD and Hegseth's Leadership," *The Megan Kelly Show*, episode 1058, April 26, 2025.
57 **podcast:** "In the Fight—Scaling AI/ML in Defense with Colin Carroll," *Acquisition Talk* podcast, December 8, 2022.
57 **Big Lebowski:** *The Big Lebowski*, directed by Joel and Ethan Coen (1998).
58 **"general asshole":** Brock Briggs, "Building a Winning Defense Tech Company with Colin Carroll," episode 99, *The Scuttlebutt Podcast*, December 16, 2023.
58 **"competitive as shit":** Ibid.
58 **fired twice from the Pentagon:** *In the Fight*; Daniel Lippman and Jack Detsch, "Third Top Pentagon Official Suspended in Leak Investigation," *Politico*, April 16, 2025; "Now I have the honor of having been fired twice by a deputy secretary and a secretary of defense, you know, in a five year period," Carroll told *The Megyn Kelly Show* in April 2025, see "Exclusive: Accused Pentagon 'Leaker' Colin Carroll on Life Inside DOD).""
58 **"certain type of person":** Colin Carroll, "12 Days of Maven," LinkedIn post, December 14, 2024.
59 **Poggemeyer:** 1st Lt. George McArthur, "31st MEU Marine Goes Back to his Amphibious Roots on USS Peleliu," The United States Marine Corps, October 24, 2014.
59 **St. Petersburg:** According to people familiar with his history.
59 **award-winning paper:** LtCol Drew Cukor et al., "Operate to Know: A Proposed Operating Concept for Intelligence and Operations to Enhance Combat Effectiveness," *Marine Corps Gazette* 98, no. 4 (April 2014): 57–62; "Top Marine Leaders from the 2014 Marine Intelligence Community Recognized at 5th Annual Marine Corps Association & Foundation Intelligence Awards Dinner," PRWeb, September 24, 2015.
59 **Mike Rhoads:** Bill Monroe, "Oregon Veterans of Several Wars Share Memories, Go Fishing Memorial Day," *Oregon Live*, May 26, 2012; Cpl. Kenneth Jasik, "Corpsman Rescue Wounded Marine During Firefight," *Defense Visual Information Distribution Service*, May 21, 2012; Rhoads declined comment on the author's reporting. Colin Carroll says Rhoads was among the first members, see *Building a Winning Defense Tech Company with Colin Carroll*.
59 **became Cukor's unofficial chief of staff:** According to Larson interview with author.
60 **primary contractor:** Jack Poulson has extensively documented companies that worked on Project Maven, including many that were not explicitly listed before, in this September 2021 report, "Easy as PAI (Publicly Available Information)," *Tech Inquiry*, September 10, 2021.
61 **emailed colleagues:** September 9, 2011, email included as part of leaked document trove, "The Global Intelligence Files: Re: Compiled Draft for Comment," WikiLeaks, on February 13, 2013.
62 **"I was not exceptional":** Details about Cukor's biography as told to author, cross-checked by public records.

63 **"will take conversions":** "The Age of Accountability: Why Am I Baptized When I Am Eight Years Old?," The Church of Jesus Christ of Latter-day Saints, February 2000.
64 **"the church has a lot of baggage":** The church permitted Black priests only in 1978, just a few years before Cukor went on his mission. Today church leaders "unequivocally condemn all racism." "Race and Priesthood," The Church of Jesus Christ of Latter-day Saints; the church also had a history of polygamy (which it called "plural marriage") in contravention of US law, until it declared an end to the practice in the US in 1890 and additionally instituted a worldwide ban in 1904 after plural marriages continued in Canada and Mexico. "Plural Marriages and Families in Early Utah," The Church of Jesus Christ of Latter-day Saints.

CHAPTER 6: RELAXED ABOUT FURY

65 **television series:** *Silicon Valley*, created by John Altschuler, Mike Judge, and Dave Krinsky (HBO, 2014).
65 **"massive egos and scores to settle":** "Silicon Valley: Season 1 with Mike Judge and Alec Berg | HBO," posted February 24, 2014, by HBO, YouTube.
65 **small bottle of water:** Recollections of the office shared by a former Mavenite.
66 **It wasn't always love:** This episode relies on accounts from multiple people.
68 **cartoon:** Three cartoons were shared with the author.
68 **"crayon-eaters":** "Marines and Crayons: The Inside Joke, Explained," USAMM.com, January 23, 2025.
68 **"concept of operations":** "Marine Corps Uniform Regulations" (The United States Marine Corps, May 1, 2018).
70 **a single tattoo:** Ibid., 1–7.
70 **facial follicle:** Ibid., 1–12.
70 **"appropriately disseminated"; "good judgment":** Ibid., 3, 1–8.
70 **"matter of tradition":** Ibid., 3.
70 **Maven take on a classic Christmas carol:** Corroborated with multiple Maven members; Carroll, "12 Days of Maven."
70 **Kelly Martin:** "Lt. Col. Kelly Martin, 92nd Operations Support Squadron Commander," Fairchild Air Force Base.
72 **Apollo program:** "There were thirty married astronauts during the Gemini and Apollo programs—all but seven marriages ended in divorce." "The Astronauts' Wives," in the "Race to the Moon" Exhibition at the Digital Public Library of America.
72 **MQ-1 Predators:** The "Q" meant that it was unmanned. For more details, see "MQ-1 Predator Unmanned Aerial Vehicle," 163d Attack Wing, May 2007; "MQ-1B Predator," United States Air Force, September 2015.
73 **MQ-9:** *MQ-9 Reaper Unmanned Aircraft System (MQ-9 Reaper)*, Selected Acquisition Report (SAR) (Department of Defense, 2016).
73 **"platform of choice":** Ibid., 6.
73 **nearly a million total flight hours:** Ibid., 7.
73 **"Gorgon Stare":** Loren Thompson, "Air Force's Secret 'Gorgon Stare' Program Leaves Terrorists Nowhere to Hide," *Forbes*, April 10, 2015.
73 **PMA-263:** "Navy and Marine Corps Small Tactical Unmanned Aircraft Systems Program," NAVAIR.com.
74 **Szymanski:** Vice Admiral (ret.) Szymanski confirmed this recollection with the author.
75 **Work signed a decree:** According to multiple people familiar with the matter.

CHAPTER 7: THE COLONEL AND THE MATH WHIZ

76 **Matt Zeiler:** Aaron Tilley, "Every AI Powerhouse Wanted This Whiz Kid. He's Taking Them On Instead," *Forbes*, July 13, 2017.

77 **undergraduate thesis:** Matthew D. Zeiler, "Learning Pigeon Behaviour Using Binary Latent Variables" (undergraduate thesis, University of Toronto—Engineering Science 2009).
77 **Jeff Dean:** "Jeffrey Dean," Google Research.
77 **swept the board:** Jia Deng et al., "Imagenet: A Large-Scale Hierarchical Image Database," *CVPR*, 2009; Olga Russakovsky et al., "ImageNet Large Scale Visual Recognition Challenge," *IJCV*, 2015.
77 **2014 paper:** Matthew D. Zeiler and Rob Fergus, "Visualizing and Understanding Convolutional Networks," in *Computer Vision—ECCV 2014: 13th European Conference, Zurich, Switzerland, September 6–12, 2014, Proceedings, Part I* (Springer, 2014), 818–33.
78 **open the windows:** Zeiler confirmed this detail in this piece, Aaron Tilley, "Every AI Powerhouse Wanted This Whiz Kid. He's Taking Them On Instead," *Forbes*, July 13, 2017.
79 **SUNet:** "Secure Unclassified Network (SUNet) Infrastructure," SAM.gov; "Evaluation of Cybersecurity Controls on the DoD's Secure Unclassified Network," Department of Defense Office of Inspector General, January 12, 2023.
79 **Pilot AI:** "Pilot AI's Jonathan Su Introduces a Deep Learning Vision Framework for Constrained Devices (Preview)," posted July 13, 2018, by Edge AI and Vision Alliance, YouTube.
79 **Xnor:** Janakiram MSV, "Apple Acquires Xnor.ai To Bolster AI at the Edge," *Forbes*, January 19, 2020.
81 **Honolulu:** "The International Conference on Information and Computer Technologies," ICICT, https://icict.org.
81 **"You Only Look Once":** Joseph Redmon et al., "You Only Look Once: Unified, Real-Time Object Detection," *Proceedings of the IEEE Conference on Computer Vision and Pattern Recognition (CVPR)*, 2016, 779–88.
81 **TEDx presentation:** Joseph Redmon, "Computers Can See. Now What?," posted June 13, 2018, by TEDx Talks, YouTube.
82 **"disastrous by-product"; "abominable deterioration of ethical standards":** Alice Calaprice, ed., *The Ultimate Quotable Einstein* (Princeton University Press, 2010), 397.
82 **"I loved the work":** Joseph Redmon (@pjreddie), "I stopped doing CV research because I saw the impact my work was having. I loved the work but the military applications and privacy concerns eventually became impossible to ignore," Twitter (now X), February 20, 2020.
82 **"The US is the largest":** Joseph Redmon (@pjreddie), "The US is the largest state sponsor of terrorism in the world. Don't work on tech for the military," X, June 21 2025.
82 **Kateryna (Katya) Volkovska:** "Kateryna Volkovska," Center for a New American Society.

CHAPTER 8: SOMALIA

86 **Somalia:** This chapter draws on interviews with multiple Mavenites and others familiar with Project Maven's work in Somalia.
86 **"twenty years":** Cukor drew on the following book for some of his assessment: James Kitfield, *Prodigal Soldiers: How the Generation of Officers Born of Vietnam Revolutionized the American Style of War* (Potomac Books, 1997).
86 **American troops:** *Vietnam War U.S. Military Fatal Casualty Statistics*, Electronic Records Reference Report (National Archives, 2008).
86 **antiwar demonstrators:** "On This Day: October 15—1969: Millions March in US Vietnam Moratorium," *BBC News: On This Day*; "Timeline: Vietnam War and Protests," *American Experience*.
86 **news-sheet:** According to a 1971 write-up, which cites Pentagon disclosure that there were about one hundred incidents of fragging—the murder or attempted murder by soldiers of their officers—in 1970. Col. Robert D. Heinl Jr., "The Collapse of the Armed Forces," *Armed Forces Journal*, June 7, 1971.
87 **"Just start":** Told to the author by someone who worked on Maven who received this advice in person.
88 **"Operate to Know":** Cukor et al., "Operate to Know."

NOTES 369

88 **John Boyd:** Dan Pilat and Dr. Sekoul Krastev, "John Boyd," TheDecisionLab.com.
88 **Hyman Rickover:** "Hyman G. Rickover," Atomic Heritage Foundation; "Science: The Man in Tempo 3," *Time*, January 11, 1954.
88 **Jimmy Doolittle:** "General James Harold Doolittle," United States Air Force.
88 **secretly advocated:** James H. Doolittle et al., *Report on the Covert Activities of the Central Intelligence Agency*, no. 15D (1976).
88 **"spurred his men":** "Science: The Man in Tempo 3."
88 **the first year:** Cheryl Pellerin, "Project Maven to Deploy Computer Algorithms to War Zone by Year's End," *DOD News*, June 21, 2017.
90 **Somalia was a good place:** Marcus Weisgerber, "The Pentagon's New Artificial Intelligence Is Already Hunting Terrorists," *Defense One*, December 21, 2017.
90 **nonpublic review:** Author reviewed 2024 document.
90 **walked the beach:** Manson, "Welcome to Mogadishu."
91 **twin bomb blast:** Hussein Mohamed, Eric Schmitt, and Mohamed Ibrahim, "Mogadishu Truck Bombings Are Deadliest Attack in Decades," *New York Times*, October 15, 2017.
92 **"Authority to Operate":** "An introduction to ATOs," Digital.gov.
93 **Baledogle:** Ty McCormick, "Exclusive: U.S. Operates Drones from Secret Bases in Somalia," *Foreign Policy*, July 2, 2015.
93 **massive assault:** Mark Olson, "New Jersey Cavalry Defeats al-Shabaab Attack," *Defense Visual Information Distribution Service*, December 27, 2022.
94 **Six times in five days:** According to a person familiar with the matter. Lt Col Garry "Pink" Floyd previously referred to updating a system six times in five days without specifiying where. See Mark Pomerleau, "What the Pentagon is Learning From its Massive Machine Learning Project," *C4ISRNet*, May 2, 2018.
95 **Justin Guzzardo:** According to multiple author interviews with Guzzardo.

CHAPTER 9: MORAL OUTRAGE

100 **"Don't be evil":** Tanya Basu, "New Google Parent Company Drops 'Don't Be Evil' Motto," *Time*, October 4, 2015; Google Inc., *Amendment No. 9 to Forms S-1*, filed with the U.S. Securities and Exchange Commission, August 18, 2004.
100 **Hassabis:** "Demis Hassabis," Center for Brain Minds + Machines.
100 **won backing:** David Rowan, "DeepMind: Inside Google's Super-Brain," *Wired*, June 22, 2015.
100 **snapped up:** Samuel Gibbs, "Google Buys UK Artificial Intelligence Startup DeepMind for £400m," *The Guardian*, January 27, 2014.
101 **2015 open letter:** "Autonomous Weapons Open Letter: AI & Robotics Researchers," *Future of Life Institute*, announced July 28, 2015, published February 9, 2016.
101 **Kalashnikov:** C. J. Chivers, *The Gun* (Simon & Schuster, 2010).
101 **signed by:** "Autonomous Weapons Open Letter: AI & Robotics Researchers—Signatories List," *Future of Life Institute*, February 9, 2016.
101 **Stop Killer Robots:** "About Us," StopKillerRobots.org.
101 **Mary Wareham:** "Mary Wareham," HRW.org (Human Rights Watch).
102 **unacknowledged CIA drone strikes:** Ken Dilanian, "CIA Drone Strike Program Targeting Pakistan Winds Down," PBS, May 29, 2014.
102 **first country:** "Press Release: Civil Society Joins Together to Encourage Regional Cooperation on Banning Killer Robots," Stop Killer Robots, December 10, 2019.
102 **Iceland's:** "On 2 June 2016, Iceland's parliament (*Alþingi*) passed a resolution expressing support for a ban on production and use of autonomous weapon systems." "Parliamentary Action," Stop Killer Robots.
102 **supported Silicon Valley:** "Defense spending has served as a monolithic catalyst, vastly accelerating California's industrial expansion and population explosion during the past decade." See James L. Clayton, "Defense Spending: Key to California's Growth," *Western Political Quarterly* 15,

no. 2 (1962): 280–93; "Secret History of Silicon Valley," posted December 4, 2008, by Computer History Museum, YouTube; "Defense Spending by State (FY 2023): California," Office of Local Defense Community Cooperation.
102 **mass spying programs:** Glenn Greenwald et al., "Edward Snowden: The Whistleblower Behind the NSA Surveillance Revelations."
102 **Ashton Carter:** "Ash Carter," US Department of Defense.
103 **two significant contracts:** According to a person familiar with the matter.
103 **Gorgon Stare:** "Sierra Nevada Corporation Achieves Milestone for USAF's Advanced Wide-Area Airborne Persistent Surveillance System—Gorgon Stare Increment 2," SNCorp, July 1, 2014.
103 **took money from DARPA:** David Hart, "On the Origins of Google," National Science Foundation, August 17, 2004; Stephen M. Griffin, "NSF/DARPA/NASA Digital Libraries Initiatives: A Program Manager's Perspective," *Graduate School of Library and Information Science, University of Illinois Urbana–Champaign*, 2000; Stephen M. Griffin, "Funding for Digital Libraries Research: Past and Present," *D-Lib Magazine* 11, no. 7/8 (2005).
103 **Burning Man:** Melia Robinson, "Here's Why Google Went to Burning Man to Find its Next CEO," *Business Insider*, August 28, 2018.
103 **tension at the heart:** "Secret History of Silicon Valley"; Gonzalez, "How Big Tech and Silicon Valley are Transforming the Military-Industrial Complex"; April Glaser, "Thousands of Contracts Highlight Quiet Ties Between Big Tech and U.S. Military," *NBC News*, July 8, 2020.
104 **"stupidest rule ever":** Peter Sagal, "Google Chairman Eric Schmidt Plays Not My Job," *Wait Wait . . . Don't Tell Me!* podcast, May 11, 2013.
104 **on-stage appearance:** Bradley Peniston, "How Will the Pentagon Create Its AIs? The Algorithmic-Warfare Team Is Charting a Path," *Defense One*, July 13, 2017; Pellerin, "Project Maven to Deploy Computer Algorithms to War Zone by Year's End."
104 **flew to California:** "Secretary of Defense Jim Mattis Visits Google Headquarters," US Department of Defense, August 11, 2017.
104 **Aurelius:** US Naval Institute Staff, "SECDEF Mattis' Speech to Cadets at Virginia Military Institute," *USNI News*, September 26, 2018.
104 **get better:** "Media Availability with Secretary Mattis at DIUx," US Department of Defense, August 10, 2017.
105 **research for this book:** According to documents reviewed by the author.
106 **AWS:** According to a person familiar with the matter. AWS did not respond for a request for comment.
106 **Fei-Fei Li:** Fei-Fei Li, "How to Make A.I. That's Good for People," *New York Times*, March 7, 2018.
106 **eventually leak:** Scott Shane, Cade Metz, and Daisuke Wakabayashi, "How a Pentagon Contract Became an Identity Crisis for Google," *New York Times*, May 30, 2018.
107 **"priceless":** Asked for comment, Black said to please make it clear she did not provide any documents and has no comment.
107 **got quietly to work:** According to people familiar with the matter.
107 **"heavily engaged":** "Scaling AI-Enabled Capabilities at the DOD: Government and Industry Perspectives," Center for Strategic & International Studies, March 28, 2024.
108 **forum:** "WEF 2018: AI Is More Profound Than Electricity or Fire," *Business Today*, January 25, 2018.
108 **read about:** Fitzgerald told me he read Jeremy Scahill, *The Assassination Complex: Inside the Government's Secret Drone Warfare Program* (Simon & Schuster, 2017).
108 **helped hide NSA leaker Edward Snowden:** "The Google Employee Who Helped Edward Snowden in Hong Kong," *The Guardian*, June 18, 2023.
108 **Whittaker:** This next section draws on author interviews with Whittaker and others.
109 **confused dogs for horses:** April Taylor, "New Google Photos Neatly Sorts Photos into Collections, But My Dogs Aren't Horses," *iTechPost*, May 31, 2015.
109 **condemnation and crisis:** James Vincent, "Google 'Fixed' Its Racist Algorithm by Removing Gorillas From Its Image-Labeling Tech," *The Verge*, January 12, 2018; "Google Apologises for Photos App's Racist Blunder," BBC, July 1, 2015.

109	**meeting at the White House:** "The Social Implications of Artificial Intelligence Technologies in the Near-Term: AI Now 2016 Primers," *AI Now*, July 7, 2016.
109	**bias and mistakes:** Jordan Pearson, "Why an AI-Judged Beauty Contest Picked Nearly All White Winners," *Vice*, September 5, 2016.
109	**2017 report:** Alex Campolo et al., *AI Now 2017 Report* (AINow, 2017).
111	**leak information:** William Fitzgerald indicated he leaked the Project Maven contract as part of a six-month campaign in this article: "Tech Workers Should Shine a Light on the Industry's Secretive Work with the Military," *MIT Technology Review*, May 10, 2024.
111	**Google's work for Project Maven:** Kate Conger, "Google Is Helping the Pentagon Build Drones," *Gizmodo*, March 6, 2018.
111	**"pilot" project:** Ibid.
111	**"save lives":** Shane and Wakabayashi, "The Business of War."
111	**She was right:** See Chapter 25: "Trump's Robots."
111	**Whittaker:** Whittaker drafted the paper on February 28, according to Nitasha Tiku, "Three Years of Misery Inside Google, the Happiest Company in Tech," *Wired*, August 13, 2019.
111	**document:** Shane and Wakabayashi, "The Business of War"; "Open Letter in Support of Google Employees and Tech Workers," International Committee for Robot Arms Control.
112	**resigned:** Kate Conger, "Google Employees Resign in Protest Against Pentagon Contract," *Gizmodo*, May 14, 2018.
112	**wouldn't renew:** Kate Conger, "Google Plans Not to Renew Its Contract for Project Maven, a Controversial Pentagon Drone AI Imaging Program," *Gizmodo*, June 1, 2018; Drew Harwell, "Google to Drop Pentagon AI Contract After Employee Objections to the 'Business of War,'" *Washington Post*, June 1, 2018.
112	**AI principles:** Sundar Pichai, "AI at Google: Our Principles," *Google: The Keyword* (blog), June 7, 2018.
112	**wouldn't help America:** Billy Mitchell, "Gen. Dunford to Meet with Google on its AI Work That 'Indirectly Benefits' China," *FedScoop*, March 21, 2019.
112	**highest moral and diplomatic authority:** "About the Role of the United Nations Secretary-General," United Nations: Secretary-General.
112	**Guterres expressed concern:** António Guterres, "Secretary-General's Address to the General Assembly," United Nations: Secretary-General, September 25, 2018.
113	**unscripted twist:** António Guterres, "Remarks at 'Web Summit,'" United Nations: Secretary-General, November 5, 2018; Axel Bugge, "U.N.'s Guterres Urges Ban on Autonomous Weapons, Reuters, November 5, 2018. Izumi Nakamitsu, UN High Representative for Disarmament Affairs, told the author in a 2025 interview that Guterres went off-script the first time he advocated the ban in 2018, adding that after that it became official policy. "Under-Secretary-General and High Representative for Disarmament Affairs," United Nations Office for Disarmament Affairs.
113	**"active participant":** Matthew Zeiler, "Why We're Part of Project Maven," Clarifai, June 13, 2018.
113	**three-page public letter:** Liz O'Sullivan, "I Quit My Job to Protest My Company's Work on Building Killer Robots," ACLU, March 6, 2019; Liz O'Sullivan, open letter to Matt Zeiler, January 2019.
113	**quietly visited:** "Google CEO Quietly Met with Military Leaders at the Pentagon, Seeking to Smooth Tensions Over Drone AI," *Washington Post*, October 5, 2018.
113	**accused Google:** Jack Poulson, a former Google employee who left the company in protest at some of its policies and now researches defense-tech, argues "the US weapons and intelligence community dramatically overreacted" to Google's decision not to renew its contract in his July 2020 paper, "Reports of a Silicon Valley/Military Divide Have Been Greatly Exaggerated," *Tech Inquiry*, July 7, 2020.
113	**"took the blows":** According to a person familiar with the matter.
114	**news story:** Ramishah Maruf, "Google Erases Promise Not to Use AI Technology for Weapons or Surveillance," CNN, February 4, 2025.

114 **fused:** Sundar Pichai, "Google DeepMind: Bringing Together Two World-Class AI Teams," *Google: The Keyword* (blog), April 20, 2023.
114 **2024 documentary:** *The Thinking Game*, directed by Greg Kohs (2024).
114 **should work together:** James Manyika and Demis Hassabis, "Responsible AI: Our 2024 Report and Ongoing Work," *Google: The Keyword* (blog), February 4, 2025; Ina Fried, "Google's Hassabis Explains Shift on Military Use of AI," *Axios*, February 14, 2025.
114 **ISR:** Intelligence, surveillance, and reconnaissance.
115 **furtively found:** According to multiple people familiar with the matter.
115 **subsequently maintained:** Devin Coldewey, "AI Acceleration Startup Xnor.ai Collects $2.6M in Funding," *TechCrunch*, February 2, 2017; Ali Farhadi, who founded Xnor, did not reply for repeated requests for comment; Apple, which bought Xnor in 2020, did not respond to outreach from the author.

CHAPTER 10: THE ALGORITHMS HAVE NO CLUE

119 **impossible bonhomie:** Based on several meetings and interviews with author and corroborated with Ward's friends, colleagues, and his mother.
120 **"Mother of All Bombs":** Sune Engel Rasmussen, " 'It Felt Like the Heavens Were Falling': Afghans Reel From MOAB Impact," *The Guardian*, April 14, 2017; Robert Burns, "U.S. Drops Largest Non-Nuclear Bomb on Isis Target in Afghanistan," PBS, April 13, 2017.
121 **"Madre":** I confirmed the details about Brian Ward's family and time growing up with Elizabeth Ward, his mother, in a phone call in September 2025. She started enthusing about Brian before my question had ended, and said that he had been awarded a "Mr. Personality" pin at school. "He's always been this way," she said.
122 **480 hours:** Details in this passage draw on a 381-page redacted report that obscures the identity of the person but relies on cross-referencing with interviews to establish the testimony as belonging to Brian Ward; see *Supplemental Information Regarding the Command Investigation Into the Friendly Fire Incident on 6 April 2011 in Regional Command-Southwest (RC-SW)*, no. 5830 (Department of Defense, 2011). Many of these details (but not Ward's identity) were first reported by the *Los Angeles Times*, which referenced the document, and secured the release of the report through a FOIA request; see David Zucchino and David S. Cloud, "U.S. Deaths in Drone Strike Due to Miscommunication, Report Says," *Los Angeles Times*, October 14, 2011, and David S. Cloud and David Zucchino, "Multiple Missteps Led to Drone Killing U.S. Troops in Afghanistan," *Los Angeles Times*, November 5, 2011.
123 **"I know their daddies":** Azmat Khan, "Two U.S. Soldiers Killed in Friendly Fire Drone Strike," *Frontline*, PBS, October 17, 2011.
125 **"needed Brians":** In September 2025, Cukor hired Ward again at a new job at TWG focused on deploying AI.
125 **Bagram:** "Afghanistan: Two Decades of the US at Bagram Base," BBC, July 2, 2021.
126 **former screener:** According to multiple interviews with author.
126 **email went back to the Pentagon:** According to people familiar with the matter.
126 **Google debacle:** Scott Shane, Cade Metz, and Daisuke Wakabayashi, "How a Pentagon Contract Became an Identity Crisis for Google," *New York Times*, May 30, 2018.
127 **30 percent in the Philippines:** "Abu Sayyaf Group (ASG)," National Counterterrorism Center, as of October 2022; John Spencer, Jayson Geroux, and Liam Collins, "Urban Warfare Project: Case Study #8—Marawi," Modern War Institute at West Point, May 23, 2024.
128 **Not every user:** This section relies on interviews with three people with knowledge of the matter.
128 **ISIS-Khorasan:** "Islamic State Khorasan (IS-K): TNT Terrorism Backgrounder," Center for Strategic & International Studies, 2018.
129 **had told me:** Katrina Manson, "Mattis in Afghanistan as US Pushes for Taliban Deal," *Financial Times*, September 7, 2018.

NOTES

129 **I accompanied him:** Katrina Manson, "US Tries to Bring Its Longest War to an End in Afghanistan," *Financial Times: News in Focus*, September 11, 2018.
129 **"Taliban Air Force":** Wesley Morgan, *The Hardest Place: The American Military Adrift in Afghanistan's Pech Valley* (Random House, 2021), 505.
129 **"just lost a war":** Manson, "Mattis in Afghanistan as US Pushes for Taliban Deal."
129 **Save the Children office:** Michael Safi, "ISIS Claims Attack on Save the Children Office in Afghanistan," *The Guardian*, January 24, 2018.
129 **rocket attack:** Associated Press in Kabul, "Rockets Fired at Afghan Presidential Palace During Eid Speech," *The Guardian*, August 21, 2018.
129 **suicide bombing:** "Afghanistan: Death Toll Soars to 68 in Suicide Bomb Attack," Al Jazeera, September 12, 2018.
129 **hasty US retreat:** Matthew Olay, "Kabul Airport Attack Review Reaffirms Initial Findings, Identifies Attacker," *DOD News*, April 15, 2024.
132 **unarmed civilians:** "Afghanistan: Weak Investigations of Civilian Airstrike Deaths," *Human Rights Watch Report*, May 16, 2018.
132 **US drone strike:** Ahmad Sultan and Abdul Qadir Sediqi, "U.S. Drone Strike Kills 30 Pine Nut Farm Workers in Afghanistan," Reuters, September 19, 2019.

CHAPTER 11: HARBINGER OF DOOM

134 **chess match:** Mike Sweeney, "Submarines Will Reign in a War with China," *Proceedings* 149, no. 3 (2023).
134 **clear superiority:** Bob Work told the author; see Katrina Manson, "Robot-Soldiers, Stealth Jets and Drone Armies: The Future of War," *Financial Times*, November 16, 2018.
134 **think tank report:** Bryan Clark and Timothy A. Walton, "Fighting into the Bastions: Getting Noisier to Sustain the US Undersea Advantage," Hudson Institute, June 2023.
135 **implications for the global order:** Alistair Gale, "The Era of Total U.S. Submarine Dominance Over China is Ending," *Wall Street Journal*, November 20, 2023.
135 **"boomers":** "Fleet Ballistic Missile Submarines—SSBN," US Navy, February 27, 2025.
137 **Woods Hole:** Catherine Musemeche, "How WHOI Helped Win World War II," *Oceanus: The Journal of Our Ocean Planet*, June 5, 2025.
138 **got both:** "Exhibit R-2, RDT&E Budget Item Justification—Digital Warfare," PB 2020 Navy, March 2019.
140 **"an immediate halt":** According to correspondence reviewed by the author.
141 **nuclear-powered attack submarine:** "Exhibit R-2, RDT&E Budget Item Justification—Digital Warfare," PB 2020 Navy.
141 **developing algorithms to rake over the data:** IBM and SAIC were initial possibilities, according to people familiar with the matter. The project settled on Blue Ridge Envisioneering, Leidos, and STR as the first three vendors, all of whose work was tested by MORSE Corp, a Massachusetts company with a suboffice by the Pentagon.
141 **SURTASS:** "An Unofficial History of the Surveillance Towed Array Sensor System (SURTASS), 1972–2015," IUSSCAA.org.
142 **IUSS:** "IUSS Mission," COMSUBPAC (Commander, Submarine Force, U.S. Pacific Fleet); Katrina Manson, "Australia Laments Bureaucratic 'Permafrost' That's Slowing Aukus Security Alliance," Bloomberg, May 23, 2023.
142 **"Underwater Great Wall":** H. I. Sutton, "China Builds Surveillance Network in South China Sea," *Forbes*, August 5, 2020.
143 **Type 096:** Ryan Chan, "China State Media Reveals New Nuclear-Armed Submarine," *Newsweek*, August 1, 2025.
143 **announced an agreement:** "Joint Leaders Statement on AUKUS," The Biden White House Archives, September 15, 2021.

143 **live onstage:** "Offset '23: International Allies & Partners | Collaboration for Mutually Beneficial Opportunities," SecondFront.com (Second Front Systems), May 30, 2023.
143 **go-ahead to sell:** Ben Felton, "Australia Cleared for $207 Million Modular SURTASS Buy," *Naval News*, October 5, 2023.
144 **lives on today:** Department of the Navy Chief Information Officer, "Navy AI Undersea Task Force Selected for Coalition Warfare Program," *Department of Navy Artificial Intelligence (AI) Newsletter*, June 2022.
144 **military parade:** Nectar Gan et al., "China Showcases Military Strength at Parade as Xi Stands Alongside Putin and Kim," CNN, September 3, 2025.
144 **passive acoustics:** Eirwen Williams, "'China Can Track Submarines Near Alaska': U.S. Declares Nationwide Alert After Alarming Low-Frequency Detector Breakthrough," *Sustainability Times*, April 20, 2025; Michael Peck, "China's Military Wants to Target US Undersea Sensor Network: Analysis," *Defense News*, August 13, 2025.

CHAPTER 12: ARMS RACE

145 **anniversary:** "The Chinese Embassy in the United States Held a Reception in Celebration of the 75th Anniversary of the Founding of the People's Republic of China," Embassy of the People's Republic of China in the United States of America, October 1, 2024.
145 **design:** Suevon Lee, "China's New Embassy in U.S. Reflects Growing Clout," *New York Times*, May 28, 2008.
145 **wrote in 2008:** Kurt M. Campbell, "McEmbassy," *The American Interest* 3, no. 5 (2008).
145 **largest holder:** "FACTBOX: China, the U.S. Treasury's Top Foreign Creditor," Reuters, February 10, 2010.
145 **second-largest economy:** Justin McCurry and Julia Kollewe, "China Overtakes Japan as World's Second-Largest Economy," *The Guardian*, February 14, 2011.
145 **overtook the US:** Naomi Xu Elegant, "China's 2020 GDP Means It Will Overtake U.S. as World's No. 1 Economy Sooner than Expected," *Fortune*, January 18, 2021.
146 **military budget:** Kathrin Hille, "China's Military Capability Set to Grow Faster Than Its Defence Budget," *Financial Times*, March 5, 2024.
146 **prioritized China:** Colin Clark, "'Mattis' *Defense* Strategy Raises China to Top Threat; Allies Feature Prominently," *Breaking Defense*, January 18, 2018.
146 **Xi Jinping's speech:** Helen Davidson, "Xi Jinping Forecasts 'Rough Seas' on 75th Anniversary of People's Republic of China," *The Guardian*, October 1, 2024.
147 **taken the stage:** "Remarks by Defense Attaché Major General Liu Zhan at the Reception Marking the 96th Anniversary of the Founding of the People's Liberation Army of China," Embassy of the People's Republic of China in the United States of America, July 25, 2023.
147 **by 2027:** Noah Robertson, "How DC Became Obsessed with a Potential 2027 Chinese Invasion of Taiwan," *Defense News*, May 7, 2024.
147 **own vision for AI:** "China's New Generation of Artificial Intelligence Development Plan," Foundation for Law and International Affairs, July 30, 2017.
147 **"military revolution":** Elsa B. Kania, "Chinese Military Innovation in Artificial Intelligence: Hearing of the U.S.-China Economic and Security Review Commission," Center for a New American Security, June 7, 2019.
147 **new AI development plan:** Arjun Kharpal, "China Wants to Be a $150 Billion World Leader in AI in Less than 15 Years," CNBC, July 21, 2017.
147 **Artificial Intelligence Summit Forum:** Ryan Fedasiuk et al., *Harnessed Lightning: How the Chinese Military Is Adopting Artificial Intelligence* (Center for Security and Emerging Technology, 2021), 27, 66n153.
148 **2023 obituary:** Cui Jia, "Young AI Pioneer Dies in Traffic Accident," *China Daily*, July 16, 2023; Minnie Chan, "Military Farewells A.I. War Games 'Pioneer,'" *South China Morning Post*, July 17, 2023.

NOTES 375

148 **Wandering Earth 2:** *The Wandering Earth 2*, directed by Frant Gwo (2023).
148 **2017 report:** Elsa B. Kania, "Battlefield Singularity Artificial Intelligence, Military Revolution, and China's Future Military Power," Center for a New American Security, November 28, 2017.
149 **footnote:** Ibid., 16n.
149 **National Security Commission:** National Security Commission Artificial Intelligence Act of 2018, 5356, 115th Congress 2 (2017); "Press Release: Stefanik Introduces Artificial Intelligence Legislation," House Armed Services Committee, March 21, 2018.
150 **publicly argue:** Eric Schmidt et al., *Final Report: National Security Commission on AI*, Technical Report (National Security Commission on Artificial Intelligence, 2021).
150 **new office:** Sydney J. Freedberg Jr., "Joint Artificial Intelligence Center Created Under DoD CIO," *Breaking Defense*, June 29, 2018; Cate Metz, "Artificial Intelligence Is Now a Pentagon Priority. Will Silicon Valley Help?," *New York Times*, August 26, 2018.
150 **JAIC:** "Labs and Facilities," Johns Hopkins University Applied Physics Laboratory.
150 **said General Joseph Dunford:** Stew Magnuson, "Dunford Slams Google for Working with China, But Not U.S. Military," *National Defense*, November 18, 2018.
150 **in briefings:** Cukor related his congressional pitch to the author.
151 **budget jumped:** Stephanie Meloni and Mark Wisinger, "Pentagon's 2019 IT Budget Appropriations Target Cyber, AI and New Organizations," *FedScoop*, February 7, 2019; Nathan Strout, "The Pentagon's AI Center Is Poised for a Breakthrough," *C4ISRNet*, August 30, 2019.
152 **early conference appearance:** Peniston, "How Will the Pentagon Create Its AIs?"
152 **national security exemption:** US Department of Defense, *Project Maven FOIA Determination*.
153 **Dome Plate:** According to those familiar with the twice-yearly conference.
153 **"Five Eyes":** Katherine Haan, "What Is the Five Eyes Alliance?," *Forbes*, June 4, 2024.
154 **The UK might not have:** The UK's Ministry of Defense did not respond to requests for comment on these details.
154 **"existential fight":** Cukor argues there is a "civilizational" contest for AI between the US and China in this podcast, *Drew Cukor—AI Adoption as a National Security Priority*.
154 **One suggested China could rush to deploy:** Elsa B. Kania, "'AI Weapons' in China's Military Innovation," *Brookings: Global China*, April 2020.
154 **another that China was using AI:** Ryan Fedasiuk et al., *Harnessed Lightning: How the Chinese Military Is Adopting Artificial Intelligence*.
154 **a third:** Jacob Stokes, *Military Artificial Intelligence, the People's Liberation Army, and U.S.-China Strategic Competition: Testimony Before the U.S.-China Economic and Security Review Commission* (Center for a New American Security, 2024).

CHAPTER 13: DADDY KARP

155 **op-ed:** Alexander C. Karp, "Our Oppenheimer Moment: The Creation of A.I. Weapons," *New York Times*, July 25, 2023.
155 **Einstein's 1939 letter:** Albert Einstein, "Letter to President Roosevelt," August 2, 1939, original at the Franklin D. Roosevelt Presidential Library and Museum.
155 **"Little Boy" and "Fat Man":** Julie Miller, "A Tale of Two Bomb Designs," *The Vault* (National Security Research Center), October 10, 2023.
156 **"one great mistake":** Deborah Nicholls-Lee, "'It Was the One Great Mistake in My Life': The Letter from Einstein That Ushered in the Age of the Atomic Bomb," BBC, August 6, 2024.
156 **overtaken:** Jason Ma, "SpaceX and Palantir Now Have Bigger Valuations than Top Aerospace-Defense Stocks as the Military Eyes Transformation," *Fortune*, December 7, 2024.
156 **national security conference:** "Reagan National Defense Forum," Reagan Foundation; "People's Perspective: Public Opinion on Defense and National Security," posted December 7, 2024, by Ronald Reagan Presidential Foundation and Institute, YouTube.
156 **Department of War:** Erica L. Green, "Trump to Sign Order Renaming the Defense Department as the Department of War," *New York Times*, September 4, 2025.

156 **outgoing speech:** "Remarks by President Biden in a Farewell Address to the Nation," The Biden White House Archives, January 15, 2025.
157 **inauguration ceremony:** Ali Swenson, "Trump, a Populist President, Is Flanked by Tech Billionaires at His Inauguration," *AP News*, January 20, 2025.
157 **would donate $1 million:** Anna Massoglia, "Million-Dollar Donors Flooded Trump's Second Inauguration," Brennan Center for Justice, July 1, 2025.
157 **S&P 500:** "Palantir Technologies, Dell Technologies, and Erie Indemnity Set to Join S&P 500," S&P Global, September 6, 2024.
157 **crested:** Ed Lin, "Palantir Chairman Peter Thiel to Sell Up to $1 Billion of Stock," *Barron's*, September 10, 2024.
157 **overtook Boeing:** Chelsey Dulaney, "The Stock-Market Winners and Losers of 2024," *Wall Street Journal*, December 31, 2024.
157 **Anduril:** "Anduril and Palantir to Accelerate AI Capabilities for National Security," Anduril, December 6, 2024.
157 **Booz Allen:** "Press Release: Booz Allen and Palantir Partner to Boost U.S. Defense—Will Drive New Capabilities for the Evolving Battlefield," Booz Allen, December 6, 2024.
157 **it banned users:** Hayden Field, "OpenAI Quietly Removes Ban on Military Use of Its AI Tools," CNBC, January 16, 2024.
157 **OpenAI announced:** "Anduril Partners with OpenAI to Advance U.S. Artificial Intelligence Leadership and Protect U.S. and Allied Forces," Anduril, December 4, 2024.
158 **$164 billion:** Bloomberg Terminal readout.
159 **Shield AI:** "Shield AI and Palantir Technologies Deepen Strategic Partnership and Announce Deployment of Warp Speed," Shield AI, December 5, 2024.
159 **to take off:** Gary Hatch and Mary Kozaitis, "SecAF Kendall Experiences VISTA of Future Flight Test at Edwards AFB," United States Air Force, May 3, 2024; Tara Copp, "An AI-Controlled Fighter Jet Took the Air Force Leader for a Historic Ride. What That Means for War," *AP News*, May 3, 2024.
159 **"They don't get tired":** Jon Harper, "Air Force's Kendall: AI Agents Had 'Roughly an Even Fight' Against Human F-16 Pilot in Recent Engagements," *Defense Scoop*, May 8, 2024.
160 **sanctions:** "China Imposes Sanctions on 13 US Firms Over Arms Sales to Taiwan," Bloomberg, December 5, 2024.
161 **"I've always had the fantasy":** Nearly three months later, Karp riffed on the theme, although he stopped suggesting he had bought his own drone company, Jawwwn (@jawwwn_), "$PLTR CEO Dr. Alex Karp 'You need a higher purpose, and I think you often need a lower purpose. @andrewrsorkin: "what's your lower purpose" Karp: "I love the idea of getting a drone and having light fentanyl-laced urine spraying on analysts that tried to screw us,"'" X, February 20, 2025.
162 **two-year low:** Trevor Jennewine, "Palantir Stock Fell Nearly 40% From Its High. History Says This Will Happen Next," *Yahoo! Finance*, April 5, 2025.
162 **wrote Edwin Dorsey:** Edwin Dorsey, "Problems at Palantir Technologies (PLTR)," *The Bear Cave*, June 1, 2023.
162 **"Daddy Karp":** "Daily $PLTR discussion thread! Come here to talk about the good, the bad and the 💎 🙌," Reddit, 2023. See Lizette Chapman, "Palantir's CEO Spars with Wall Street as Shares Keep Rising," *Bloomberg Businessweek*, September 23, 2024.
162 **"KARPGOD":** OpenGod (@Opendoor_God, formerly @Karp_God), "Changed up the profile a bit. Big plans ahead for the KARPGOD account. Stay with me," X, September 11, 2025.
162 **he told the host:** "Debating the $2.50 PLTR Bear Case: Edwin Dorsey of The Bear Cave," posted June 7, 2023, by Palantir Vision, YouTube.
162 **Karp told Bloomberg TV:** "Palantir CEO: Not Sure We Should Sell Our AI to Some Clients," interview by Ed Ludlow, June 1, 2023, Bloomberg TV.
162 **through $50 billion:** According to stock market on Bloomberg Terminal.
163 **rival paper routes:** Kenrick Cai, "Shield AI Rejected A Pivot To 'Selfie Drones'—Now Its Drones Are Being Used By the Military Overseas," *Forbes*, July 16, 2020.

NOTES 377

164 **Team 7's mantra:** Chief Petty Officer Eric Chan Naval Special Warfare Group ONE, "Looking to the Past: SEAL Team 7 Celebrates 20 Year Legacy," *U.S. Navy: Naval Special Warfare Command*, March 17, 2022.
165 **237 US forces:** 237 in 2012; "U.S. Military Casualties—Operation Enduring Freedom (OEF)," Defense Casualty Analysis System.
165 **3,000 more were wounded:** Ibid.
167 **Wi-Fi charging company:** The company was WiPower; see Anthony Clark, "Fortune 500 Company Acquires Startup," *Gainesville Sun*, September 17, 2010.
167 **seven other companies' aircrafts:** According to interview with author. *Forbes* reported the company also had mishaps, which a Shield AI representative later confirmed to me, without confirming all the details in David Jeans, "How a Grisly Injury Threw a $5 Billion Drone Startup Off Course," *Forbes*, May 13, 2025.
167 **Doug Philippone:** "Investing Profile: Doug Philippone," SignalNFX.com; "Doug Philippone: Partner & Co-Founder," Snowpoint.com.
168 **"founding father":** Karp, "National Security Innovation Forum," November 20, 2024.

CHAPTER 14: PALANTIR, PALANTIR, PALANTIR

169 **Palantir bagged:** Carley Welch, "Army Consolidates Dozens of Palantir Software Contracts into One Deal Worth up to $10 Billion," *Breaking Defense*, August 1, 2025.
169 **$411 billion:** Ty Roush, "Palantir Jumps 8% To Record High After Revenue Boosted By 'Astonishing' AI Impact," *Forbes*, August 5, 2025.
169 **top twenty-five:** Samantha Subin, "Palantir Joins List of 20 Most Valuable U.S. Companies, with Stock More than Doubling in 2025," CNBC, July 25, 2025.
170 **Special Operations Forces conference:** Matthew Olay, "Experts Say Special Ops Has Made Good AI Progress, But There's Still Room to Grow," *DOD News*, May 7, 2025; Photo by Senior Airman Sterling Sutton, "SOF Week 2025: Artificial Intelligence Innovation, Integration in National Security," *Defense Visual Information Distribution Service*, May 6, 2025.
171 **Robert Imig:** "Robert Imig," Krach Institute for Tech Diplomacy at Purdue.
171 **Anand Gupta:** Morgan Brennan, "Manifest Space: Sorting Space Data with Palantir's Katie Ward & Anand Gupta," episode 6, *Squawk on the Street*, March 15, 2022.
171 **unorthodox fellowship:** Matt Lynley, "Peter Thiel Is Paying These 20 Entrepreneurs Who Can't Even Drink Yet $100,000 to Drop Out of College," *Business Insider*, June 12, 2012.
173 **Supermicro:** Maven team members told me they were aware of Bloomberg's 2018 report that China had compromised Supermicro circuit boards, which some US officials contested, and that the risk was deemed too small to act upon. See Jordan Robertson and Michael Riley, "The Big Hack," Bloomberg, October 4, 2018 and "The Long Hack," Bloomberg, February 12, 2021.
173 **Minotaur:** "Press Release: HII Awarded $244 Million Contract to Integrate Minotaur Software Products into Maritime Platforms," HII, October 17, 2023.
173 **Novetta:** Novetta, a company that described itself as an advanced analytics company with leading edge capabilities in machine learning, cyber, and cloud engineering supporting national security missions across the federal government, was bought by Accenture in 2021, see "Novetta Is Acquired by AFS," *RWBaird*, June 2021; Novetta, "Novetta Full-Spectrum Mission Command Solutions," LinkedIn post, May 14, 2021.
174 **NRO:** "Members of the IC (Intelligence Community)," Office of the Director of National Intelligence.
174 **FADE; MIST:** "Multi-INT Spatial Temporal (MIST) Toolsuite," CACI, 2021.
175 **crash and burn:** Steve Lohr, "What Ever Happened to IBM's Watson?," *New York Times*, July 16, 2021.
175 **Jock Padgett:** Ben Debow, "Jock Padgett, Chief Data Officer of the XVIII Airborne Corps," episode 7, *Tech[e]Valuation* podcast, April 14, 2023.
175 **Cukor flew to Palo Alto:** According to multiple people familiar with the matter.

177 **transformer models:** Cory Stryker and Dave Bergmann, "What Is a Transformer Model?" *IBM Think*, n.d.
177 **first described in a 2017 paper:** Ashish Vaswani et al., "Attention Is All You Need," *31st Conference on Neural Information Processing Systems (NIPS)*, 2017.
177 **Gotham:** "Palantir Gotham," Software One; *Palantir Platform: Gotham—Service Definition Document* (Palantir Technologies UK, 2024).
179 **Gaia:** *Palantir Platform: Gotham—Service Definition Document*.
181 **DEVGRU:** Michele Metych, "SEAL Team 6, Also Known as: DEVGRU, Naval Special Warfare Development Group," *Encyclopedia Brittanica* (2025); Mark Mazzetti et al., "SEAL Team 6: A Secret History of Quiet Killings and Blurred Lines," *New York Times*, June 6, 2015.
182 **Djibouti:** Katrina Manson, "Jostling for Djibouti," *Financial Times*, April 1, 2016.
182 **suicide bombing:** "Al Shabaab Claims Responsibility for Djibouti Suicide Attack," Reuters, May 27, 2014.
183 **"g-wot":** Manson, "Jostling for Djibouti."

CHAPTER 15: PALANTIR SPLITS THE TEAM

184 **Donahue:** "Commander General Christopher Donahue," NATO: Allied Land Command.
184 **special operations:** Jellen Koch and Ward Vanhaute, "NATO Special Operations Forces in Afghanistan: Lessons Learned from a Belgian Perspective," *Defense Institute*, June 2020.
186 **Carroll thought:** Carroll's position as explained in this paragraph relies on documentation reviewed by the author and interviews with people familiar with Carroll's thinking.
187 **private email:** According to correspondence reviewed by the author.
188 **reasoning models:** Dave Bergmann, "What Is a Reasoning Model?," *IBM Think*, n.d.
188 **develop reasoning models:** According to documentation reviewed by the author.
188 **"hated it":** Interview with author.
189 **laptops:** According to author interviews with people familiar with the matter.
190 **a visit:** According to people familiar with the matter.
191 **went wrong:** According to people familiar with the matter.
192 **emailed Cukor:** According to correspondence reviewed by the author.
192 **"Some Program Thoughts":** According to correspondence reviewed by the author.
193 **Aiyer:** Krishnan Aiyer did not respond to multiple outreaches from the author.

CHAPTER 16: A STRIKING OPERATION

194 **"died like a dog":** "Remarks by President Trump on the Death of ISIS Leader Abu Bakr Al-Baghdadi," The Trump White House Archives, October 27, 2019.
194 **Jackpot:** NBC reported that the special operations commander said on the radio, "100 percent Jackpot, over;" see Ken Dilanian, "Inside the Secret U.S. Mission That Took al-Baghdadi Out," *NBC News*, October 27, 2019.
195 **ISIS:** "Islamic State of Iraq and Ash-Sham (ISIS)," National Counterterrorism Center, as of September 2022.
195 **Abu Ghraib:** Joshua Eaton, "U.S. Military Now Says ISIS Leader Was Held in Notorious Abu Ghraib Prison," *The Intercept*, August 25, 2016.
195 **Camp Bucca:** Andrew K. Woods, "ISIS Was Born in an American Detention Facility (And It Wasn't Gitmo)," *Lawfare*, February 3, 2016.
195 **Yazidis:** *"They Came to Destroy": ISIS Crimes Against the Yazidis* (Human Rights Council: 32nd Session, 2016).
195 **set about destroying:** *History: Combined Joint Task Force—Operation Inherent Resolve, 2014–2023* (Operation Inherent Resolve).
195 **"Where's al-Baghdadi?":** "Remarks by President Trump on the Death of ISIS Leader Abu Bakr Al-Baghdadi."

195	**"completely confident":** "Department of Defense Press Briefing by Assistant to the Secretary of Defense for Public Affairs Jonathan Rath Hoffman and General Kenneth F. McKenzie, Jr., Commander, United States Central Command," Department of Defense, October 30, 2019.
196	**"just death":** "Remarks by President Trump on the Death of ISIS Leader Abu Bakr Al-Baghdadi."
196	**35,000 air strikes:** Jared Morgan, "Coalition Aircraft Launched Almost 35,000 Strikes on ISIS Targets over Six Years," *Air Force Times*, November 5, 2020; Garrett Nada, "The U.S. and the Aftermath of ISIS," Wilson Center, December 17, 2020.
196	**"These people are amazing":** "Remarks by President Trump on the Death of ISIS Leader Abu Bakr Al-Baghdadi."
196	**ten Delta Force commandos:** Seen on video later released by CENTCOM, see Shawn Snow, "CENTCOM Commander Releases Video of Raid on Baghdadi Compound, Which Now Looks like a 'Parking Lot with Large Potholes,'" *Military Times*, October 30, 2019; Eyal Tsir Cohen and Eliora Katz, "What We Can Learn about US Intelligence from the Baghdadi Raid," Brookings, November 6, 2019.
197	**"By the time":** "Remarks by President Trump on the Death of ISIS Leader Abu Bakr Al-Baghdadi."
197	**the end of May:** According to a person familiar with the timeline of the request.
197	**pinged:** According to a person familiar with the details of Maven's use during the raid.
197	**"hostile intent":** "Department of Defense Press Briefing by Assistant to the Secretary of Defense for Public Affairs Jonathan Rath Hoffman and General Kenneth F. McKenzie, Jr., Commander, United States Central Command."
197	**Take it out:** According to the person familiar with details of Maven's use during the raid.
198	**Barakat Ahmad Barakat:** Daniel Estrin, "Pentagon Files Reveal Flaws in U.S. Claims about Syrian Casualties in Baghdadi Raid," *NPR Investigation: The Baghdadi Raid Files*, July 28, 2023.
198	**another interview:** Daniel Estrin and Lama Al-Arian, "Syrians Say U.S. Helicopter Fire Killed Civilians During the Raid on Baghdadi," *NPR Investigation: The Baghdadi Raid Files*, December 3, 2019.
199	**acknowledged:** "Operation Inherent Resolve Civilian Casualty," Operation Inherent Resolve, March 10, 2022.
199	**questioned:** Azmat Khan et al., "The Civilian Casualty Files," *New York Times*, December 18, 2021.
199	**second investigation:** Daniel Estrin, "The Pentagon Is Reinvestigating If Troops Killed Civilians in Its 2019 Baghdadi Raid," *NPR Investigation: The Baghdadi Raid Files*, August 7, 2024.
199	**"moral and ethical thing to do":** *Annual Report on Civilian Casualties in Connection with United States Military Operations* (Department of Defense, 2018), 3.
199	**report:** United States Government Accountability Office, *Civilian Harm: DOD Should Take Actions to Enhance Its Plan for Mitigation and Response Efforts*, Report to Congressional Committees (2024).
199	**media reporting:** See, for example, Azmat Kahn, "Hidden Pentagon Records Reveal Patterns of Failure in Deadly Airstrikes," *New York Times*, December 18, 2021. In a May 2022 Pentagon press briefing, the Pentagon press secretary, in congratulating the *New York Times* for the Pulitzer Prize it had won for coverage of civilian casualties caused by the US military and military operations, stated that the DoD knew the department had made mistakes but that the *New York Times*' reporting had reinforced those concerns and, in some cases, identified additional concerns; see "Pentagon Press Secretary John F. Kirby Holds a Press Briefing," Department of Defense, May 10, 2022.
200	**as a much younger man:** Jen Judson, "Gen. James McConville Reflects on His Tenure as Army Chief of Staff," *Army Times*, August 2, 2023.
200	**machine bias:** James Holdsworth, "What Is AI Bias?," *IBM.com*, December 22, 2023.
201	**"pounding the table":** "Lt. Gen. Jack Shanahan Media Briefing on A.I.-Related Initiatives within the Department of Defense," Department of Defense, August 30, 2019.

201 **"I'm never going to say":** Ibid.
201 **Civilian Protection Center of Excellence:** *Report on the Civilian Protection Center of Excellence* (Department of Defense, 2023); Alex Horton et al., "Pentagon Moves to Gut Operations Focused on Reducing Civilian Harm," *Washington Post*, March 4, 2024.
202 **detail far greater numbers:** Azmat Khan, "Hidden Pentagon Records Reveal Patterns of Failure in Deadly Airstrikes," *New York Times*, December 18, 2021; Dave Philipps et al., "Civilian Deaths Mounted as Secret Unit Pounded Isis," *New York Times*, December 21, 2021.
202 **post-action review:** "Operation Desert Storm: Limits on the Role and Performance of B-52 Bombers in Conventional Conflicts," General Accounting Office, draft (Washington, DC), 3, 57, quoted in *Gulf War: Air Power Survey—Planning and Command and Control (Volume 1)* (Department of Defense, 1993), 293n100.
202 **four hundred civilians:** Sofia Barbarani, "Amiriyah Bombing 30 Years on: 'No One Remembers' the Victims," Al Jazeera, February 13, 2021.
202 **"series of errors and omissions":** Thomas Pickering, "Oral Presentation to the Chinese Government Regarding the Accidental Bombing of the P.R.C. Embassy in Belgrade," U.S. Department of State Archive, July 6, 1999.
202 **Chinese embassy in Belgrade:** Kevin Ponniah and Lazara Marinkovic, "The Night the US Bombed a Chinese Embassy," *BBC News*, May 6, 2019.
202 **intended target:** Three Chinese nationals were killed, and twenty embassy staff were wounded. The embassy had been clearly marked from the ground—a red flag, a plaque, and a long fence. Several Chinese officials suggested the attack was intentional, and US and UK newspaper articles investigated whether it might be, especially since it was the sole strike in the war for which the CIA, rather than the Pentagon, was responsible. Thomas Pickering, a senior State Department official who flew to Beijing to hand deliver a letter from President Bill Clinton after President Jiang Zemin refused his calls, described it as a blunder.
202 **failed to update:** None of the military and intelligence databases used to verify target information contained the correct location of the embassy, and the review process failed to detect either mistake. The embassy's correct address was not included on the "no hit list," and no one who made the list checked it with anyone with local knowledge of the city. One intelligence officer who worked on the targeting package even thought he'd spotted a discrepancy between the intended target and the location set to be bombed, but "missed phone calls and lack of follow-up" meant his doubts were never aired at a command level soon enough to stop the attack. If this were all true, something as simple—and as complex—as database management and effective communications could have averted the attack.
202 **unheard:** A few months earlier, NATO mistakenly bombed a train of Albanian refugees thinking they were Serbian military vehicles. Peter G. Gosselin and John-Thor Dahlburg, "NATO Concedes Civilians Killed in Bomb Attack," *Los Angeles Times*, May 16, 1999.
202 **forty-two were killed:** "On 3 October 2015, US Airstrikes Destroyed Our Trauma Hospital in Kunduz, Afghanistan, Killing 42 People," Medecins Sans Frontieres.
203 **"the middle of this field":** Oriana Pawlyk, "Pilots, Crew of AC-130 Gunship Acknowledged Confusion before Striking Kunduz Hospital," *Air Force Times*, April 29, 2016.
203 **"combination of human errors":** "Press Briefing by Army General Joseph Votel, Commander, U.S. Central Command," Department of Defense, April 29, 2016.
203 **"a tragic mistake":** C. Todd Lopez, "DoD: August 29 Strike in Kabul 'Tragic Mistake,' Kills 10 Civilians," *DOD News*, September 17, 2021.

CHAPTER 17: DATA HELL

204 **thirty-eight classes of object:** Cheryl Pellerin, "Project Maven to Deploy Computer Algorithms to War Zone by Year's End," *DOD News*, July 21, 2017.
205 **Figure Eight:** Figure Eight's involvement in Project Maven was previously reported by Lee Fang,

"Google Hired Gig Economy Workers to Improve Artificial Intelligence in Controversial Drone-Targeting Project," *The Intercept*, February 4, 2019.
206 **10,000 images:** According to author interviews with Maven algorithm vendors.
206 **Jane Pinelis:** "Jane Pinelis," Center for Strategic & International Studies.
206 **2023 interview:** Parts of this interview appeared in the following article, Katrina Manson, "AI Warfare Is Already Here," *Bloomberg Businessweek*, February 28, 2024.
207 **"responsible":** Dana Deasy and Lieutenant General Jack Shanahan, "Department of Defense Press Briefing on the Adoption of Ethical Principles for Artificial Intelligence," Department of Defense, February 24, 2020.
207 **I was told:** Author interviews with people familiar with the matter. When I asked Carroll about this, he said he was way more scared of nonperformant AI than unethical AI, and recounted to me that his exact line was: "Those who can, build and test performant AI; those who cannot, talk about AI ethics." His delivery of this line during a Pentagon lunch prompted at least one AI ethics specialist to get up and walk away, he said.
207 **thirty static frames:** According to author interviews with algorithm vendors.
209 **when Apple bought them:** Alan Boyle et al., "Exclusive: Apple Acquires Xnor.ai, Edge AI Spin-out from Paul Allen's AI2, for Price in $200M Range," *GeekWire*, January 15, 2020.
209 **wasted away:** Xnor co-founder Ali Farhadi did not respond to repeated author requests for comment. Apple did not respond to a request for comment.
209 **"Garbage in, garbage out,":** According to author interviews with people familiar with the matter.
209 **Aberdeen Proving Ground:** "Aberdeen Proving Ground," US Army Installation Management Command.
209 **I'd visited Aberdeen:** Manson, "Robot-Soldiers, Stealth Jets and Drone Armies: The Future of War."
209 **at Aberdeen:** According to author interview with member of Maven team who was part of this visit.
210 **One analysis:** Reviewed by the author.
212 **third tack:** According to author interviews with multiple people familiar with the matter.
213 **roommates:** Cory Weinberg, "Fame, Feud and Fortune: Inside Billionaire Alexandr Wang's Relentless Rise in Silicon Valley," *The Information*, June 28, 2024.
213 **preaching to Congress:** "Statement by Alexandr Wang Founder and Chief Executive Officer Scale AI Before the House Committee on Energy and Commerce," April 9, 2025; "Full Committee: 'Future of AI Technology, Human Discovery, and American Global Competitiveness,'" streamed live April 9, 2025, by House Committee on Energy and Commerce, YouTube.
213 **youngest self-made billionaire:** Cole Horton, "The New Youngest Self-Made Billionaire in the World Is a 25-Year-Old College Dropout," *Forbes*, May 25, 2022; Krystal Hu et al., "Meta Poaches 28-Year-Old Scale AI CEO after Taking Multibillion Dollar Stake in Startup," Reuters, June 13, 2025.
213 **startup I visited:** Katrina Manson, "Neurodiversity Emerges as a Skill in Artificial Intelligence Work," Bloomberg, October 18, 2022.
214 **100 million:** According to author interviews with people familiar with the matter; also cited by Colin Carroll in "In the Fight—Scaling AI/ML in Defense," *Acquisition Talk*.
214 **"pathfinder project":** "Lt. Gen. Jack Shanahan Media Briefing on A.I.-Related Initiatives within the Department of Defense," Department of Defense, August 30, 2019.
214 **"warfighting-focused":** Ibid.
214 **"non-MIP":** For a primer on MIP Funding, see Miachel E. DeVine, *Defense Primer: Budgeting for National and Defense Intelligence*, no. IF10524 (Congressional Research Service, 2025).
215 **Smart Sensor:** "GA-ASI Awarded Smart Sensor Contract," General Atomics, November 24, 2020.
216 **"liberate data":** "Lt. Gen. Jack Shanahan Media Briefing on A.I.-Related Initiatives within the Department of Defense."
218 **June 2020:** *Audit of Governance and Protection of Department of Defense Artificial Intelligence Data and Technology* (Department of Defense Office of Inspector General, 2020).

218 **"threaten the safety":** Ibid., ii.
219 **"Maven's biggest failure":** *In the Fight—Scaling AI/ML in Defense with Colin Carroll.*
219 **still raw and angry:** The pair would later reconcile after a fashion. Gen (ret.) Shanahan told me that after they were both retired and some of the discombobulation of Covid ebbed, he sent a simple greeting text message to Cukor, who replied as if nothing had ever gone wrong between them. Cukor invited Shanahan to give a talk on AI at J.P. Morgan in spring 2023, and he accepted.

CHAPTER 18: WE'LL FIND IT AND WE'LL STRIKE IT

224 ***The Princess Bride***: *The Princess Bride*, directed by Rob Reiner (1987); "Princess Bride Clips— Most Likely Kill You In The Morning," posted May 25, 2020, by Kelly L. Call, YouTube.
224 **Kurilla:** "Commander, General Michael E. Kurilla," United States Central Command.
224 **picture of Kurilla:** According to author interviews with people familiar with the matter.
225 **lieutenant competitions:** According to Anthony Pfaff, a retired US army colonel. Pfaff told me the pair attended French jump school together. (That was nerve-wracking; the French jump out of planes differently from the Americans—US personnel take a position in the door of the aircraft and then leap; the French simply run straight out the door, he said. And unlike the Americans, the French pull their reserve parachute at the same time as their main parachute.)
225 **every single year:** "Commander, General Michael E. Kurilla."
225 ***Relentless Strike***: Sean Naylor, *Relentless Strike: The Secret History of Joint Special Operations Command* (Macmillan, 2015).
225 **"the finest soldier":** "2015—Brigadier General (Promotable) Michael 'Erik' Kurilla at the Children of Fallen Patriots Gala," posted November 25, 2015, by Children of Fallen Patriots Foundation, YouTube. A representative for Gen (ret.) Kurilla did not respond to requests for comment.
225 **speech:** "2015—Brigadier General (Promotable) Michael 'Erik' Kurilla at the Children of Fallen Patriots Gala."
225 **"like my dad":** Quoted in Michael Yon, "The Battle for Mosul," May 14, 2005.
225 **bullet in each leg:** Michael Yon, "Gates of Fire," August 31, 2005.
225 **"a little shooting":** Ibid.
226 **Nick Villarruel:** Interview with author.
226 **Kurilla wanted to understand:** According to author interview with Anthony Pfaff, who discussed these issues with General Kurilla during a visit to the 18th.
226 **strike on Soleimani:** "Statement by the Department of Defense," Department of Defense, January 2, 2020; Katrina Manson et al., "Iran's Top Military Leader Qassem Soleimani Killed in US Air Strike," *Financial Times*, January 2, 2020.
226 **Qasem Soleimani:** also spelled Quassem Soleimani, Quassim Suleimani, Quasim Suleimani, Quasim Soleimani.
229 **three-year-old:** Colin Clark, "Air Combat Commander Doesn't Trust Project Maven's Artificial Intelligence—Yet," *Breaking Defense*, August 21, 2019.
229 **"dazzling":** *Deception*, MCTP 3-32F (US Marine Corps, 2024), 2–8.
229 **"art of the possible":** Manson, "AI Warfare Is Already Here."
229 **Scarlet Dragon:** Spc. Osvaldo Fuentes, "Data Centric Exercise Showcases Joint Capabilities and Lethality during Scarlet Dragon Oasis," Defense Visual Information Distribution Service, February 3, 2023.
230 **"Rainman":** According to author interviews with multiple people familiar.
230 **biggest bump:** According to internal document reviewed by the author.
232 **July 2021 white paper:** Joe O'Callaghan, "White Papers: Maven Smart System," United States Field Artillery Association, July 29, 2021.
232 **ignominious withdrawal:** "Remarks by President Biden on Evacuations in Afghanistan," The Biden White House Archives, August 20, 2021; Congressman Michael McCaul, *House Republican Interim Report: A "Strategic Failure": Assessing the Administration's Afghanistan Withdrawal* (US House of Representatives, 117th Congress, 2022).

NOTES   383

233 **120,000 people:** Justin Vallejo, "Read the Full Transcript of Biden's Remarks on the US Withdrawal from Afghanistan War," *The Independent*, August 31, 2021.
233 **dangerous conditions:** Nick Paton Walsh and Mick Krever, "Exclusive: New Evidence Challenges the Pentagon's Account of a Horrific Attack as the US Withdrew from Afghanistan," CNN, April 24, 2024.
233 **using Maven Smart System:** According to people familiar with the matter.
233 **Abbey Gate:** Joshua Kaplan et al., "Hell at Abbey Gate: Chaos, Confusion and Death in the Final Days of the War in Afghanistan," *ProPublica*, April 2, 2022; "Read U.S. Central Command's Investigation Into Botched Aug. 29, 2021 Kabul Drone Strike," *New York Times*, January 6, 2023; Olay, "Kabul Airport Attack Review Reaffirms Initial Findings, Identifies Attacker."
234 **Joe Biden warned:** "Statement by President Joe Biden on the Evacuation Mission in Kabul," The Biden White House Archives, August 28, 2021.
234 **"righteous":** Alex Horton et al., "Examining a 'Righteous' Strike," *Washington Post*, September 10, 2021.
234 **later report:** Charlie Savage et al., "Newly Declassified Video Shows U.S. Killing of 10 Civilians in Drone Strike," *New York Times*, January 19, 2022.
234 **briefly visible:** "General Kenneth F. McKenzie Jr. Commander of U.S. Central Command and Pentagon Press Secretary John F. Kirby Hold a Press Briefing," Department of Defense, September 17, 2021.
235 **told him to remove the post:** According to two people familiar with the matter.
236 **from Sudan:** Emma Bowman et al., "The U.S. Evacuates Some 1,000 Americans from Sudan," *NPR News*, April 30, 2023.
236 **reduced the human role:** Unclassified slides shared with the author depict these changes.

CHAPTER 19: NOBODY KNOWS TARGETING BETTER THAN TREY

238 **Whitworth:** In 2020, Whitworth was a rear admiral. He was later promoted to vice admiral; see "Vice Admiral Frank Whitworth—Director of Intelligence, J-2, Joint Staff," United States Navy, September 20, 2021; "Vice Admiral Frank Whitworth: Director of the National Geospatial-Intelligence Agency," National Geospatial-Intelligence Agency. He retired in November 2025.
238 **J2:** "J2 Joint Staff Intelligence," Joint Chiefs of Staff.
239 **joint targeting cycle:** The 2013 version of the doctrine is public. See *Joint Targeting*, no. 3-60, Joint Chiefs of Staff, 2013. A 2018 version was made public in late 2020 following a 2019 FOIA request. See John Greenewald, "Joint Publication (JP) 3-60, Joint Targeting," *The Black Vault*, November 11, 2020.
239 **"distorted":** Major Matthew P. McKeon, "Joint Targeting: What's Still Broke?" (School of Advanced Airpower Studies at Air University, 1999), 3.
239 **"excessive focus":** John Patch, "Obstacles to Effective Joint Targeting," *Joint Force Quarterly*, April 2007.
239 **communication problems:** GR Brown, "Maximizing Joint Targeting Synergy within the USINDOPACOM AOR," n.d.
239 **STAR targets:** For more on the Sensitive Target Approval and Review (STAR) process, see Joint Staff, *(U) Sensitive Target Approval and Review (STAR) Process* (2018).
239 **"distinction":** "Rule 1. The Principle of Distinction between Civilians and Combatants," International Humanitarian Law Databases.
240 **language of doctrine:** Joint Staff, *(U) Sensitive Target Approval and Review (STAR) Process*.
240 **Whitworth visited:** This retelling is based on separate author interviews with multiple people present.
243 **Shanahan's retirement:** The JAIC, "JAIC Director Retires, Bids Farewell to Partners in Government & Private Sector," *CHIPS: The Department of the Navy's Information Technology Magazine*, June 2, 2020.

243 **CDAO:** "About: Our Mission," CDAO.
243 **halting success:** The following IG evaluation of CDAO's AI Governance and Acquisition Plan found its implementation plan for the Defense Department's AI Adoption Strategy and AI policy were overdue. It also found that as a result of delays in issuing key foundational documents, the DoD "may not be effectively implementing or achieving the full benefits of AI." See *Evaluation of the Effectiveness of the Chief Digital and Artificial Intelligence Office's Artificial Intelligence Governance and Acquisition Process* (Department of Defense Office of Inspector General, 2024).
243 **new management:** Brandi Vincent, "Feinberg Orders Major Shakeup in Pentagon's AI Enterprise," *Defense Scoop*, August 15, 2025.
243 **National Geospatial-Intelligence Agency:** "About the National Geospatial-Intelligence Agency," National Geospatial-Intelligence Agency.
243 **hurriedly drawn:** Based on a history of the NGA shared with the author.
244 **Five Guys:** "President Obama Stops for Lunch at Five Guys," posted May 29, 2009, by C-SPAN, YouTube.
244 **greater renown:** Marc Ambinder, "The Little-Known Agency That Helped Kill Bin Laden," *The Atlantic*, May 5, 2011.
244 **model replica:** "Artifacts—Model of Abbottabad Compound," Central Intelligence Agency.
244 **2011 Navy SEAL raid:** "Digital Exhibitions: Operation Neptune Spear," 9/11 Memorial & Museum.
244 **memo:** The Joint Requirements Oversight Council Memorandum transferring Maven to NGA is referenced in the following partly redacted January 2022 document; see *(U) Evaluation of Contract Monitoring and Management for Project Maven*, DODIG-2022-049 (Department of Defense Office of Inspector General, 2022). The author reviewed a non-redacted version of the same report, which gave the date as September 3, 2021. It stipulates that Maven's AI training and geospatial data lines of effort would go to NGA.

CHAPTER 20: KILL CHAIN

248 **one classified report:** *Evaluation of the Algorithmic Warfare Cross-Functional Team (Project Maven)*, DODIG-2020-025 (Department of Defense Office of Inspector General, 2019).
248 **saying in September 2025:** "Secretary of War Pete Hegseth Addresses General and Flag Officers at Quantico, Virginia."
248 **IG:** "About DoD Office of Inspector General," Department of Defense Office of Inspector General.
250 **told colleagues:** According to documents reviewed by the author.
254 **Poggemeyer left Project Maven:** According to contemporaneous records relied on by the author, Cukor announced on July 28, 2020, that Poggemeyer would be leaving.
254 **note card:** According to multiple people; Christ could not be reached for comment.
254 **jumped right back into work:** According to multiple people. Cukor did not dispute this version of events.
254 **ran a story:** Cade Metz et al., "What's a Palantir? The Tech Industry's Next Big I.P.O.," *New York Times*, August 26, 2020.
255 **Was any current or future financial benefit accruing to Cukor:** Cukor and Palantir deny this. The IG report into Maven's contracting practices until spring 2020 found no irregularities. Cukor turned down job offers from Palantir made after he left the Pentagon, according to Cukor and Palantir's Aki Jain. Cukor has continued working with Palantir in his private sector work, forging new partnerships with the company. Palantir says it attracts customers because of the quality of its products and service.
256 **public post:** Colin Carroll, "On 21 August 2021, DSD Kathleen Hicks asked Michael Groen for my resignation as the Joint AI Center COO," LinkedIn post, January 24, 2025.
256 **review of Maven:** A redacted version of the report dated January 6, 2022, is available publicly, see *Evaluation of Contract Monitoring and Management for Project Maven*,

(DODIG-2022-049) (Department of Defense Office of Inspector General, 2022). The author has also reviewed the unredacted version. The period under review spanned September 2018 to February 2020, and omits the arrival of Palantir and Scale AI as prime contractors on Project Maven alongside ECS Federal after September 2020. The Inspector General's Office concluded that Cukor spent "a considerable amount of time" monitoring contract performance because the research and development required for AI development and machine learning is not widely understood across the DoD "and requires a heavy degree of oversight and monitoring," the report said. He conducted regular in-depth meetings with contractor project managers to review deliverables, financials, and potential project risks and a detailed weekly review of all contract financial reporting, the report added. The report recommended Maven formalize and document its processes, which Cukor then did as the project prepared to transition to NGA.

257 **genuinely and deeply upset:** The environment "worked for some but not for others," according to an internal March 2024 document about Maven that I reviewed. It noted high turnover and burnout rates. Maven faced "antibodies" to changes in the status quo, it added, saying the project relearned the lesson to integrate user interaction and "applied it more aggressively."
259 **farewell dinner:** According to contemporaneous records shared with the author.
259 **gold-framed farewell:** Reviewed by the author.
260 **photograph:** Original provided to the author.

CHAPTER 21: UKRAINE FIGHTS BACK

261 **Ukraine Fights Back:** This chapter draws on multiple author interviews with multiple people who visited Wiesbaden or have knowledge of conditions and operations of the US mission in support of Ukraine from 2022 and its impact, plus correspondence and other documentation reviewed by the author. Conversations are reconstructed based on email and memories of those present.
261 **looking up at the television:** "Defense Department Briefing," C-SPAN, February 2, 2022; Natasha Bertrand et al., "US Troops to Deploy to Eastern Europe amid Ukraine Crisis," CNN, February 2, 2022.
261 **on a military plane:** Beth Hutson, "18th Airborne Corps Soldiers Arrive in Germany amid Russian Threat to Ukraine," *Fayetteville Observer*, February 5, 2022.
261 **ensconced in a US military base:** John Vandiver, "Fort Bragg Troops Arrive in Germany and Poland as Russian Buildup Continues near Ukraine," *Stars and Stripes*, February 4, 2022.
262 **"The Pit":** David E. Sanger previously reported the name in *New Cold Wars*; Sanger also describes some of Project Maven's broad use in supporting Ukraine in the book and the following excerpt, David E. Sanger, "In Ukraine, New American Technology Won the Day. Until It Was Overwhelmed," *New York Times*, April 23, 2024.
262 **US intelligence indicated:** Shane Harris and Paul Sonne, "Russia Planning Massive Military Offensive against Ukraine Involving 175,000 Troops, U.S. Intelligence Warns," *Washington Post*, December 3, 2021.
262 **Russian force assembled on Ukraine's border:** Max Seddon et al., "Ukraine: What Does Vladimir Putin Want?," *Financial Times*, December 10, 2021.
263 **the next revolution:** Hearing to Consider the Nomination of Lieutenant General Michael E. Kurilla, USA to Be General and Commander, United States Central Command: Hearing before the Committee on Armed Services (US Senate, 2022).
263 **Command Post Computing Environment (CPCE):** "Command Post Computing Environment (CPCE)," Office of the Director, Operational Test and Evaluation. The Army noted "deficiencies" in the CPCE, including that, "when under stress, CPCE can discontinue generating outbound server data for logged in users."
264 **Putin's face beamed out:** Andrew Osborn and Polina Nikolskaya, "Russia's Putin Authorises 'Special Military Operation' Against Ukraine," Reuters, February 24, 2022.

264 **Satellite internet connections started going out:** Katrina Manson, "The Satellite Hack Everyone Is Finally Talking About," Bloomberg, March 1, 2023; "U.S. Government Attributes Cyberattacks on SATCOM Networks to Russian State-Sponsored Malicious Cyber Actors," Cybersecurity & Infrastructure Security Agency, May 10, 2022.

264 **invasion:** For more on the war, see "Russia & Ukraine," Institute for the Study of War.

264 **pre-written list:** See Christopher R. Bolton and Matthew R. Prescott, "Commander's Critical Information Requirements: Crucial for Decisionmaking and Joint Synchronization," National Defense University Press, July 19, 2024.

264 **gruff voice answered:** According to a person familiar with the matter. Representatives for Gen (ret.) Milley did not respond to requests for comment.

265 **Kyiv was nearly overrun:** Liam Collins and John Spencer, "Urban Warfare Project: Case Study #12—Kyiv," Modern War Institute at West Point, February 21, 2025.

265 **Ukrainians pushed back:** James Marson, "Putin Thought Ukraine Would Fall Quickly. An Airport Battle Proved Him Wrong," Wall Street Journal, March 3, 2022.

265 **heavyweight war leader:** John Yang and Ryan Connelly Holmes, "Volodymyr Zelensky's Improbable Rise from Comedian to Wartime Leader of a Defiant Nation," PBS News, March 3, 2022.

265 **Hundreds were dying:** "Ukraine: Civilian Casualties as of 15 March 2022," United Nations Human Rights: Office of the High Commissioner, March 15, 2022.

265 **"needs to go to Poland":** Aamer Madhani, "Top Biden Aide Says Ukraine Invasion Could Come 'Any Day,'" AP News, February 6, 2022; "Press Release—USAREUR-AF Repositions Troops in Poland," US Army Europe and Africa, April 7, 2025.

265 **fired into Poland:** Steve Inskeep and Greg Myre, "Poland Says a Missile That Crashed on Its Territory Was Friendly Fire from Ukraine," NPR News, November 17, 2022.

268 **last US soldier to leave Afghanistan:** Jeff Schogol, "Why a 2-Star General Was the Last American Service Member to Leave Afghanistan," Task & Purpose, September 1, 2021.

268 **night-vision photograph:** US Central Command (@CENTCOM), "The last American Soldier leaves AfghanistanMajor [sic] General Chris Donahue, commander of the U.S. Army 82nd Airborne Division, @18airbornecorps," X, August 30, 2021.

269 **snarled Ben Sasse:** Steve Inskeep, "Ukrainians Aren't Getting U.S. Intelligence on Russia Fast Enough, Sasse Says," Morning Edition, March 1, 2022.

269 **Mark Warner:** Ken Dilanian, "Biden Administration Walks Fine Line on Intelligence-Sharing with Ukraine," NBC News, March 4, 2022.

269 **Adam Smith . . . worried out loud:** "Russian Military 'Bogged Down' but Has the Advantage: Armed Services Committee Member," Morning Joe, MSNBC, March 3, 2022.

270 **met with Jim Caggy:** According to documentation reviewed by the author.

270 **got in a room at AWS:** According to correspondence reviewed by the author.

270 **war doesn't heed weekends:** According to a person familiar with the matter.

271 **"the cornerstone":** Amy Walker et al., "Global Network Super Highway Postures Army for Multi-Domain Operations," US Army, July 7, 2021.

271 **"hit the ground":** "Regional Hub Nodes / Global Agile Integrated Transport," PEO C3N.

271 **blown up:** Colin Demarest, "Data Centers Are Physical and Digital Targets, Says Pentagon's Eoyang," C4ISRNet, November 17, 2022.

271 **Elon Musk answered an appeal:** Elon Musk (@elonmusk), "Starlink service is now active in Ukraine. More terminals en route.," Twitter (now X), February 26, 2022.

272 **"point of interest" package:** The following investigation explores the role of points of interest and broader US military support to Ukraine. See Adam Entous, "The Partnership: The Secret History of the War in Ukraine," New York Times, March 29, 2025, and Adam Entous, "Key Takeaways from America's Secret Military Partnership with Ukraine," New York Times, March 30, 2025.

273 **Delta:** Kateryna Bondar, "Does Ukraine Already Have Functional CJADC2 Technology?," Center for Strategic & International Studies, December 11, 2024.

273 **Kropyva:** Bruno Maçães, "How Palantir Is Shaping the Future of Warfare," Time, July 10, 2023.

NOTES

CHAPTER 22: TENS OF THOUSANDS OF TARGETS

275 **Tens of Thousands of Targets:** This chapter is based on several interviews with multiple people familiar with the episodes described. Conversations are reconstructed based on those present.

276 **June 2022 decision:** Steve Holland and Doina Chiacu, "Biden Announces New $700 Million in Military Aid for Ukraine," Reuters, June 1, 2022.

276 **expand nearly tenfold:** "Immediate Release: Fact Sheet on U.S. Security Assistance to Ukraine," Department of Defense, December 30, 2024.

276 **Turkish Bayraktar drones:** Stijn Mitzer et al., "A Monument of Victory: The Bayraktar TB2 Kill List," *Oryx*, February 23, 2022.

276 **Izyum:** Giulia Carbonaro, "HIMARS Have Killed These Members of Russia's Military Elite—Full List," *Newsweek*, July 25, 2022; "Ukraine Is Already Using U.S.-Supplied Rocket Systems in Conflict, Top General Says," Reuters, June 25, 2022.

277 **tweet:** Oleksii Reznikov (@oleksiireznikov), Twitter (now X), July 9, 2022.

277 **hundred would win the war:** Patrick Tucker, "Ukraine Says It Needs at Least 100 HIMARS and Longer-Range Rockets," *Defense One*, July 19, 2022.

278 **Maxar:** Maxar has confirmed working on Project Maven. Colin Demarest and Courtney Albon, "Q&A: Maxar Execs Discuss US Army Simulation, Project Maven," *C4ISRNet*, June 5, 2023.

279 **$275 million on Maven:** According to internal document reviewed by the author.

279 **sixteen HIMARS:** Caitlin Doornbos, "US to Send Another $550M in Ammunition for Rocket Systems, Howitzers to Ukraine," *Stars and Stripes*, August 1, 2022.

281 **estimates from the British government:** George Allison, "The UK Ministry of Defence Estimates over 350,000 Russian Personnel Casualties, and Significant Losses of Military Assets, Including 2,600 Tanks and 4,900 Armoured Vehicles, in Ukraine since February 2022," *UK Defence Journal*, February 1, 2024.

281 **NETCOM said:** Solutions span commercial cloud transport, leased circuits from telecommunications providers, and microwave line-of-sight or satellite communications when fiber is unavailable at the point of need, NETCOM added.

282 **arrival of railroads:** A. J. P. Taylor makes this point in *The First World War: An Illustrated History* (Perigee, 1972).

282 **accredited by the National Security Agency:** Michael Peck, "ViaSat [sic] Unveils High-Speed Encryptor," *C4ISRNet*, August 22, 2016.

282 **KG-142:** "Datasheet: KG-142," Viasat, 2017. Viasat did not respond to request for comment.

282 **"devastating effect":** "Secretary of Defense Lloyd J. Austin III and Chairman of the Joint Chiefs of Staff Army General Mark A. Milley Hold a Press Conference Following the Ukraine Defense Contact Group Meeting, Ramstein Air Force Base, Germany," DoD, September 8, 2022.

283 **lunch meeting:** According to people familiar with the briefing.

283 **AI had no morals:** Judson, "Gen. James McConville Reflects on His Tenure."

283 **Signal:** Meredith Whittaker, "A Message from Signal's New President," *Signal*, September 6, 2022.

285 **military's own trust, or lack of it:** Paul Lushenko has studied the extent of trust for such systems, arguing for humans and machines to work together rather than for the primacy of machines, Paul Lushenko, "Trust but Verify: US Troops, Artificial Intelligence, and an Uneasy Partnership," Modern War Institute at West Point, January 16, 2024.

286 **"Your adversaries"; "already happening":** Manson, "AI Warfare Is Already Here."

CHAPTER 23: WE'VE DRUNK THE KOOL-AID

287 **stood face-to-face:** According to people familiar with the matter.

288 **coming out party:** See "GEOINT Artificial Intelligence," National Geospatial-Intelligence Agency.

288 **Maven ATR:** According to people familiar with the matter.

288 **in public:** "Remarks as Delivered by VADM Frank 'Trey' Whitworth, USN," GEOINT 2025 Symposium, *NGA News*, May 21, 2025.

288 **customer event:** "AIPCON 5 September 2024 | Livestream," streamed live September 12, 2024, by Palantir, YouTube.
288 **narrating the event:** Ibid., 1 hour, 21 min.
288 **$480 million ceiling:** Jon Harper, "Palantir Lands $480M Army Contract for Maven Artificial Intelligence Tech," *Defense Scoop*, May 29, 2024.
288 **$100 million:** Katrina Manson, "Palantir Wins $100 Million US Contract for AI Targeting Tech," Bloomberg, September 19, 2024.
288 **$1.3 billion:** Brandi Vincent, "'Growing Demand' Sparks DOD to Raise Palantir's Maven Contract to More Than $1B," *Defense Scoop*, May 23, 2024.
289 **GBP 750 million deal:** Larisa Brown, "Trump Visit Brings £750m Palantir Deal for Military AI Technology," *Times* (London), September 17, 2025.
289 **contract for data labeling:** "NGA Announces $708M Data Labeling RFP," National Geospatial-Intelligence Agency, September 30, 2024.
289 **serial number:** NGA did not provide a specific reason for the request. A former NSA official told me it would be possible to hack a voice recorder that connects to an internet-connected device, such as a computer, knowing only its make, model, and serial number, but adamantly argued it would not have been for the US intelligence community to intercept my communications and that it was most likely a way for a spy agency to check if the tool had been infiltrated by Chinese or Russian hackers.
290 **new mantra:** "USGIF GEOINT Symposium 2023—From Seabed to Space Video," posted February 10, 2025, by National Geospatial-Intelligence Agency, YouTube; "Remarks as Prepared for Vice Adm. Frank Whitworth, Director, National Geospatial-Intelligence Agency For 2023 USGIF GEOINT Symposium," NGA press release, distributed by Public Now, May 22, 2023.
290 **"program of record":** Office of Inspector General, *Semiannual Report to Congress*.
291 **rang combatant commanders:** Author interview with Admiral Whitworth in 2023.
291 **former Google AI expert:** "Press Release: U.S. Central Command Hires Former Google AI Cloud Director Dr. Andrew Moore, One of World's Leading Experts on Artificial Intelligence and Machine Learning," US Central Command, April 19, 2023.
291 **"a pretty seamless shift":** Katrina Manson, "US Says It's Using AI for Targeting Help in Mideast Airstrikes," Bloomberg, February 26, 2024.
291 **Schuyler Moore:** Ibid.
291 **"off to the races":** "Data-Centric Warfighting with Brigadier General John P. Cogbill," *Second Front* podcast.
291 **helped narrow down more than eighty-five targets:** "Press Release: CENTCOM Statement on U.S. Strikes in Iraq and Syria," US Central Command, February 2, 2024.
291 **death of three US service members:** C. Todd Lopez, "3 U.S. Service Members Killed, Others Injured in Jordan Following Drone Attack," *DOD News*, January 29, 2024.
291 **confirmation that the US military was using AI to identify enemy systems:** Chief Warrant Officer 4 Joseph P. Lyddane, "138th Field Artillery Brigade Incorporates Artificial Intelligence," Defense Visual Information Distribution Service, February 29, 2024.
291 **its own weapons to strike:** Manson, "US Says It's Using AI for Targeting Help in Mideast Airstrikes."
292 **told a podcast:** "Data-Centric Warfighting with Brigadier General John P. Cogbill," *Second Front* podcast.
292 **Admiral Hulin:** He spoke in place of VADM Brad Cooper at an event attended by the author, see Center for Security and Emerging Technology, "Building the Tech Coalition."
292 **Target Workbench:** "Target Workbench," Palantir Technologies Inc., 2023.
292 **"Shortening kill chains is universally good":** Lauren Bedula and Hondo Geurts, "Digital Combatant Command Coordination, with BG John Cogbill," episode 47, *Building the Base* podcast, April 23, 2024.

NOTES

292 **topic of choice:** According to a person familiar with the matter.
292 **"C" grade:** Evan Lynch, "18th Airborne Corps' 'Tricky' AI Journey," *AFCEA: Signal Media*, August 21, 2024.
292 **Kurilla told Congress:** "A Strategic Window in the Central Region: Statement for the Record General Michael 'Erik' Kurilla Commander, US Central Command Before the Senate Committee on Armed Services on the Posture of US Central Command," United States Central Command, June 10, 2025.
293 **twelve-day Iran-Israel war:** The information in this paragraph relies on people familiar with the matter.
294 **Chinese spy balloon:** Humeyra Pamuk et al., "U.S. Briefed 40 Nations on China Spy Balloon Incident, Diplomats and Official Say," Reuters, February 9, 2023.
294 **other aircraft coming near the US:** "Press Release: NORAD Detects and Tracks Russian Aircraft Operating in the Alaskan Air Defense Identification Zone," North American Aerospace Defense Command, August 26, 2025.
295 **"armed conflict":** Charlie Savage and Eric Schmitt, "Trump 'Determined' the U.S. Is Now in a War with Drug Cartels, Congress Is Told," *New York Times*, October 2, 2025.
295 **That was the same month:** Charlie Savage, "U.S. Military Attacked Boat Off Venezuela, Killing Four Men, Hegseth Says," *New York Times*, October 3, 2025.
295 **"I don't give a shit what you call it":** JD Vance (@JDVance), "I don't give a shit what you call it," X, September 6, 2025.
295 **killing eighty-seven people:** "Tracking U.S. Military Killings in Boat Attacks," Lazaro Gamio, *New York Times*, December 5, 2025.
295 **imperial boomerang:** Aimé Césaire, *Discours sur le colonialisme (Discourse on Colonialism)* (Éditions Réclame, 1950); other writers see an example of imperial boomerang in the wake of techniques used in Gaza, and what they see as the expansion of authoritarianism at home. See Noura Erakat, "The Boomerang Comes Back," *Boston Review*, Winter 2025.
296 **"optimised for domestic warfare":** Jacquelyn Schneider, "Trump's War on the Enemy within May Reward the Enemy without," *Financial Times*, October 15, 2025.
296 **"WOPR":** *WarGames*, directed by John Badham (1983). Framing it in context of the WOPR, Michael Hirsh explores how AI is being integrated into US strategic defenses in this article, "The AI Doomsday Machine Is Closer to Reality Than You Think," *Politico*, September 2, 2025.
296 **Unit X:** Christopher Kirchoff and Raj M. Shah, *Unit X: How the Pentagon and Silicon Valley Are Transforming the Future of War* (Scribner, 2024).
297 **Defense Department's policy on autonomy:** "DOD Directive 3000.09: Autonomy in Weapon Systems," Department of Defense, January 25, 2023; Michael C. Horowitz, "Autonomous Weapon Systems: No Human-in-the-Loop Required, and Other Myths Dispelled," *War on the Rocks*, May 22, 2025.
297 **"appropriate levels of human judgment over the use of force":** Ibid., 3.
297 **Probasco advocates for scaling up:** Probasco has extensively studied Maven and the integration of AI into command systems and drones. See Emelia Probasco and Minji Jang, "Military AI: Angel of Our Better Nature or Tool of Control?," *War on the Rocks*, May 9, 2025.
297 **erroneously shot down:** *Formal Investigation into the Circumstances Surrounding the Downing of Iran Air Flight 655 on 3 July 1988* (Department of Defense, 1988); Fred Kaplan, "America's Flight 17," *Slate*, July 23, 2014.
298 **agentic AI:** Corey Stryker, "What Is Agentic AI?," *IBM Think*, n.d.
298 **ChatGPT-3:** Kyle Wiggers et al., "ChatGPT: Everything Released from the AI-Powered Chatbot in 2022," *Tech Crunch*, December 31, 2022.
299 **"the next great sea change":** Alexandr Wang, "A Letter from Our CEO," Scale AI, May 10, 2023.
299 **Thunderforge:** "DIU's Thunderforge Project to Integrate Commercial AI-Powered Decision-Making for Operational and Theater-Level Planning," Defense Innovation Unit, March 5, 2025.
300 **Craig Martell:** "DEF CON 31—Shall We Play a Game—Craig Martell," posted September 15, 2023, by DEFCONConference, YouTube.

300 **Mistakes I came across:** Katrina Manson, "Hackers Trick AI with 'Bad Math' to Expose Flaws and Biases," *Bloomberg*, August 12, 2023.
300 **trigger nuclear war:** Max Lamparth and Jacquelyn Schneider, "Why the Military Can't Trust AI," *Foreign Affairs*, April 29, 2024.
300 **maintain human control:** Jarrett Renshaw and Trevor Hunnicutt, "Biden, Xi Agree That Humans, Not AI, Should Control Nuclear Arms," *Reuters*, November 16, 2024.
301 **Defense Llama:** Brandi Vincent, "Scale AI Unveils 'Defense Llama' Large Language Model for National Security Users," *Defense Scoop*, November 4, 2024.
301 **April 2025:** This was at the "2025 Modern Conflict Summit," Vanderbilt Institute of National Security: Summit on Modern Conflict and Emerging Threats, April 10, 2025.
301 **Sam Altman:** Katrina Manson and Jamie Tarabay, "OpenAI's Altman Won't Rule Out Helping Pentagon on AI Weapons," *Bloomberg*, April 10, 2025.
301 **"frenemy":** According to former Mavenites, a characterization that Miller laughed at but did not dispute.
302 **were finding:** Adam Tauman Kalai et al., "Why Language Models Hallucinate," OpenAI, September 4, 2025.
304 **huge fan of Claude:** According to people familiar with the matter.
304 **US military officials in Hawaii:** "2025 Honolulu Defense Forum," Pacific Forum, February 12, 2025.
304 **"drinking the Kool-Aid":** Patrick J. Kiger, "Why Did Hundreds of Americans 'Drink the Kool-Aid' at Jonestown?," *How Stuff Works*.

CHAPTER 24: MACHINES SHOULDN'T KILL PEOPLE

309 **bronze sculpture:** Carl Fredrik Reuterswärd, *Non-Violence*, June 11, 1989, United Nations Gifts.
310 **long-running discussion:** "Convention on Certain Conventional Weapons—Group of Governmental Experts on Lethal Autonomous Weapons Systems," United Nations Office for Disarmament Affairs, 2025.
310 **lethal autonomous weapons systems:** "UNODA Science, Technology and International Security Unit—Meeting," United Nations Office for Disarmament Affairs, 2024.
310 **"no place for lethal autonomous weapon systems":** "'Politically Unacceptable, Morally Repugnant': UN Chief Calls for Global Ban on 'Killer Robots,'" *UN News*, May 14, 2025.
310 **AI in the military domain:** "Artificial Intelligence in the Military Domain," United Nations Office for Disarmament Affairs; *Resolution Adopted: Artificial Intelligence in the Military Domain and Its Implications for International Peace and Security* (United Nations General Assembly: 79th Session, 2024).
311 **NATO had bought Maven Smart System:** "NATO Acquires Ai-Enabled Warfighting System," NATO: Allied Land Command, April 14, 2025.
311 **lunchtime event:** "Side Events: A Hazard to Human Rights: Autonomous Weapons Systems and Digital Decision-Making," posted on May 12, 2025, UN Web TV.
312 **child soldier:** "ILC Ideas | Timothy Musa Kabba," Yale Jackson School of Global Affairs, September 27, 2023.
312 **thanks to the Geneva Conventions:** "The Geneva Conventions: 'Rules of War' Saved Me, Says Former Child Soldier," *UN News*, August 26, 2024.
312 **"gold standard":** "Resolution 79/239 'Artificial Intelligence in the Military Domain and Its Implications for International Peace and Security,'" United Nations Office for Disarmament Affairs, 2025.
312 **responsible AI in the military domain:** David Vergun, "U.S. Endorses Responsible AI Measures for Global Militaries," *DOD News*, November 22, 2023; "Political Declaration on Responsible Military Use of Artificial Intelligence and Autonomy," US Department of State, November 27, 2024.

312 **"appropriate levels"; "meaningful human control":** "U.S. Delegation Statement on 'Appropriate Levels of Human Judgment,'" US Mission to International Organizations in Geneva, April 12, 2016.
312 **incorrect claims:** Michael Horowitz, an advocate for the development of autonomous systems, describes these as based on a "pernicious myth." See Horowitz, "Autonomous Weapon Systems: No Human-in-the-Loop Required, and Other Myths Dispelled."
312 **derided "meaningful human control":** "Resolution 79/239 'Artificial Intelligence in the Military Domain and Its Implications for International Peace and Security.'"
313 **2015 open letter:** "Research Priorities for Robust and Beneficial Artificial Intelligence: An Open Letter," Future of Life Institute, October 28, 2015.
313 *Slaughterbots*: "Slaughterbots," posted November 13, 2017, by Future of Life Institute, YouTube.
313 **robot goats:** Lance Cpl. Justin Marty, *U.S. Marines Test Fire the M72 LAW with a Robotic Goat*, September 9, 2023, Photograph, Defense Visual Information Distribution Service.
313 **robotic dog:** Joseph Trevithick, "Marines Test Fire Robot Dog Armed with Rocket Launcher," *The War Zone*, October 18, 2023.
314 **China was reportedly on a similar path:** Emma Helfrich and Tyler Rogoway, "China Pairs Armed Robot Dogs with Drones that Can Drop Them Anywhere," *The War Zone*, October 5, 2022.
314 **looking at the question since 2014:** *Expert Meeting—Autonomous Weapon Systems: Technical, Military, Legal and Humanitarian Aspects* (International Committee of the Red Cross, 2014).
316 **UN published its report:** *Artificial Intelligence in the Military Domain and Its Implications for International Peace and Security*, Report of the Secretary-General (United Nations General Assembly: 80th Session, 2025).
316 **Trump administration's four-page submission:** "Resolution 79/239 'Artificial Intelligence in the Military Domain and Its Implications for International Peace and Security,' Submission by the United States of America."
317 **China's alarmed submission:** *Artificial Intelligence in the Military Domain and Its Implications for International Peace and Security*, 21–23.
317 **Russia:** Ibid., 72–76.
317 **accuse of intentionally striking:** "UN Commission Concludes that Russian Armed Forces' Drone Attacks against Civilians in Kherson Province Amount to Crimes against Humanity of Murder," OHCHR, May 28, 2025, and "Ukraine: Russian Using Drones to Attack Civilians," Human Rights Watch, June 3, 2025.
318 **"AI Red Lines":** "We Urgently Call for International Red Lines to Prevent Unacceptable AI Risks," AI Red Lines.
318 **said the International Committee of the Red Cross:** "Submission to the United Nations Secretary-General on Artificial Intelligence in the Military Domain Re: Oda/2025-00029/Aimd," ICRC, April 2025; Open to Debate, "Wartime Kill Switch: Human or AI?," Council on Foreign Relations, September 15, 2025.

CHAPTER 25: TRUMP'S ROBOTS

319 **Trump's Robots:** The opening episodes are based on documentation reviewed by the author and author interviews with people familiar with the matter. Vice Admiral Okano, through a representative, declined comment.
319 **Replicator:** "Replicator," Defense Innovation Unit, https://www.diu.mil/replicator.
319 **Overmatch:** In 2023, Admiral Michael Gilday, chief of naval operations for the US Navy, said the Navy would comprise 40 percent unmanned surface fleet by the early 2040s. It wasn't clear how the Navy planned to get there or if it was allocating funds to do so. See Diego Laje, "Improving Unmanned Capabilities Ensures Effective Deterrence," *AFCEA: Signal Media*, February 16, 2023; Elisha Gamboa, "Project Overmatch Achieves Historic Milestone with Five Eyes Agreement," *NavWar*, February 26, 2025.
320 **Hegseth had just testified to Congress:** Pete Hegseth, "FY26 Written Posture Statement," House Appropriations Committee—Defense, June 10, 2025.

321 **and capsize:** Brandi Vincent ("Navy Experiment Cut Short after Unmanned Vessel Flipped a Support Boat," *DefenseScoop*, July 1, 2025) previously reported the support boat was upturned due to an inadvertent command; David Jeans ("The US Navy Is Building a Drone Fleet to Take On China. It's Not Going Well," Reuters, August 20, 2025) previously reported the captain was thrown into the water.

322 **did not work at L3Harris:** According to a spokesperson for L3Harris.

322 **software:** The L3Harris autonomy software is called Amorphous Collaborative Autonomy.

322 **errant ENTER:** This is also known as executing a zero command.

322 **none of the unmanned surface vessels had ever carried a weapon:** According to interviews with people familiar with the matter. See also Lt.j.g. Drew Verbis, "Increase in Small Boat Activity, Around Naval Base Ventura County, Local Harbors," Defense Visual Information Distribution Service, April 14, 2025.

322 **reportedly:** David Jeans, "The US Navy Is Building a Drone Fleet to Take On China. It's Not Going Well."

322 **gone awry:** In a statement, a Navy spokesperson said: "Safety is always our top priority, and for any test event there are multiple mitigation measures and backup systems in place to prevent danger to participants and observers. There is inherent risk in testing new capabilities, but the lessons we learn during these events drive improvements in the systems we ultimately provide the Fleet and the warfighter. In order to meet the pressing challenges to our Nation's security we are taking a 'fail fast' approach and moving quickly, but not at the expense of safety."

323 **"effective immediately":** Asked for comment, a spokesperson for L3Harris said the company successfully delivered a prototype and the company's discontinuation on Replicator had nothing to do with safety. The spokesperson said the company is committed to autonomy at scale and that its autonomy command and control software has demonstrated its ability to control a mix of uncrewed platforms, payloads, and commercial technologies even if they were produced by different manufacturers. A spokesperson for the Navy declined comment.

323 **faced the axe:** A spokesperson for BlackSea Technologies, which produces GARCs, said the company has sold more than 300 GARCs to the US Navy, including more than 250 that are operating on the water in every combatant command's area of responsibility. GARCs have integrated more than seven autonomy software stacks during more than 15,000 hours of operation by the Navy.

323 **quietly involved in Replicator's effort:** The next two paragraphs rely on people familiar with the matter and documents reviewed by the author. In May 2025 testimony to Congress, Doug Beck flicked at Maven's role to deliver "needed levels" of ATR for unmanned platforms that use autonomy. See "Prepared Remarks by Douglas A. Beck, Director, Defense Innovation Unit Before the United States House of Representatives Armed Services Committee Subcommittee on Tactical Air and Land Forces on Small UAS and Counter-Small UAS: Gaps, Requirements, and Projected Capabilities," 119th Cong., 1st Sess., May 1, 2025.

324 **"I love killer robots":** "The AI Arsenal That Could Stop World War III | Palmer Luckey | TED," posted April 25, 2025, by TED, YouTube.

324 **four thousand drones a day:** "Prepared Remarks by Douglas A. Beck, Director, Defense Innovation Unit."

324 **China made more than a hundred:** Hegseth, "FY26 Written Posture Statement," 12.

325 **$13.4 billion:** "Background Briefing on FY 2026 Defense Budget," Department of Defense, June 26, 2025.

325 **demos they viewed:** None responded to request for comment.

325 **share data or models with commercial AI companies:** According to people familiar with the matter. See also "Scaling AI-Enabled Capabilities at the DOD: Government and Industry Perspectives," streamed live March 26, 2024, by Center for Strategic & International Studies, YouTube.

325 **when she announced the initiative:** Joseph Clark, "Hicks Underscores U.S. Innovation in Unveiling Strategy to Counter China's Military Buildup," *DOD News*, August 28, 2023.

326 **"harder to plan for":** "Deputy Secretary of Defense Kathleen Hicks' Remarks: 'Unpacking the

Replicator Initiative' at the Defense News Conference (As Delivered)," Department of Defense, September 6, 2023.

327 **wouldn't necessarily be accurate:** Lucy Suchman, "Algorithmic Warfare and the Reinvention of Accuracy," *Critical Studies on Security* 8, no. 2 (2020); Ingvild Bode and Tom Watts, *Loitering Munitions and Unpredictability: Autonomy in Weapon Systems and Challenges to Human Control* (Center of War Studies, 2023).

329 **"Hedge":** Michael Brown et al., "A Hedge Strategy to Strengthen Defense Capabilities," Defense Innovation Unit, July 2022; Michael Brown and Rear Adm. Lorin Selby, "Revisiting the Hedge Strategy with Renewed Urgency," *War on the Rocks*, September 7, 2023.

329 **bagging an extra $1 billion:** "Deputy Secretary of Defense Kathleen H. Hicks and Vice Chairman of the Joint Chiefs of Staff Admiral Christopher W. Grady Opening Remarks on the Department of Defense F.Y. 2025 Budget Request," Department of Defense, March 11, 2024; Courtney Albon and Noah Robertson, "Pentagon Says $1 Billion Planned for First Two Years of Replicator," *Defense News*, March 11, 2024.

329 **"Hellscape":** At the same event where Hicks launched Replicator, the commander of Indo-Pacific Command, Admiral John Aquilino, said in 2023 that he wanted to be able to hit a thousand targets in twenty-four hours. "There is a term—'hellscape'—that we use."

329 **"unmanned hellscape":** Josh Rogin, "The U.S. Military Plans a 'Hellscape' to Deter China from Attacking Taiwan," *Washington Post*, June 10, 2024.

329 **wanted INDOPACOM to lead the way:** Sydney J. Freedberg Jr., "'Who Should Shoot Who?': INDOPACOM Getting 'Combat Representative' JFN This Year," *Breaking Defense*, July 17, 2024; Mikayla Easley, "Air Force Standing Up Integrated Program Office for Joint Fires Network on Oct. 1," *Defense Scoop*, September 23, 2025.

330 **"It's no longer training":** Katrina Manson, "US Commander Warns of China, Russia, North Korea 'Troublemakers,'" Bloomberg, February 14, 2025.

330 **more than 2,400:** "2025 HASC Testimony [of Frank D. Whitworth] on National Security Space Programs," National Geospatial-Intelligence Agency, May 14, 2025.

330 **"profoundly" adding:** Manson, "US Commander Warns of China, Russia, North Korea 'Troublemakers.'"

330 **into the hands of sailors:** Sam Lagrone, "First Replicator Initiative Capability on Track for August, Officials Say," *USNI News*, January 28, 2025.

330 **Goalkeeper:** Multiple Navy budget documents refer to Goalkeeper, see "Department of Defense Fiscal Year (FY) 2024 Budget Estimates: Justification Book Volume 1 of 1," Department of the Navy, March 2023, and "Department of Defense Fiscal Year (FY) 2024 Budget Estimates: Justification Book Volume 3 of 5," Department of the Navy, March 2023. David Hambling, a defense technology journalist, has also explored the emergence of Goalkeeper here, "The U.S. Navy's Billion-Dollar Mystery 'Kamikaze Drones,'" *19fortyfive*, May 31, 2023.

330 **budget documents:** "Department of Defense Fiscal Year (FY) 2024 Budget Estimates: Justification Book Volume 1 of 1"; "Department of Defense Fiscal Year (FY) 2025 Budget Estimates: Justification Book Volume 1 of 1," Department of the Navy, March 2024.

331 **"our stagnant defense industrial base":** Hegseth, "FY26 Written Posture Statement," 8.

331 **ONR worked closely with Georgia Tech:** "During this past fiscal year, ONR completed 22 grants here at GTRI worth $23.6 million, and Georgia Tech currently has 72 active contracts and grants with the Navy worth $216 million," Navy Secretary Carlos del Toro said in October 2024. He said four of the five selected "Replicator" systems came out of the Department of the Navy's innovation ecosystem. "Secretary Del Toro As-Written Remarks at the Georgia Tech Research Institute," US Navy, October 23, 2024.

331 **AI computer vision software:** Known as SPOTR-Edge; "AV Unveils New AI Capability and Autonomy Kit for Unmanned Systems," AV (AeroVironment), April 23, 2024. In May 2025, AeroVironment launched Red Dragon, which it described as a fully autonomous capable, software-defined unmanned aircraft system designed for one-way attack missions in high-threat, GPS-denied, and communications-degraded environments. AeroVironment, Inc., also

known as AV, Inc., declined to comment on this and work for ONR and in Ukraine, citing customer sensitivities.

331 **2015 low-cost effort:** I was told LOCUST program drones, intended to defend against enemy drones, emerged into a new idea for drones to strike at land and sea targets under the Goalkeeper program. The ONR announced the LOCUST program in 2015, David Smalley "LOCUST: Autonomous Swarming UAVs Fly Into the Future," Office of Naval Research, April 14, 2015. A YouTube video depicting LOCUST, cited by ONR in 2015, has since been turned to Private. Journalist David Hambling previously explored LOCUST here, "U.S. Navy Destroys Target with Drone Swarm—And Sends a Message to China," *Forbes*, April 30, 2021. ONR did not respond to author outreach by phone and email.

331 **"never been done before":** Smalley "LOCUST: Autonomous Swarming UAVs Fly Into the Future."

331 **small expeditionary teams:** "The system will be expeditionary and deployable by small teams to support operations in various environments." Department of Defense Fiscal Year (FY) 2024 Budget Estimates.

331 **3000.09:** "DOD Directive 3000.09: Autonomy in Weapon Systems."

331 **"exercise appropriate levels of human judgment over the use of force":** Ibid., 10.

331 **"senior review":** Ibid., 5, 15, 18.

332 **Whiplash:** "The Expeditionary Loitering Munitions Portfolio consists of two Navy loitering munition programs (GOALKEEPER and WHIPLASH) designed to meet an urgent Geographic Combatant Command Requirement. Projects GOALKEEPER (GK) and WHIPLASH (WL) will leverage and continue prior development efforts including an Office of Naval Research (ONR) Innovative Naval Prototype (INP). The Navy is pursuing commercial-off-the-shelf solutions with autonomous government-provided software and a government-provided launcher systems [sic]. The systems will be expeditionary and deployable by small teams to support operations in various environments." Department of Defense Fiscal Year (FY) 2025 Budget Estimates.

332 **China shipbuilding capacity:** Hegseth, "FY26 Written Posture Statement," 10.

332 **unidentified jet ski:** Thomas Newdick, "Explosives-Packed Jet Ski Drone Appears Off Turkish Coast," *The War Zone*, July 25, 2024. The CIA did not respond to a request for comment. Adam Entous, "The Partnership: The Secret History of the War in Ukraine," reports several elements of the CIA's support to Ukraine.

332 **disappeared from budget documents:** "Department of Defense Fiscal Year (FY) 2026 Budget Estimates: Justification Book Volume 1 of 1, Procurement of Ammo, Navy & MC," Department of the Navy, June 2025.

333 **DAWG:** According to multiple people familiar with the matter. The *Wall Street Journal* reported the name here, Shelby Holliday, "U.S. Military Is Struggling to Deploy AI Weapons," September 26, 2025.

333 **defense of Taiwan:** See Katrina Manson, "US Elite Forces Ill-Equipped for Cold War with China," *Financial Times*, May 16, 2020.

333 **"the GWOT ship, has sailed":** Sean Carberry, "JUST IN: SOCOM Needs New Tech, Old Approaches, Senior Leader Says," *National Defense*, August 25, 2023.

333 **"back in the boats":** Ibid.

333 **Replicator test event:** Replicator did field hundreds of drones within two years; "Prepared Remarks by Douglas A. Beck, Director, Defense Innovation Unit Before the United States House of Representatives Armed Services Committee Subcommittee"; CITI Hearing: "Science, Technology, and Innovation Posture": Hearing before the House Armed Services Committee (2025) [the video on this page has since been deleted].

CHAPTER 26: THE WINCHESTER HOUSE

335 **didn't like war one bit:** The opening section is based on interviews during a visit to Drew Cukor at his office on April 25, 2025, the Cukors' home on April 26, 2025, and follow-up interviews.

335 **turned down:** According to Jain and confirmed by Cukor.

336 **Thomas Tull:** Natalie Robehmed, "Box Office Billionaire: How Legendary's Thomas Tull Used Comics, China and a Secret Formula to Remake Hollywood," *Forbes*, February 24, 2016; Alyssa Hillwig, "'I Have a Very Different Day-to-Day Experience than Most Folks,'" *Shale Oracle*, May 24, 2022.
335 **national security forum:** This was the "National Security Innovation Forum" on November 20, 2024.
336 **"into the fabric":** "Embedding Intelligence into the Future of Finances," TWG Global, accessed October 7, 2025, https://www.twgglobal.com/palantir-xai-partnership/.
336 **"develop and deploy":** "xAI, TWG Global and Palantir Unite to Redefine Financial Services through Enterprise AI," Palantir, May 6, 2025.
336 **"don't have to trust your taste on AI":** "Palantir CEO Alex Karp and TWG Global Co-Chairman Thomas Tull Talk New Partnership on CNBC," posted May 6, 2025, by Palantir, YouTube.
339 **The Winchester family:** Pamela Haag, "The Heiress to a Gun Empire Built a Mansion Forever Haunted by the Blood Money That Built It," *Smithsonian*, July 7, 2016; "History," Winchester Mystery House; "The Complete History of Winchester Repeating Arms," Winchester Guns.
340 ***Winchester*:** *Winchester*, directed by Peter and Michael Spierig (2018).
340 **places that were important to him:** These cities in Afghanistan and Iraq were at one time the epicenter for US Marine Corps combat operations.
340 **"All of us carry PTSD":** By some estimates, more than one-quarter of veterans from the most recent conflicts have symptoms of PTSD, depression, or traumatic brain injury. More than 40 percent of veteran offenders who served in combat zones developed PTSD, according to this study, Glenn R. Schmitt and Amanda Kerbel, *Federal Offenders Who Served in the Armed Forces* (United States Sentencing Commission, 2021).
341 **moral harm:** "PTSD: National Center for PTSD," US Department of Veterans Affairs, March 26, 2025.
342 **Just War:** James Turner Johnson, "Just War, As It Was and Is," *First Things*, January 1, 2005.
342 **moral injury:** "The reality is that the high level of guilt symptoms among veteran populations is significant, and that many counselors and clinicians have not developed proven techniques to help." See "Moral Injury & Moral Resilience," Human Performance Office, TECOM; William Nash, a former US Marine Corps director of psychological health and leading researcher on moral injury, wrote the foreword to this book, Robert Emmet Meagher and Douglas A. Pryer, eds., *War and Moral Injury: A Reader* (Cascade, 2018); this series explores moral injury, David Wood, "'I'm A Good Person and Yet I've Done Bad Things': A Warrior's Moral Dilemma," *Huffington Post*, March 18, 2014. Gen (ret.) Jim Mattis argued in a 2014 speech to which some veterans have attributed a life-saving impact that there was no room for self-pity, cynicism or for veterans to see themselves as victims "even if so many of our countrymen are prone to relish that role." He argued for unit leadership based on compassion, trust and affection, and for "post-traumatic growth," which he described as emerging from trauma feeling kinder to one's fellow man and woman. "There's one misperception of our veterans and that is they are somehow damaged goods," he said. See "2014 Salute to Iraq & Afghanistan Veterans – General James Mattis, USMC (Ret.) – Full Version," YouTube, May 1, 2014.
342 **"War may be a human activity":** C. Anthony Pfaff, "Respect for Persons and the Ethics of Autonomous Weapons and Decision Support Systems," *Strategy Bridge*, March 4, 2019.
342 **2019 paper:** C. Anthony Pfaff, "The Ethics of Acquiring Disruptive Technologies: Artificial Intelligence, Autonomous Weapon and Decisions Support Systems," Simons Center Special Report: Ethical Implications of Large Scale Combat Operations, 2019.
345 **not reflected in our doctrine:** Steve Leonard, a former senior military strategist, quotes a Soviet observation about the relevance of US military doctrine during the Cold War: "A serious problem in planning against American doctrine is that the Americans do not read their manuals, nor do they feel any obligation to follow their doctrine." Steve Leonard, "Broken and Unreadable: Our Unbearable Aversion to Doctrine," *Modern War Institute at West Point*, May 18, 2017.
345 **AI targeting against Gazans:** Marissa Newman, "Israel Quietly Embeds AI Systems in Deadly Military Operations," Bloomberg, July 16, 2023.
345 **"AI-assisted genocide":** Marc Owen Jones, quoted in "'AI-Assisted Genocide': Israel Reportedly

Used Database for Gaza Kill Lists," Al Jazeera, April 4, 2024. See also Sam Mednick et al., "How US Tech Giants Supplied Israel with AI Models, Raising Questions about Tech's Role in Warfare," *AP News*, February 18, 2025; "Press Release: Situation in the State of Palestine: ICC Pre-Trial Chamber I Rejects the State of Israel's Challenges to Jurisdiction and Issues Warrants of Arrest for Benjamin Netanyahu and Yoav Gallant," International Criminal Court, November 21, 2024; "Press Release: Israel Has Committed Genocide in the Gaza Strip, UN Commission Finds," United Nations, September 16, 2025.

345 **majority of them civilians:** *Legal Analysis of the Conduct of Israel in Gaza Pursuant to the Convention on the Prevention and Punishment of the Crime of Genocide* (Human Rights Council: 60th Session, 2025); the WHO updates Palestinian Casualties here, "Palestinian Casualties," World Health Organization, accessed October 6, 2025.

345 **"No part of life and death":** António Guterres, "Secretary-General's Press Encounter on Gaza," United Nations: Secretary-General, April 5, 2024.

345 **"lawful," "responsible":** "Press Release: The IDF's Use of Data Technologies in Intelligence Processing," Israel Defense Forces, June 18, 2024.

345 **"proportionate":** "We ended up, I believe, with about twenty-five thousand Hamas killed and twenty-five thousand civilians," a former political leader told me earlier this year. "This is a better proportion than was ever achieved by a modern military." Dexter Filkins, "Is the U.S. Ready for the Next War?," *The New Yorker*, July 14, 2025.

345 **Mark Milley:** Mark A. Milley and Eric Schmidt, "America Isn't Ready for the Wars of the Future and They're Already Here," *Foreign Affairs*, August 5, 2024.

345 **cited reporting from +972 Magazine:** Yuval Abraham, "'Lavender': The AI Machine Directing Israel's Bombing Spree in Gaza," *+972 Magazine*, April 3, 2024. The magazine also published the first reports about the use of AI targeting in 2023, Yuval Abraham, "'A Mass Assassination Factory': Inside Israel's Calculated Bombing of Gaza," *+972 Magazine*, November 30, 2023.

345 **talk of his I attended:** "Retired Gen. Mark Milley Calls on a New Generation to Confront Changing Era of Security Threats at Washington, D.C. Launch Event for the Institute of National Security," Vanderbilt Institute of National Security, October 21, 2024.

346 **human safari:** *"They Are Hunting Us": Systematic Drone Attacks Targeting Civilians in Kherson* (Human Rights Council: 59th session, 2025); Zarina Zabrisky, "'Human Safari'—Kherson Civilians Hunted Down by Russian Drones," *Kyiv Independent*, October 2, 2024.

346 **Michael Horowitz ... makes this argument:** "'The civilian casualties in Gaza were not an A.I. issue—they appear to be a rules-of-engagement issue,' Michael Horowitz, a Deputy Assistant Secretary of Defense in the Biden Administration, said." See Filkins, "Is the U.S. Ready for the Next War?"

346 **controversial even inside the IDF:** Elizabeth Dwoskin, "Israel Built an 'AI-Factory' for War. It Unleashed It in Gaza," *Washington Post*, December 29, 2024.

347 **criticism:** Fitzgerald, "Tech Workers Should Shine a Light on the Industry's Secretive Work with the Military."

347 **resulting report:** Gaza Assessment Task Force, *The October 7 War: Observations, October 2023–May 2024* (Jewish Institute for National Security of America, 2024).

348 **"operationally critical operations":** "Palantir Technologies | Q4 2023 Earnings Webcast," streamed live February 5, 2024, by Palantir, YouTube. Alex Koller, "Palantir CEO Says His Outspoken Pro-Israel Views Have Caused Employees to Leave Company," CNBC, March 13, 2024.

348 **strategic partnership:** Marissa Newman, "Palantir Supplying Israel with New Tools Since Hamas War Started," Bloomberg, January 10, 2024; Marissa Newman, "Thiel's Palantir, Israel Agree Strategic Partnership for Battle Tech," Bloomberg, January 12, 2024; Caroline Haskins, "'I'm the New Oppenheimer!': My Soul-Destroying Day at Palantir's First-Ever AI Warfare Conference," *The Guardian*, May 17, 2024.

348 **"to scare enemies and on occasion kill them":** Karp quoted in Michael Eby, "Palantir's Idea of Peace," *The Nation*, May 27, 2025.

348 **"the innate superiority" of the West:** "Palantir Technologies | Q4 2024 Earnings Webcast,"

NOTES

streamed live February 3, 2025, by Palantir, YouTube; Jack McCordick, "Alex Karp's War for the West," *New Republic*, February 28, 2025.

348 **Google:** Billy Perrigo, "Exclusive: Google Workers Revolt Over $1.2 Billion Contract with Israel," *Time*, April 10, 2024; Kelvin Chan and Wyatte Grantham-Philips, "Google Fires More Workers Who Protested its Deal with Israel," *AP News*, April 23, 2024.

348 **Microsoft:** "Microsoft Statement on the Issues Relating to Technology Services in Israel and Gaza," Microsoft, May 15, 2025; Brad Smith, "Update on Ongoing Microsoft Review," Microsoft Corporate Blogs, September 25, 2025; Bruna Horvath, "Pro-Palestinian Protestor Interrupts Microsoft AI CEO During Live Event," *NBC News*, April 4, 2025.

348 **Palantir:** Jason Sutich, "'Jews Say Let Gaza Live': Protestors Rally Against Palantir Over Alleged Role in Gaza Conflict, ICE Operations," *My Northwest*, July 14, 2025; "'Purge Palantir': Day of Action Protests Firm's Role in Gov't Surveillance, ICE & Genocide in Gaza," *Democracy Now*, July 15, 2025.

348 ***Guardian* reporting:** Harry Davies and Yuval Abraham, "'A Million Calls an Hour': Israel Relying on Microsoft Cloud for Expansive Surveillance of Palestinians," *The Guardian*, August 6, 2025.

348 **lost employees and investors:** "Palantir CEO on Generative AI and Competition," posted March 13, 2024, by CNBC Television, YouTube; Stefania Spezzati and Gwladys Fouche, "Thiel's Palantir Dumped by Norwegian Investor Over Work for Israel," Reuters, October 25, 2024.

348 **"mostly terrorists":** "Palantir Technology CEO at Hill and Valley Forum," posted April 30, 2025, by C-SPAN.

348 **Geneva Conventions:** "The Geneva Conventions and Their Commentaries," ICRC.

348 **distinguish between combatants and civilians:** "Practice Relating to Rule 7. The Principle of Distinction between Civilian Objects and Military Objectives," International Humanitarian Law Databases.

349 **knowingly kill "excessive" numbers of civilians:** "Practice Relating to Rule 14. Proportionality in Attack," International Humanitarian Law Databases.

349 **"Even wars have rules":** "Frequently Asked Questions: Rules of War," ICRC, July 20, 2023.

349 **2024 book:** Pete Hegseth, *The War on Warriors: Behind the Betrayal of the Men Who Keep Us Free* (Broadside, 2024).

349 **questioned the Geneva Conventions:** As Hegseth went through a testy confirmation process, his lawyer told Congress that he supported the Law of Armed Conflict. "He is not in any way advocating that anybody not follow the Law of Armed Conflict," Parlatore said. "He is saying that the way that has been interpreted at the local level is overly restrictive." Dan De Luce and Courtney Kube, "Some Military Officials Worry that Pete Hegseth Could Turn a Blind Eye to U.S. War Crimes," *NBC News*, January 13, 2025.

349 **"Our boys should not fight by rules":** Hegseth, *The War on Warriors*, quoted in Parker Yesko, "Turning a Blind Eye to War Crimes," *The New Yorker*, November 22, 2024.

349 **"AI in everything":** "Secretary of War Pete Hegseth Addresses General and Flag Officers at Quantico, Virginia."

EPILOGUE

353 **it wasn't sufficient to change:** "I'm certainly questioning my original premise that the fundamental nature of war will not change," Defense Secretary Jim Mattis told me when I asked him about the impact of AI on the character and nature of war in February 2018. "I—you've to question that now. I just don't have the answers yet." "Press Gaggle by Secretary Mattis En Route to Washington, D.C.," Department of Defense, February 17, 2018.

353 **nature of war:** von Clausewitz, *On War*, 4.

353 **the Marine Corps:** "Announcement of Maven Smart System Licensing for Marine Corps," US Marine Corps, September 11, 2025.

353 **halfway through:** Among those who did not respond to requests for comment on aspects of this book and whose responses are not already logged elsewhere, are ECS Federal, i2 Group, and the Australian Government Department of Defence. AWS, Anduril Industries, Inc., Google, Microsoft, and Pilot AI did not provide comment; the UK's Ministry of Defence declined comment.

INDEX

Page numbers in *italics* refer to figures. Page numbers after 356 refer to notes.

Abbottabad, Pakistan, 244
Aberdeen Proving Ground, MD, 209–10
Adkins, Winfield, 251
Advanced Field Artillery Tactical Data System (AFATDS), 231–32
AEGIS weapons system, 297
Aerosonde drones, 72
AeroVironment, 323, 331, *393*
Afghanistan, 8–9
  Bagram Air Base, 19, 125, 184–85, 188–89
  Camp Delaram, 39
  Camp Fenty, 130
  Camp Leatherneck, 30, 39–40
  Camp Rhino, 14–15, 17
  Jalalabad, 129, 130, 359
  Kabul, 19, 34, 129–31, 232–35, 359
  Kandahar, 13–14, 16–17, 60, 97, 130, 164, 218
  the "Mother of All Bombs," 120
  the problem of computer vision and drone deployment in, 48–54
  Project Maven in, 48–54, 59, 61, 65–72, *69, 71*, 97, 105, 122–33, 338–39, *339*, 344

Soviet invasion of, 17, 43, 45, 93, 101, 260, 296, 395
  the UK in, 8–9
  US invasion of, 13–19, 26–28, 30, 33, 37–42, 164–165, 202–3, 359
  US withdrawal from, 176, 183, 184–85, 188–89, 202–3, 232–35, 268–69
Africa, 8, 48, 74, 83–85, 91, 97, 294, 182, 311
AI Now Institute, 109
Air Force. *See* US Air Force
Aiyer, Krishnan, 193
"Algorithmic Warfare Cross-Functional Team," 51. *See* Project Maven
Alion Science and Technology (defense contractor), 173
Allbirds, 68
Alphabet, 56. *See also* Google
AlphaGo, 1–2, 89, 100–101, 114, 358
al-Qaeda, 13, 15–16, 30–31, 48, 1, 183, 190
al-Shabaab, 91, 93, 183
Altman, Sam, 213, 301
Amazon Web Services (AWS), 4, 103, 106, 114, 141–42, 149, 212, 268, 270–71, 278, 281, 285, 304, 352
Amazon, 47, 156–57

Anderson, Jeff, 259–60
Anduril Industries, 30, 58, 157, 177–78, 299, 320, 324–25, 352
Anthropic, 5, 304
Apple, 5, 77, 80, 209, 372, 381
Army. *See* US Army
Atropos Group, 215
AUKUS deal, 143–44
Austin, Lloyd, 233
Australia, 7, 143–44, 153, 294
Automated Information Discovery Environment (AIDE), 173, 175, 177

al-Baghdadi, Abu Bakr, 194–99, 201, 226
Bagram Air Base, Afghanistan, 19, 125, 184–85, 188–89
Bahrain, 294
Baledogle Military Airfield, Somalia, 93–94, 97
Barakat, Barakat Ahmad, 198–99
Barchick, Matt, 87, 185, 188–90
Barisha village, Syria, 196–98
Batir, Sean, 303–4
Beale Air Force Base, CA, 95–97, 107
Beck, Doug, 324, 392
Belgrade, Serbia, 202, 380

INDEX 399

Biden, Joe, 58, 156–57, 233–34, 276, 300, 319, 328
bin Laden, Osama, 156, 244
Black, Aileen, 104–7
Bloch, Jan, 8
Bloomberg, 8, 162
Boeing, 49, 72, 139, 157
Boggess, James, 344
Booz Allen, 49, 75, 157
Bosnia, 224
Boston Consulting Group, 59
Bowden, Mark, 156
Boyd, John, 88–89, 260
Breakfast Club, 43–44, 48
Broad Area Search for Targeting (BAS-T), 232
Bryant, Wes, 201–3
Burkina Faso, 195
Burry, Michael, 352
Bush, Vannevar, 159

CACI, 174
Caggy, Jim, 270–71, 352
Caine, Dan, 325
Cambodia, 294
Camp Bucca detention center, Iraq, 195
Camp Delaram, Afghanistan, 39
Camp Fenty, Afghanistan, 130
Camp Leatherneck, Afghanistan, 30, 39–40
Camp Lejeune, NC, 26
Camp Lemonnier, Djibouti, 182–83
Camp Rhino, Afghanistan, 14–15, 17
Camp Titan, Djibouti, 183
Campbell, Kurt, 145, 146
Canada, 153, 294
Cape Cod, MA, 137–38
Carley, Kathleen, 31
Carroll, Colin, 57–61, 66–68, 71, 90–95, 123, 125, 172, 180, 186–88, 191–93, 207, 215–19, 249–60, 352, 366, 381
Carter, Ashton, 102
CCW-GGE (Convention on Certain Conventional Weapons—Group of Governmental Experts), 314–15

Central Command (CENTCOM), 41, 201, 245, 291–92, 301
Césaire, Aimé, 295, 389
ChatGPT, 177, 188, 298
Chief Digital and Artificial Intelligence Office (CDAO), 243, 245, 290–91, 384
China, 2, 4, 6, 8–9, 23–25, 324–32, 333, 351
  the AI arms race with, 23–25, 43–45, 57, 84, 134–36, 137, 140, 142–44, 145–54, 158, 160, 218, 324–32
  autonomous weaponry and global pushback, 310, 315, 316–17
  centralized data labeling, 206
  defense spending in, 145–46
  facial recognition, 213
  Google and, 112, 113, 149
  Navy, 142–43
  People's Liberation Army (PLA), 24, 146–51, 323
  potential conflict over Taiwan, 7, 9, 24, 44, 147, 151, 154, 160, 170, 283, 320, 323–30, 333–34, 351
  spy balloon incident, 294
  technology sharing with Russia, 142–43
  US bombing of the Chinese embassy in Belgrade, 202, 380
  use of landmines, 112
Chomsky, Noam, 108
Christ, Gregory "Jesus," 71, 217, 254, 260
CIA, 5, 31–33, 88, 37, 102, 106, 124, 149, 157, 202, 325, 332, 362, 380
Cisco, 171
Clarifai, 49, 77, 79–80, 83, 85, 113, 207–8, 212–13, 278, 285, 323
Clarke, Richard, 74
Clark, Shannon, 253
Clausewitz, Carl von, 16, 359
Coast Guard, 295

Coddington, Jaim, 68, 69, 227–29
Cogbill, John, 291–92
Combating Terrorism Technical Support Office (CTTSO), 38
Command Post Computing Environment (CPCE), 263
Congress
  casualties of war, 199, 359
  designated software "programs of record," 36, 40, 263, 288, 290, 357
  House of Representatives, 23, 269
  National Security Commission on Artificial Intelligence, 149–50
  Project Maven funding, 5, 24–25, 38, 41, 51–52, 87, 95, 149–51, 213–15, 226, 230, 240–42, 247, 255
  the Replicator initiative and beyond, 319–34, 352
  Senate, 23, 269
  various AI deployments, 294–95
  and the war in Ukraine, 269–70, 290–92
Cornelius, Andrew, 165–67
Costa Rica, 64, 313
Covid, 213, 247, 251, 254, 381
Crawford, Kate, 109
CrowdFlower (now Figure Eight), 205, 207–8, 381
Cukor, Drew, 1–2, 3–9, 351–54
  building the Project Maven team, 55–64, 71
  early life and career, 19–21, 43–44, 62–64
  early service in Afghanistan 2001, 13–19
  early service in Iraq, [to include references to 2003 and Project Legacy and after]
  enter Palantir, 29–42
  fear of China, 23–25
  on the human element, 25–26

Cukor, Drew (*continued*)
intelligence work at Marine HQ, 29–30
at J.P. Morgan, 3, 255, 259, 335–36, 382
leadership style, 5–6
"Marine Ground Intelligence Reform," 21, 22–25
master's thesis, 21, 22–23
as a Mormon in Panama, 63–64
"Operate to Know," 48, 87–89
Project Maven after his retirement, 256–60, 287–98, 335–49
work ethic of, 60–62
See also Palantir; Project Maven
Cukor, Kirsten, 20–21, 26, 61, 63, 255–56, 259, 335, 337–41
Customs and Border Protection, 295
Cyber Command, 283–84, 301, 347

Dam Neck, VA, 181
DCGS-A (Distributed Common Ground System—Army), 35, 38, 40, 362
Dean, Jeff, 77, 102–30
DeepMind, 1, 100–102, 113–14
DEF CON hacker conference, 299–300
Defense Advanced Research Projects Agency (DARPA), 103, 352
Defense Autonomous Warfare Group (DAWG), 332–34
Defense Department
Breakfast Club, 43–44, 48
Central Command (CENTCOM), 41, 201, 245, 291–92, 301
Civilian Protection Center of Excellence, 201
Combating Terrorism Technical Support Office (CTTSO), 38
Cyber Command, 283–84, 301, 347

Defense Criminal Investigative Service, 255
European Command (EUCOM), 245, 270, 286, 296, 299
Indo-Pacific Command (INDOPACOM), 142–43, 215, 283, 291, 294, 296, 299, 303–4, 329–30, 353, 393
Intelligence Support to Targeting, 240
Joint Artificial Intelligence Center (JAIC), 150, 187, 214–19, 243–45, 250, 256, 290–91
Network Enterprise Technology Command (NETCOM), 281
Northern Command (NORTHCOM), 294
Southern Command (SOUTHCOM), 295
Standards of Conduct Office, 249
See also Special Operations Command (SOCOM); individual branches of the military; specific theaters of war
Defense Innovation Unit (DIU), 102, 296, 324, 333, 352
Defense Intelligence Agency, 138
Delaram FOB, Nimruz, Afghanistan, 39
Delta battle management system, 273
Delta Force, 196
Dimon, Jamie, 3
Dixon, Peter, 36–38, 40
Djibouti, 74, 98, 122, 182–83, 294
Dome Plate conferences, 153
Donahue, Christopher, 184, 190, 225, 233, 268–69, 272, 275–79, 283–86, 296–97, 344, 353
Donovan, Frank, 333–34
Doolittle, Jimmy, 88
Dorchak, Rich, 260, 365

Dorsey, Edwin, 162
Dover Air Force Base, DE, 130
Dunford, Joseph, 113, 150
Duong, Ninh, 143

ECS Federal, 60, 62, 66, 82, 97–98, 106, 172–73, 186, 192, 210–13, 215, 248, 252–55, 385
Edwards Air Force Base, CA, 159
18th Airborne Corps, 224–32, 235–36, 240, 242, 261, 268, 270, 275–77, 280, 283, 299, 342–44
82nd Airborne Division, 233, 271
Einstein, Albert, 82, 155–56
Eisenhower, Dwight D., 109, 111
Enabled Intelligence (startup), 213, 289
Erbil, Iraq, 97, 195, 197, 224, 226
European Command (EUCOM), 245, 270, 286, 296, 299

Facebook (now Meta), 5, 77, 158, 213, 301
FADE: Fusion Analysis and Development Effort, 174
Farhadi, Ali, 80, 372
Feinberg, Stephen, 352
Fenty FOB, Jalalabad, Afghanistan, 130
Fergus, Rob, 77
Figure Eight (formerly CrowdFlower), 205, 207–8, 381
Financial Times, 7–8
Finish, The (Bowden), 156
Fitzgerald, William, 108–11, 371
"Five Eyes" intelligence partners, 153–54
Floyd, Garry "Pink," 55, 66–67, 70, *71*, 139, 152, 369
Flynn, Michael, 37, 38
Fort A.P. Hill training center, VA, 93
Fort Belvoir Army Base, VA, 289

France, 143
Freedom of Information Act
  (FOIA), 152–53, 357, 372
Future of Life Institute, 313

GARCs (global autonomous
  reconnaissance crafts),
  320–23, 392
Gaza, 345–46, 348, 396, 398
General Atomics, 72–73
General Dynamics, 281–82
Geneva Conventions of 1949,
  7, 310–12, 314, 348–49,
  397
GEOINT (geospatial intelli-
  gence), 289
Georgia Tech Research Insti-
  tute, 331, 393
Germany, 137, 261, 264–65,
  269–71, 278, 281–82, 294,
  314–15
Goalkeeper program, 318,
  330–34, 393, 394
Google, 37, 47, 54, 76–77,
  100–115, 121, 126, 130,
  175–77, 357, 362, 371
  data quality and, 207
  employee protest at, 5, 149–
    53, 217
  Israel and, 348
  Trump and, 156–57
Google Advanced Solutions
  Lab, 107
Google Brain, 77, 102–3, 114
Google Cloud, 103, 104, 106–8,
  111, 207
Google DeepMind, 1, 100–102,
  113–14
Google Earth, 14, 15, 32, 103,
  175
Gordon, Susan, 32
"Gorgon Stare" sensor, 73, 96,
  103, 105, 130–31
Government Accountability
  Office, 199
Grant, Greg, 43–44, 48–53
Greene, Diane, 104–5
Griffin, Mike, 149
Group 10 (Naval Special War-
  fare), 74–75
Gulf War of 1991, 22–23, 360
Gupta, Anand, 171–72

Guterres, António, 112–13,
  310–11, 317, 345
Guzzardo, Justin, 95–99

Hamas, 291, 345, 396
Harbinger project, 137–44
Hassabis, Demis, 100–102, 114
Hatan village, Syria, 198
Hegseth, Pete, 320, 324, 349
Hicks, Kathleen, 256, 325–26,
  329, 331, 393
HIMARS (high mobility artil-
  lery rocket systems), 229,
  276–77, 279
Hindu Kush, 131
Hinton, Geoffrey, 77, 103
Hoffman, Frank, 147
Holmes, Mike, 228
Horowitz, Michael, 346, 391,
  396
House of Representatives, 23,
  269
Hulin, Liam M., 292
Human Rights Watch, 101,
  314
Hunter, Mike, 138–41
Hurst, Jay (Hurst, Jules, III),
  138–39, 352

i2 (software company), 33–36
IA (information assurance), 92
IBM, 33, 46–47, 106, 175, 188,
  373
Iceland, 102, 369
IDenTV, 50, 80, 365
IEDs (improvised explosive
  devices), 29, 37, 96, 165,
  166, 359, 363
ImageNet computer vision
  competition, 77
Imig, Robert, 171–72, 179–80,
  189–90, 352
Indo-Pacific Command
  (INDOPACOM), 142–
  43, 215, 283, 291, 294,
  296, 299, 303–4, 329–30,
  353, 393
In-Q-Tel, 32, 37, 149, 362
Integrated Undersea Surveil-
  lance System (IUSS), 142
Intelligence Support to Targeting
  (DOD), 240

Intelligence Systems Support
  Office (ISSO), 50
International Committee for
  Robot Arms Control, 318
International Committee of the
  Red Cross (ICRC), 311,
  315, 318, 349, 359
Iran, 226
Iran-Israel war, 293, 297
Iraq
  Abu Ghraib and other
    detention centers in, 195
  civilian casualties in, 199,
    202
  Erbil, 97, 195, 197, 224, 226
  Gulf War of 1991, 22–23, 360
  Mosul, 34, 49, 195, 225
  US invasion in 2003, 26–28,
    30, 34, 45, 48, 128, 195,
    223–25, 344
  weapons of mass destruc-
    tion in, 262
  Yazidi people of, 195, 340
Iraqi Army, 22, 26–27, 202, 225
Is War Now Impossible?
  (Bloch), 8
ISIS, 48, 51, 57, 91, 128–32,
  183, 195–99, 203, 224,
  233–34
Israel, 112, 156, 291–93, 310,
  316, 345–48
Israel Defense Forces (IDF),
  345–48

J.P. Morgan, 3, 255, 259, 335–
  36, 382
Jain, Akash "Aki," 32–33,
  169–70, 175–79, 190–91,
  335, 384
Jalalabad, Afghanistan, 129,
  130, 359
Japan, 88, 145, 155, 294, 334
Jassy, Andy, 114
Jevons Paradox, 343–44
Johns Hopkins University
  Applied Physics Labo-
  ratory (JHU APL), 173,
  177–78, 180, 192, 249, 253
Joint Artificial Intelligence
  Center (JAIC), 150, 187,
  214–19, 243–45, 250, 256,
  290–91

402 INDEX

Joint Base Langley-Eustis, VA, 228
Joint Special Operations Command (JSOC), 4, 85, 90, 93, 120, 128–30, 138–39, 175, 183, 198, 201, 224–26, 333, 344
Joint Special Operations Task Force in Iraq, 26
Jordan, 115, 122, 172, 291–92, 294
Judson, Jen, 200

Kabba, Timothy Musa, 311–13
Kabul, Afghanistan, 19, 34, 129–31, 232–35, 359
Kaiser, Carl, 138
Kalashnikovs, Mikhail, 101
Kania, Elsa, 148–49
Kant, Peter, 213
Karp, Alex, 5, 32–33, 155–63, 168–69, 175–76, 288, 336, 348
Karzai, Hamid, 16–17
Kazakhstan, 294
Kelly, John, 28, 29, 105
Kelly, Robert Michael, 28, 29, 39, 105
Kendall, Frank, 49, 159
Kernan, Joe, 190, 246–47, 250–51, 254–56, 259
KG-142 Ethernet Data Encryptor, 282
Koehler, Steve, 330
Korea, Democratic People's Republic of, 144, 294
Korea, Republic of, 135, 294
Kropyva artillery system (Ukraine), 273
Kukreja, Amit, 288
Kurilla, Michael "Erik," 130, 224–26, 263–66, 268, 291–93
Kuwait, 224

L3Harris Technologies, 320, 322, 323, 392
Larson, Joe, 30–35, 51–52, 57–62, 66–67, 70, 71, 73–74, 79
alleged conflicts of interest, 186, 249, 365

as chief of Project Maven
after Cukor, 270, 281, 295, 298–99, 300
at OpenAI, 352, 365
personal life, 253
Law of Armed Conflict, 123, 199, 235, 347, 397
LeCun, Yann, 77
Lee Sedol, 1, 89, 100, 358
Legacy project, 27–28, 38
Lethal autonomous weapons systems (also known as "killer robots"), 310–334
Li, Fei-Fei, 106–7
LinkedIn, 152, 227, 235, 335
Liu Guozhi, 147–48, 154
Liu Zhan, 146–47
LLMs (large language models), 7, 114, 168, 188, 243, 298–302, 303, 314, 317
Lockheed Martin, 49, 75, 76, 143
Luckey, Palmer, 324
Lymberopoulos, Dimitrios, 208, 211–14, 228

Mabe, Brady, 311
Mali, 195
Marasco, Daniel, 212–14
Marcinko, Richard, 247
Marine Corps. *See* US Marine Corps
Martell, Craig, 300
Martin, Kelly, 70, 260
Martin, Rachael, 244, 304
Massachusetts, 137–38
Mastroianni, Lee, 331
Mattis, Jim, 13–14, 16, 41, 104, 129, 146–47, 152, 358, 395, 397
Maven. *See* Project Maven
Maxar Technologies, 227, 244, 268, 278, 378
McConville, James, 200, 283
McGunnigle, John, 137–38
McKenzie, Kenneth F., Jr., 197, 379
McKnight, Matt, 27, 34, 41
McRaven, William, 225
Médecins Sans Frontières, 202–3
Meta, 5, 77, 158, 213, 301

Microsoft
Algorithm development and, XXXX
Azure cloud, 103, 209, 348
in the field, 4, 16, 50, 106, 109, 114, 127, 180, 208–9, 211–14, 215, 228, 268, 278, 285, 299, 304, 323, 348, 352
Israel and, 348
in politics, 149, 156–57
Trump and, 156–57
Military Intelligence Program (MIP), 214
Miller, Alex, 301–2, 390
Milley, Mark, 233–34, 264–66, 282, 345–46
Minotaur software suite, 173, 177–80
MIST (Multi-Intelligence Spatial-Temporal) tool suite, 174
Mistral (company), 304
Moore, Andrew, 291, 301
Moore, Schuyler, 291
Mormonism, 63–64
MORSE Corp, 373
Mosul, Iraq, 34, 49, 195, 225
Mozambique, 195
MQ-1 "Predator" drones, 72, 367
MQ-1B "Predator" drones, 123, 367
MQ-9 "Reaper" drones, 79, 93–94, 96, 103, 108, 126, 130, 286, 215–16, 234, 292
Mulchandani, Nand, 216, 249–50
Musk, Elon, 61, 100–101, 107, 156–57, 271, 313, 332–33, 336, 389
Myanmar, 294

Nadella, Satya, 114
Nakasone, Paul, 283–84, 301
Nash, William, 342, 395
National Defense University, 292
National Geospatial-Intelligence Agency (NGA), 7, 243–45, 259, 287–97, 302–5, 323–25, 330, 357, 385

National Military Command Center (NMCC), 233
National Reconnaissance Office (NRO), 174–75, 183
National Security Agency (NSA), 35, 55, 102, 109, 156, 170, 282, 290, 301, 347, 388
National Security Commission on Artificial Intelligence, 149–50
NATO, 7, 184, 262–63, 265, 288–89, 292, 304, 311, 353, 380
Navigator project, 37, 363
Navy. *See* US Navy
Naylor, Sean, 225
Netflix, 179
Network Enterprise Technology Command (NETCOM), 281
New York Times, 56, 111, 145, 155, 202, 234, 254, 357, 379
New Zealand, 153
9/11, 13, 16, 24, 30–31, 164, 361–62
Ng, Andrew, 102
Nolan, Laura, 110–11
NORAD, 294, 296
Noriega, Manuel, 63–64
North Korea, 144, 294
Northern Command (NORTHCOM), 294
Northrop Grumman, 75, 188
Novetta, 175, 177, 179, 189, 377
NPR, 198–99
Nvidia, 5, 134, 173

O'Callaghan, Joseph, 223–32, 236–37, 261, 280–81, 283, 293, 304–5
Obama, Barack, 36, 195, 244, 300
Okano, Seiko, 319–20
OpenAI, 5, 157–58, 213, 300–301, 352
"Operate to Know" (Cukor), 48, 87–89
Operation Desert Storm, 23, 202, 360

Operation Enduring Freedom, 359
"Operation Swift Freedom," 16
Osborn, Phil, 153
Overmatch project, 319, 391

Padgett, Jock, 175–76
Pakistan, 14, 101–2, 108, 244, 294
Palantir Technologies, 5, 29–41, 155–68, 352–53
 Alex Karp from, 5, 32–33, 155–63, 168–69, 175–76, 288, 336, 348
 enter Drew Cukor, 29–42
 growing Defense Department contracts, 169–72
 Israel and, 348
 Jain, Akash "Aki" and, 32–33, 169–70, 175–79, 190–91, 335, 384
 protests against, 348
 representatives from, 230, 232–33, 236, 268
 Shield AI and, 159–68
 Sankar, Shyam and, 35, 158–59, 163 [should we include here too?]
 See also Project Maven
Palestinians, 345
Panama, 63–64, 224
Paparo, Samuel, 283, 304, 325, 329–30, 351
Payne, Jason, 31–34
Perry, William, 46
Pfaff, Anthony, 342–44, 382
Philippines, 122, 127–28, 195
Philippone, Doug, 167–68
Pichai, Sundar, 108, 113
Pilot AI, 80
Pinelis, Jane, 206, 211, 249, 327
PMA-263 drone program, 73–74
Poggemeyer, Sy, 59–60, 68, 71, 191–92, 249–54, 352, 384
Poland, 265, 271, 294
Port Hueneme Naval Base, CA, 320
Prabhakar, Arati, 300
Probasco, Emelia, 297–98, 389
Project Atropos, 215
Project Legacy, 27–28, 38

Project Maven, 1–2, 3–9, 351–54
 in Afghanistan, 48–54, 59, 61, 65–72, 69, 71, 97, 105, 122–33, 338–39, 344
 agentic AI and, 298–303
 AI and the conduct of war, 199–203
 and AI technology among allies, 153–54
 in the AI arms race with China, 23–25, 43–45, 57, 84, 134–36, 137, 140, 142–44, 145–54, 158, 160, 218, 324–32
 arms control and, 309–18
 building the team, 55–64, 71
 after Cukor's retirement, 256–60, 287–98, 335–49
 data access and, 214–19
 data quality and, 204–9
 developing autonomous weaponry, 319–34
 divisions over Palantir, 184–93
 ECS Maven Integration Lab, Fairfax, VA, 210–13
 embedded in the kill chain, 246–60
 ephemera from, 68, 69, 205–6, 206, 338, 339
 Google and, 100–15
 the hunt for al-Baghdadi, 194–99, 201, 226
 the hunt for bin Laden, 156, 244
 the hunt for Soleimani, 226–27
 in the lab vs. in the field, 30, 39, 72, 87–89, 107, 150, 170–72, 180, 202–3, 210–13, 229–31, 266–67, 300, 322, 330, 352
 last holdouts against, 303–4
 linking AI and drone technology, 72–75
 Maven appliances (MAPPs), 172–73
 Maven ATR (automatic target recognition), 288, 325, 392

Project Maven (*continued*)
  Maven Smart System (MSS), 7, 181–83, 184–86, 188, 190–91, 197, 202, 226–27, 231–33, 265–66, 270, 273, 275–78, 288–304, 311, 316–17, 323, 343, 351–53
  Maven Target Workbench, 289, 292, 297–98
  onboarding AI for the military, 76–85
  other "test" deployments, 172–83
  "The Reboot" of, 209–14
  under scrutiny, 238–57
  Shield AI and, 159–68
  in Somalia, 90–95
  surveillance tests at the Aberdeen Proving Ground, 209–10
  survivability of, 87–90
  tapping targeting expertise, 95–99
  in Ukraine, 261–74, 275–83, 285–87
  for undersea surveillance, 134–44
  usability and user adoption issues with, 232–37
  work environment at, 65–72, 69
  See also Cukor, Drew; Palantir; Special Operations Command (SOCOM)
Project Navigator, 37, 363
Project Overmatch, 319, 391
PTSD (post-traumatic stress disorder), 340–41, 359, 395

Qatar, 294
Quds Force (Iran), 226
Qurmo, Khaled Abdel Majid, 198
Qurmo, Khaled Mustafa, 198

Rafferty, John, 286
Ramstein Air Base, Germany, 282
Rangers (US Army), 14, 124, 128–30, 133, 224–25, 291
Rastegari, Mohammad, 80

Raytheon, 49, 103, 156, 159, 331
Reaper drones, 79, 93–94, 96, 103, 108, 126, 130, 286, 215–16, 234, 292
Red Cross, 311, 315, 318, 349, 359
Redmon, Joe, 81–82, 368
Relentless Strike (Naylor), 225
Replicator initiative, 319–34, 352
Reserve Officers' Training Corps (ROTC), 63
Reznikov, Oleksii, 277
Rhino FOB, Afghanistan, 14–15, 17
Rhoads, Mike, 59–60, 366
Rickover, Hyman, 88–89
Rizzo, James, 294–96
Rogers, Michael, 347
Rooker, Kelly, 249
Roosevelt, Franklin D., 155–56
Roper, Will, 50–51, 364, 365
Rosas, Ann Marie, 30, 253
Rubinstein, Anna, 303
Russia
  espionage on, 226–28, 253
  invasion of Ukraine, 9, 142–43, 159, 162, 261–86, 294, 310, 315–17, 324, 328, 332, 346
  military power, 43, 45, 86–87, 395
  in Somalia, 93
  Soviet invasion of Afghanistan, 17, 43, 45, 93, 101, 260, 296, 395
  technology sharing with China, 142–43
  undersea surveillance by, 134–36, 142–43
  use of landmines, 112

SAIC, 373
Sankar, Shyam, 35, 158–59, 163
Saronic, 322–32, 325
Sasse, Ben, 269
Save the Children, 129
Scale AI, 5, 212–13, 289, 299–301, 385
Scale Donovan, 299

ScanEagle, 72, 93, 98
Scarlet Dragon, 229, 231, 343
Schmidt, Eric, 56, 103–4, 108, 149
2nd Marine Division, 39
Senate, 23, 269
75th Ranger Regiment, 124, 128, 224, 291
Shanaberger, Joy, 326
Shanahan, Jack, 44, 51–53, 55, 60–61, *71*, 108, 113, 138, 150–55, 201, 207, 214–19, 238, 243, 382
Shanahan, Patrick, 152
Sharp, Robert, 244–45
Shield AI, 159–68, 336, 377
Shihadah, Mohammad and Amro, 50
Sierra Leone, 8, 311–12
SIGINT (signals intelligence), 275, 280–81
Signal app, 110, 283–84, 329
SIPRNet classified network, 140
Skynet, 51, 125, 148, 223, 338
Slant count, 181–82
Smith, Adam, 269
Smith, Jeremy, 123
Smith, Jerry, 123
Snowden, Edward, 35, 109
Soleimani, Qasem, 226–27
Somalia, 8, 74, 90–95, 97–98, 115, 125–26, 149, 172, 182–83, 195
South Korea, 135, 294
Southern Command (SOUTHCOM), 295
Soviet Union. *See* Russia
Space and Naval Warfare Systems Center, Newport, RI, 140
Special Operations Command (SOCOM), 17, 50, 72, 74, 90, 125, 170, 176, 201, 294, 333
  Army Rangers, 14, 124, 128–30, 133, 224–25, 291
  hunting al-Baghdadi in Syria, 194–99, 201, 226
  hunting Soleimani in Iraq, 226–27

INDEX

Joint Special Operations Command (JSOC), 4, 85, 90, 93, 120, 128–30, 138–39, 175, 183, 198, 201, 224–26, 333, 344
Joint Special Operations Task Force in Iraq, 26
Navy SEALs, 74, 124, 138–39, 142, 164–67, 180–82, 244, 246–47, 331, 333–34
See also Project Maven
Spirk, Dave, 13–19, 49–50, 59–62, 73–75, 80, 218, 352, 359
Stanley, Cameron, 295
Starlink, 142, 244, 271
Stewart, Vince, 36, 39–40
Stop Killer Robots campaign, 101, 314–15, 369
Su, Jonathan, 80
Substack, 162, 357
Sudan, 236
Sun Tzu, 23–24
SUNet, 79, 216
Supermicro chips, 173, 377
SURTASS ships, 141, 143
Syria, 122, 128, 194–95, 199, 224, 291
Szymanski, Tim, 74, 205

Tactical Imagery Production System (TIPS), 91–93
Taiwan, 7, 9, 24, 44, 147, 151, 154, 160, 170, 283, 320, 323–30, 333–34, 351
Taliban, 13–18, 48, 123, 128–29, 164–66, 233, 246, 257
"Task Force 58," 13
Task Force Champion, Stuttgart, Germany, 270
Task Force Dragon, Wiesbaden, Germany, 262, 272, 280
Tegmark, Max, 313–14
Temple, Joey, 232–33, 236, 261, 285
Terre Haute Air National Guard Base, IN, 122
Tesla, 49
Thiel, Peter, 32, 100, 172
Thomas, Tony, 56, 125
Thunderforge, 299

TikTok, 276
ToChip, Elston, 249
Toolan, John, 39
Trump, Donald
  first term of, 8, 108, 120, 146, 194–96, 226
  second term of, 57, 156–57, 243, 289, 294–95, 301, 316, 324, 331–33, 349, 360
Tseng, Brandon, 160, 163–67, 336
Tseng, Ryan, 160, 163, 167
Tull, Thomas, 336
Turkey, 276, 332
Twentynine Palms Marine training base, CA, 19
26th Marine Expeditionary Unit (Special Operations Capable), 14–17, 26
TWG Global, 336, 372
Twitter, 276, 368, 386

UAE, 115, 172
Uber, 49
Ukraine, 9, 142–43, 159, 162, 261–86, 294, 310, 315–17, 324, 328, 332, 346
Undersea Warfighting Development Center, Groton, CT, 137, 141–42
Unit X (Kirchoff and Shah), 296, 357
United Kingdom
  in Afghanistan, 8–9
  intelligence sharing, 7, 153–54, 273
  Project Maven and, 288–89, 291, 294, 304, 326
  Special Air Service, 154
  technology sharing, 143–44
United Nations (UN), 91, 102, 112–13, 309–18, 345, 375
United States
  bombing the Chinese embassy in Belgrade, 202, 380
  Djibouti, 74, 98, 122, 182–83, 294
  in Iraq, 26–28, 30, 34, 45, 48, 128, 195, 223–25, 344

  in Somalia, 8, 74, 90–95, 97–98, 115, 125–26, 149, 172, 182–83, 195
  Taiwan and Sino-American relations, 7, 9, 24, 44, 147, 151, 154, 160, 170, 283, 320, 323–30, 333–34, 351
  US invasion of Afghanistan, 13–19, 26–28, 30, 33, 37–42, 164–65, 202–3, 359
  US withdrawal from Afghanistan, 176, 183, 184–85, 188–89, 202–3, 232–35, 268–69
  use of landmines, 112
US Air Force, 44–46, 50, 51, 55, 66–67, 70–73, 88–89, 329
  in Afghanistan, 14
  AI and the, 173, 228–29
  Air Combat Command (ACC), 228–29
  Air Force Academy, 104
  Beale AFB, 95–97, 107
  bombing accuracy in Desert Storm, 202
  Dover AFB, 130
  drone warfare and the, 70–73, 101, 191–93
  Edwards AFB, 159
  GPS satellites, 45
  Intelligence Systems Support Office (ISSO), 50
  National Military Command Center (NMCC), 233
  Palantir and the, 176
  target approval, 239
US Army, 19–20, 25, 46, 67
  Army Rangers, 14, 124, 128–30, 133, 224–25, 291
  Army Research Laboratory, 39
  Command Post Computing Environment (CPCE), 263
  data security concerns, 283–84
  DCGS-A (Distributed Common Ground System—Army), 35, 38, 40, 362

US Army (continued)
18th Airborne Corps, 224–32, 235–36, 240, 242, 261, 268, 270, 275–77, 280, 283, 299, 342–44
82nd Airborne Division, 233, 271
Fort A.P. Hill training center, VA, 93
Fort Belvoir Army Base, VA, 289
Palantir and the, 37–41, 156, 165, 169–70, 175, 225–26, 288
technology issues, 230–31, 271–72
US Army Garrison in Wiesbaden, 262–73, 277–79, 281
US Coast Guard, 295
US Marine Corps
in Afghanistan, 13–17, 122–24, 233–34
Cukor's service in the, 13–19, 29–30, 256–60
Cukor's thesis "Marine Ground Intelligence Reform," 21, 22–25
deception by "dazzling," 229
in the Gulf War (1991), 22, 24–26
intelligence and the global war on terror, 29–34
in the Iraq War (2003), 26–28
Marine Aircraft Wing, 39
Palantir and Project Maven, 29–41, 55–60, 68–70, 69, 73, 89–90, 176, 188, 191, 215, 249, 353
2nd Marine Division, 39
Twentynine Palms Marine training base, CA, 19
26th Marine Expeditionary Unit (Special Operations Capable), 14–17, 26
the Van Riper plan, 23, 360
US Naval Academy, Annapolis, MD, 59, 90

US Naval Submarine Base at Groton, 137, 141–42
US Navy, 19–20, 25
Group 10 (Naval Special Warfare), 74–75
Naval Criminal Investigative Service, 255
Naval Postgraduate School, Monterey, CA, 21
Naval Special Warfare Command, 74, 205, 331
Naval Special Warfare Development Group (DEVGRU), 181
Navy Information Warfare Systems Command (NAVWAR), 319
nuclear-powered submarines and undersea surveillance, 88, 135–41, 144
Office of Naval Research (ONR), 137, 330–31, 393–94
Palantir and the, 176–81
Port Hueneme Naval Base, CA, 320
Seabees, 17
SEALs, 74, 124, 138–39, 142, 164–67, 180–82, 244, 246–47, 331, 333–34
SIPRNet, 140
Space and Naval Warfare Systems Center, Newport, RI, 140
Submarine Base at Groton, 137, 141–42
US Naval Academy, Annapolis, MD, 59, 90
USS *Cole* bombing, 182
Uzbekistan, 294

Van Riper, Paul, 23, 360
Vance, JD, 295, 389
Venezuela, 295
Viasat, 282
Vietnam, 86–87, 294
Villarruel, Nick, 226, 227
Volkovska, Kateryna (Katya), 82, 172, 192, 215, 252–53, 256, 352

Waguespack, Nicholas, 352
WAMI (wide-area motion imagery), 73, 103–8, 128, 130, 209, 215, 365
Wang, Alexandr, 212–13, 299
Ward, Brian, 62, 71, 119–27, 169, 204, 206, 234–35, 242, 260, 265, 352
Wareham, Mary, 101–2, 111–12, 314–15, 330
Warner, Mark, 269
Waymo, 49
Wheeler, Stuart "Death," 71
Whiplash program, 332–34, 394
Whipps, Gene, 192, 249
Whittaker, Meredith, 108–12, 284
Whitworth, Frank "Trey," 238–45, 287–94, 303, 325, 383
Winchell, Stephen, 352
Winchester, Sarah, 339–40
Woods Hole Oceanographic Institution, 137–38
"WOPR," 296, 389
Work, Robert, 44–53, 75, 84, 89, 134, 149, 152, 243, 365
World Trade Organization, 145

xAI, 336
Xi Jinping, 2, 146–47, 300, 326, 328
Xnor, 80, 81, 115, 209, 372

Yazidi people in Iraq, 195, 340
Yemen, 182, 291, 294
Yon, Michael, 225
YouTube, 162, 179

Zabul province, Afghanistan, 164, 165
Zeiler, Matt, 76–80, 83–85, 103, 113, 207–8, 386
Zelenskyy, Volodymyr, 265
Zuckerberg, Mark, 77